*Springer* **M**onographs *in* **M**athematics

Springer
*Berlin*
*Heidelberg*
*New York*
*Barcelona*
*Hong Kong*
*London*
*Milan*
*Paris*
*Singapore*
*Tokyo*

Tonny A. Springer · Ferdinand D. Veldkamp

# Octonions,
# Jordan Algebras
# and Exceptional Groups

Springer

*Tonny A. Springer*
Mathematisch Instituut
Budapestlaan 6
3584 CD Utrecht, The Netherlands
e-mail: springer@math.uu.nl


*Ferdinand D. Veldkamp* †

Library of Congress Cataloging-in-Publication Data

Springer, T. A. (Tonny Albert), 1926-
  Octonions, Jordan algebras and exceptional groups / Tonny A. Springer, Ferdinand D. Veldkamp.
  p. cm. -- (Springer monographs in mathematics)
  "[Revised] English version of the original German notes"--Pref.
  Includes bibliographical references and index.
  ISBN 3540663371 (alk. paper)
  1. Jordan algebras. 2. Alternative rings. 3. Linear algebraic groups.  I.Veldkamp,
Ferdinand D., 1931-1999. II. Title. III. Series.

  QA252.5 .S69 2000
  512'.24--dc21                                          00-021741

---

Mathematics Subject Classification (1991): 17 C 50, 17 D 05,  20 G 15

---

ISBN 3-540-66337-1 Springer-Verlag Berlin Heidelberg New York

Springer-Verlag is a company in the BertelsmannSpringer publishing group.
© Springer-Verlag Berlin Heidelberg 2000
Printed in Germany

Cover design: *Erich Kirchner, Heidelberg*
Typesetting by the author using a Springer TEX macro package
Printed on acid-free paper    SPIN 10651162        41/3143AT-5 4 3 2 1 0

# Preface

In the summer of 1963, T.A. Springer gave a course of lectures at the Mathematical Institute of Göttingen University on the theory of octonion and Jordan algebras and some exceptional groups related to these. Notes were written by Peter Eysenbach and these were published in mimeographed form by the Göttingen Institute under the title *"Oktaven, Jordan-Algebren und Ausnahmegruppen"* ([Sp 63] in the references at the end of this book). For a considerable part the results exposed in these lectures were not new, but they were brought in greater generality, with new proofs or with a different approach. New features were the introduction of twisted composition algebras, and their use in the description of exceptional Jordan division algebras.

The Göttingen notes shared the fate of so many mimeographed lecture notes: to become of difficult access after some time. But they are still being referred to in recent publications, so that they have some actual value. Therefore a new edition does not seem out of place.

The present volume contains an English version of the original German notes. The text has been completely revised: the order of exposition has been changed at several places and proofs have been rewritten. Also, the notes have been expanded: proofs are included for results that originally were mentioned without proof, additional results that fit in the framework are included and more recent developments are discussed.

The initiative to publish the old notes was due to Martin Kneser. We thank him heartily for his past and present interest. Also, we are indebted to Peter Eysenbach for writing the original German notes.
We most gratefully acknowledge the help of Joseph C. Ferrar, who very carefully read the whole text of the present version. He saved us from several errors and offered numerous suggestions for improvement. We are also very

grateful to Markus Rost for his comments, mathematical and typographical. Finally, we thank Springer-Verlag for the interest in the publication of this new version of old notes.

T. A. Springer
F. D. Veldkamp

When the manuscript of this book was almost completed, F. D. Veldkamp fell ill. He passed away on August 3, 1999.

T. A. S.

# Contents

# 1. Composition Algebras

In this chapter we present the basic theory of composition algebras and determine their structure. Since these are (not necessarily associative) algebras with quadratic norms, we need some fundamental parts of the theory of quadratic forms, which we therefore recall in the first section. It will be shown that the norm on a composition algebra already determines the algebra up to isomorphism. This leads to a more or less explicit determination of all composition algebras over some special fields, using the classification of quadratic forms over these fields.

The theory in this chapter, and in the following two chapters, will in general be developed over arbitrary fields. In particular, there will be no restriction on the characteristic, though occasionally in characteristic two the situation may be slightly different from other characteristics.

## 1.1 Quadratic and Bilinear Forms

In this section we present some basic material from the theory of quadratic and bilinear forms. For proofs and more information about these forms, we refer the reader to [Che 54, Ch. 1] or [Dieu, Ch. I, § 11 and § 16]; for forms over fields of characteristic $\neq 2$, one may consult textbooks like [Ja 74, Ch. 6], [Lam], [Lang, Ch. XIV], [O'M, § 42F] or [Schar, Ch. 1]. For the basic definitions and most elementary results no restriction on the dimension of vector spaces is needed, but sometimes the dimension must be finite, which we will make mention of in each case. For quadratic forms in infinite-dimensional spaces in particular, we refer to [Gro].

To begin with, we recall that a *quadratic form* on a vector space $V$ over a field $k$ is a mapping $N : V \to k$ with the properties:

(i) $N(\lambda x) = \lambda^2 N(x)$ $(\lambda \in k, x \in V)$;

(ii) The mapping $\langle \, , \, \rangle : V \times V \to k$ defined by

$$\langle x, y \rangle = N(x + y) - N(x) - N(y)$$

is bilinear, i.e., it is linear in each of $x$ and $y$ separately.

One calls $\langle \, , \, \rangle$ the *bilinear form associated with $N$*; as one sees, it is *symmetric*. If there is danger of confusion, we may use the notation $N( \, , \, )$

for the bilinear form associated with a quadratic form $N$, so $N(x,y)$ instead of $\langle x,y \rangle$. From the conditions (i) and (ii) it follows that

$$\langle x,x \rangle = N(x{+}x){-}N(x){-}N(x) = N(2x){-}2N(x) = 4N(x){-}2N(x) = 2N(x).$$

So if $\mathrm{char}(k) \neq 2$, then $N(x) = \frac{1}{2}\langle x,x \rangle$. But if $\mathrm{char}(k) = 2$, then always $\langle x,x \rangle = 0$. The case $\mathrm{char}(k) = 2$ will sometimes require a separate treatment in the sequel.

A vector $x \in V$ is called *isotropic* if $N(x) = 0$ and *anisotropic* if $N(x) \neq 0$. The quadratic form $N$ is said to be *isotropic* if there exist nonzero isotropic vectors in $V$ and *anistropic* otherwise. Finally, a subspace $W$ of $V$ is said to be *anisotropic* or *isotropic* if the restriction of $N$ to $W$ is anisotropic or isotropic, respectively, and *totally isotropic* if $N(x) = 0$ for all $x \in W$.

The *radical $R$* of $N$ is defined by

$$R = \{\, r \in V \mid \langle r,x \rangle = 0 \text{ for all } x \in V,\ N(r) = 0 \,\}.$$

If $\mathrm{char}(k) \neq 2$, $\langle r,x \rangle = 0$ for all $x \in V$ implies already that $N(r) = \frac{1}{2}\langle r,r \rangle = 0$, so in that case

$$R = \{\, r \in V \mid \langle r,x \rangle = 0 \text{ for all } x \in V \,\}.$$

Consider a bilinear form $\langle\ ,\ \rangle : V \times V \to k$. The vectors $x,y \in V$ are said to be *orthogonal* if $\langle x,y \rangle = 0$; notation: $x \perp y$. Two subspaces $P$ and $Q$ of $V$ are called *orthogonal*, $P \perp Q$, if $x \perp y$ for all $x \in P$ and $y \in Q$. The *orthogonal complement* of a subspace $P$ of $V$ is

$$P^{\perp} = \{\, x \in V \mid x \perp y \text{ for all } y \in P \,\}.$$

$P^{\perp}$ is a subspace of $V$. For $a \in V$ we usually write $a^{\perp}$ instead of $(ka)^{\perp}$.

The form $\langle\ ,\ \rangle$ is said to be *nondegenerate* if $V^{\perp} = 0$, i.e., if

$$\langle x,y \rangle = 0 \text{ for all } y \in V \implies x = 0.$$

From this condition we infer: If $\langle a,y \rangle = \langle b,y \rangle$ for all $y \in V$, then $a = b$. A linear subspace $W$ of $V$ such that the restriction of $\langle\ ,\ \rangle$ to $W \times W$ is nondegenerate, is called a *nonsingular* subspace. If $W$ is a nonsingular subspace of finite dimension, then $V = W \oplus W^{\perp}$, a direct sum decomposition, and $W^{\perp}$ is nonsingular, too.

A quadratic form $N$ on a vector space $V$ over $k$ is called *nondegenerate*, if the associated bilinear form $\langle\ ,\ \rangle$ is nondegenerate (see Rem. 1.2.2 below). If $N$ is a quadratic form on $V$ and the subspace $W$ of $V$ is nonsingular with respect to the associated bilinear form, then $W$ may be considered as a vector space with a nondegenerate quadratic form, viz., the restriction of $N$ to $W$.

If $\mathrm{char}(k) = 2$, the fact that $N$ is nondegenerate on a finite-dimensional space $V$ means that $V$ has even dimension and $N$ is nondefective in the sense of [Dieu, Ch. I, § 16].

With respect to a quadratic form $N$ on $V$, all maximal totally isotropic subspaces of $V$ have the same dimension, which is called the *Witt index* (or *index* for short) of $N$. If $N$ is nondegenerate and $\dim V$ is finite, the index is at most equal to $\frac{1}{2}\dim V$.

Let $V_i$ $(i = 1, 2)$ be vector spaces over fields $k_i$, with nondegenerate quadratic forms $N_i$ and associated bilinear forms $\langle\ ,\ \rangle_i$. A $\sigma$-*similarity* $t$ of $V_1$ onto $V_2$, where $\sigma$ is an isomorphism of $k_1$ onto $k_2$, is a surjective $\sigma$-linear mapping such that

$$N_2(t(x)) = n(t)\sigma(N_1(x)) \qquad (x \in V_1)$$

for some $n(t) \in k^*$, called the *multiplier* of $t$. Clearly, $\langle t(x), t(y)\rangle_2 = n(t)\sigma(\langle x, y\rangle_1)$, from which it follows that $t$ is also injective and hence bijective. If $\sigma = \mathrm{id}$, one calls $t$ a *similarity*; if $n(t) = 1$, then $t$ is said to be a $\sigma$-*isometry*; if both $n(t) = 1$ and $\sigma = \mathrm{id}$, then $t$ is an *isometry*. If there exists a $\sigma$-similarity or a similarity of $V_1$ onto $V_2$, one calls the quadratic forms $N_1$ and $N_2$ $\sigma$-*similar* or *similar*, respectively; in case of $\sigma$-isometries one speaks of $\sigma$-*isometric* forms and in case of isometries of *isometric* or rather of *equivalent* forms.

A key result about isometries is Witt's Theorem: If $V_1$ and $V_2$ have finite dimension and the nondegenerate forms $N_1$ on $V_1$ and $N_2$ on $V_2$ are $\sigma$-isometric, then every $\sigma$-isometry of a subspace of $V_1$ onto a subspace of $V_2$ can be extended to a $\sigma$-isometry of $V_1$ onto $V_2$.

In the case of one vector space $V$ with a nondegenerate quadratic form $N$, an isometry of $V$ onto itself is called an *orthogonal transformation* in $V$ with respect to $N$. These orthogonal transformations form a group, the *orthogonal group* in $V$, denoted by $O(N)$. Orthogonal transformations in finite dimension have determinant $\pm 1$, and if $\mathrm{char}(k) \neq 2$, the orthogonal transformations having determinant 1 form a subgroup $SO(N)$ of index 2 in $O(N)$, called the *special orthogonal group* or *rotation group* with respect to $N$; its elements are called *rotations*. The definition of $SO(N)$ in characteristic 2 is different; see [Dieu, Ch. II, § 10].

Consider again an arbitrary quadratic form $N$ on a vector space $V$ over $k$. A special type of orthogonal transformations which are not rotations are the following. For $a \in V$ with $N(a) \neq 0$ we define $s_a$ by

$$s_a(x) = x - N(a)^{-1}\langle x, a\rangle a \qquad (x \in V). \tag{1.1}$$

If $\mathrm{char}(k) \neq 2$, this is the *reflection* in the hyperplane orthogonal to $a$. If $\mathrm{char}(k) = 2$, it is the *orthogonal transvection* with *center* $a$. We will use the word "reflection" to also mean an *orthogonal transvection* in characteristic 2.

By a *hyperbolic plane* we understand a two-dimensional subspace $H = ka \oplus kb$ with $N(a) = N(b) = 0$, $\langle a, b\rangle = 1$. If $k$ has characteristic $\neq 2$, this is equivalent to requiring that $H = kc \oplus kd$ with $N(c) = 1$, $N(d) = -1$ and $\langle c, d\rangle = 0$ (take $c = a + b$, $d = a - b$, and conversely $a = \frac{1}{2}(c + d)$, $b = \frac{1}{2}(c - d)$). Every nonzero isotropic vector not contained in the radical is contained in a hyperbolic plane. On a hyperbolic plane, the quadratic form $N$ takes all values since $N(\xi a + \eta b) = \xi\eta$. In particular, a nondegenerate isotropic quadratic form takes all values. If $H_1$ and $H_2$ are hyperbolic planes in $V$, the restrictions of $N$ to $H_1$ and $H_2$, respectively, are equivalent, so by Witt's Theorem the same holds for their orthogonal complements $H_1^\perp$ and $H_2^\perp$, provided $N$ is nondegenerate and $\dim V$ is finite.

## 1.2 Composition Algebras. The Minimum Equation

After the above preparations, we start with composition algebras.

**Definition 1.2.1** A *composition algebra* $C$ over a field $k$ is a not necessarily associative algebra over $k$ with identity element $e$ such that there exists a nondegenerate quadratic form $N$ on $C$ which *permits composition*, i.e., such that

$$N(xy) = N(x)N(y) \qquad (x, y \in C).$$

The quadratic form $N$ is often referred to as the *norm* on $C$, and the associated bilinear form $\langle\ ,\ \rangle$ is called the *inner product*. A linear subspace of $C$ is said to be *nonsingular*, if it is nonsingular with respect to the inner product.

By a *subalgebra* of a composition algebra $C$, or *composition subalgebra*, we understand a nonsingular linear subspace $D$ which is closed under multiplication and contains the identity element $e$ of $C$.

Let $C_i$ be a composition algebra over $k_i$ with norm $N_i$ for $i = 1, 2$, and let $\sigma$ denote an isomorphism of $k_1$ onto $k_2$. A bijective $\sigma$-linear transformation $t : C_1 \to C_2$ is called a $\sigma$-*isomorphism* if

$$t(xy) = t(x)t(y) \qquad (x, y \in C_1).$$

We call $t$ a *linear isomorphism*, or *isomorphism* for short, if $k_1 = k_2$ and $\sigma = \mathrm{id}$.

In Cor. 1.2.4 we will see that the norm $N$ on a composition algebra is already determined by the algebra structure, i.e., the structure of vector space with a product, and that a $\sigma$-isomorphism is automatically a $\sigma$-isometry. Conversely, it will be shown in Th. 1.7.1 that the existence of a $\sigma$-similarity between two composition algebras implies that they are $\sigma$-isomorphic. So the algebra structure and the metric structure on a composition algebra mutually determine each other.

In the above definition, no restriction on the dimension of $C$ over $k$ is made. We will see, however, that this dimension must be finite and can, in fact, only be 1, 2, 4 or 8 (see Th. 1.6.2).

**Remark 1.2.2** In the literature, one also calls a quadratic form $N$ *nondegenerate* if $N(a+x) = N(x)$ for all $x$ implies $a = 0$, or equivalently, $N(a) = 0$ and $\langle a, x \rangle = 0$ for all $x$ implies $a = 0$. In characteristic $\neq 2$ this is equivalent to the definition we have given in § 1.1, since then $\langle a, x \rangle = 0$ for all $x$ implies $N(a) = 0$. In characteristic 2, the alternative definition would lead to some more composition algebras, viz., $k$ itself (as in all other characteristics, see Th. 1.6.2) and certain purely inseparable field extensions of $k$; cf. [BlSp 59, p. 408]. Since the latter case is uninteresting, we have preferred to call a quadratic form nondegenerate if the associated bilinear form is nondegenerate; this makes certain proofs simpler.

From Def. 1.2.1 we will derive a number of equations for the norm and the associated bilinear form on a composition algebra. From

$$N(x) = N(ex) = N(e)N(x) \qquad (x \in C)$$

we deduce

$$N(e) = 1. \tag{1.2}$$

Further, we have

$$N(x_1 y + x_2 y) = N(x_1 y) + N(x_2 y) + \langle x_1 y, x_2 y \rangle$$
$$= N(x_1)N(y) + N(x_2)N(y) + \langle x_1 y, x_2 y \rangle,$$

and on the other hand

$$N(x_1 y + x_2 y) = N((x_1 + x_2)y) = N(x_1 + x_2)N(y) =$$
$$N(x_1) + N(x_2) + \langle x_1, x_2 \rangle)N(y),$$

hence

$$\langle x_1 y, x_2 y \rangle = \langle x_1, x_2 \rangle N(y) \qquad (x_1, x_2, y \in C). \tag{1.3}$$

In a similar way we find

$$\langle xy_1, xy_2 \rangle = N(x)\langle y_1, y_2 \rangle \qquad (x, y_1, y_2 \in C). \tag{1.4}$$

If we replace $y$ by $y_1 + y_2$ in equation (1.3) and subtract from both sides the terms with $y_1$ only or $y_2$ only, we obtain

$$\langle x_1 y_1, x_2 y_2 \rangle + \langle x_1 y_2, x_2 y_1 \rangle = \langle x_1, x_2 \rangle \langle y_1, y_2 \rangle \qquad (x_1, x_2, y_1, y_2 \in C). \tag{1.5}$$

We say that this equation is obtained from (1.3) by *linearizing* it with respect to the variable $y$. A special case of (1.5) is

$$\langle x, yx \rangle + \langle x^2, y \rangle = \langle x, y \rangle \langle e, x \rangle \qquad (x, y \in C), \tag{1.6}$$

which we will use in the proof of the proposition that follows. This proposition says that every element of a composition algebra satisfies a quadratic equation, which is its minimum equation if the element is not a scalar multiple of the identity.

**Proposition 1.2.3** *Every element $x$ of a composition algebra $C$ satisfies*

$$x^2 - \langle x, e \rangle x + N(x)e = 0. \qquad (1.7)$$

*For $x, y \in C$ we have*

$$xy + yx - \langle x, e \rangle y - \langle y, e \rangle x + \langle x, y \rangle e = 0. \qquad (1.8)$$

*If the subspace $ke \oplus kx$ is two-dimensional and nonsingular, it is a composition algebra.*

Proof. Form the inner product of the left hand side of equation (1.7) with an arbitrary element $y$ of $C$:

$$
\begin{aligned}
\langle x^2 - \langle x, e \rangle x + N(x)e, y \rangle &= \langle x^2, y \rangle - \langle x, e \rangle \langle x, y \rangle + N(x) \langle e, y \rangle \\
&= \langle x^2, y \rangle - \langle x, e \rangle \langle x, y \rangle + \langle x, yx \rangle \qquad \text{(by (1.3))} \\
&= 0 \qquad \text{(by (1.6))}.
\end{aligned}
$$

Since this holds for all $y \in C$, (1.7) follows. The other equation is proved by linearizing, i.e., by replacing $x$ by $x + y$ in (1.7) and subtracting from the expression thus obtained the equations (1.7) for $x$ and $y$, respectively.

To prove the last point observe that $W = ke \oplus kx$ is a commutative and associative subalgebra of $C$ by (1.7). If $W$ is two-dimensional, $N(\alpha e + \beta x) = \alpha^2 + \alpha \beta \langle x, e \rangle + \beta^2 N(x)$ is the determinant of the linear map of $W$ defined by multiplication with $\alpha e + \beta x$. It follows that the restriction of $N$ permits composition and that $W$ is a composition subalgebra if that restriction is nonsingular. □

Formula (1.8) is convenient for changing the order in certain products; it implies, in particular, that $xy = -yx$ if $x, y \in e^\perp$ with $x \perp y$.

**Corollary 1.2.4** *The norm $N$ on a composition algebra is uniquely determined by its algebra structure (vector space with multiplication). A $\sigma$-isomorphism is necessarily a $\sigma$-isometry.*

Proof. $N(\lambda e) = \lambda^2$, and for $x \notin ke$, (1.7) is the minimum equation of $x$, which is unique. The second statement follows by a similar argument. □

The *powers* of an element $x$ of any nonassociative algebra $A$ are defined inductively by

$$x^1 = x \qquad \text{and} \qquad x^{i+1} = x^i x \quad (i \geq 1).$$

(If $A$ has an identity element $e$, one starts off with $x^0 = e$.) $A$ is said to be *power associative* if $x^i x^j = x^{i+j}$ $(i, j \geq 1)$ for all $x \in A$, or equivalently if the subalgebra $k[x]$ generated by $x$ is associative for every $x \in A$. (The word "subalgebra" here means a linear subspace closed under multiplication, as is usual in the theory of general nonassociative algebras; it is also assumed to contain the identity element, if there exists one in the whole algebra. For a "composition subalgebra" we require in addition that it is a nonsingular subspace, see Def. 1.2.1.) From (1.7) it is immediate that in a composition algebra $xx^2 = x^2x$ for all $x$, and from this it follows by a straightforward computation that for any $x \in C$ the one- or two-dimensional algebra $k[x]$ is associative. Thus we have:

**Corollary 1.2.5** *Composition algebras are power associative.*

## 1.3 Conjugation. Inverses

We now introduce *conjugation* in a composition algebra $C$, which is the mapping $^- : C \to C$ defined by

$$\bar{x} = \langle x, e \rangle e - x = -s_e(x) \qquad (x \in C), \tag{1.9}$$

where $s_e$ is the reflection in $e^\perp$. One calls $\bar{x}$ the *conjugate* of $x$.

**Lemma 1.3.1** *The following rules hold for conjugation in a composition algebra.*

(i) $x\bar{x} = \bar{x}x = N(x)e$;
(ii) $\overline{xy} = \bar{y}\bar{x}$;
(iii) $\bar{\bar{x}} = x$;
(iv) $\overline{x+y} = \bar{x} + \bar{y}$;
(v) $N(\bar{x}) = N(x)$;
(vi) $\langle \bar{x}, \bar{y} \rangle = \langle x, y \rangle$.

Proof. (i) Immediate from (1.7).
(ii) This is proved by the following computation.

$$\begin{aligned}
\bar{y}\bar{x} &= (\langle y, e \rangle e - y)(\langle x, e \rangle e - x) \\
&= \langle x, e \rangle \langle y, e \rangle e - \langle x, e \rangle y - \langle y, e \rangle x + yx \\
&= \langle x, e \rangle \langle y, e \rangle e - xy - \langle x, y \rangle e \qquad \text{(by (1.8))} \\
&= \langle xy, e \rangle e - xy \qquad \text{(by (1.5))} \\
&= \overline{xy}.
\end{aligned}$$

(iii) This follows from the fact that $x \mapsto -\bar{x}$ is a reflection, or by a straightforward computation.
(iv) Immediate.
(v) $N(\bar{x})e = \bar{\bar{x}}\bar{x} = x\bar{x} = N(x)e$.
(vi) Immediate from (v) and (iv). $\qquad\qquad\qquad\qquad\qquad\qquad\qquad\square$

Other useful identities are given in the following lemma.

**Lemma 1.3.2** *For* $x, y, z \in C$,

$$\langle xy, z \rangle = \langle y, \bar{x}z \rangle, \tag{1.10}$$
$$\langle xy, z \rangle = \langle x, z\bar{y} \rangle, \tag{1.11}$$
$$\langle xy, \bar{z} \rangle = \langle yz, \bar{x} \rangle. \tag{1.12}$$

*The last equation shows that* $\langle xy, \bar{z} \rangle$ *is invariant under cyclic permutations.*

Proof. We prove the first equation; the other ones then easily follow by using Lemma 1.3.1.

$$\begin{aligned}
\langle y, \bar{x}z \rangle &= \langle y, (\langle x, e \rangle e - x)z \rangle = \langle x, e \rangle \langle y, z \rangle - \langle y, xz \rangle \\
&= \langle xy, z \rangle + \langle xz, y \rangle - \langle y, xz \rangle \quad \text{(by (1.5))} \\
&= \langle xy, z \rangle.
\end{aligned}$$

$\square$

From the above lemma we derive some more important formulas.

**Lemma 1.3.3** *For all* $x, y, z \in C$,

*(i)* $x(\bar{x}y) = N(x)y$;
*(ii)* $(x\bar{y})y = N(y)x$;
*(iii)* $x(\bar{y}z) + y(\bar{x}z) = \langle x, y \rangle z$;
*(iv)* $(x\bar{y})z + (x\bar{z})y = \langle y, z \rangle x$.

Proof. To prove (i), we take the inner product of $x(\bar{x}y)$ with any $z \in C$:

$$\begin{aligned}
\langle x(\bar{x}y), z \rangle &= \langle \bar{x}y, \bar{x}z \rangle \quad \text{(by the previous lemma)} \\
&= N(x)\langle y, z \rangle \quad \text{(by (1.4) and Lemma 1.3.1 (v))} \\
&= \langle N(x)y, z \rangle.
\end{aligned}$$

Since this holds for all $z \in C$, we get the result. By conjugating (i) we get (ii), and (iii) and (iv) follow by linearizing (i) and (ii). $\square$

The first two statements of this lemma can be interpreted as associativity results; indeed, since $x\bar{x} = N(x)$, we can read (i) as

$$x(\bar{x}y) = (x\bar{x})y$$

and similarly for (ii). Recall that we have not assumed associativity in Def. 1.2.1. We will see in Th. 1.6.2 that there are, indeed, composition algebras that are not associative.

In a not necessarily associative algebra $A$ an element $a$ is said to have $b$ as an *inverse* if

$$a(bx) = b(ax) = (xa)b = (xb)a = x \quad (x \in A).$$

In other words: $a$ has $b$ as an inverse if and only if the left multiplication $l_a : A \to A$, $x \mapsto ax$, and the right multiplication $r_a : A \to A$, $x \mapsto xa$, are invertible and have $l_b$ and $r_b$ as inverses, respectively. If $A$ has an identity element $e$, then $ab = ba = e$; in that case the inverse $b$ is uniquely determined by $a$, viz., $b = l_a^{-1}e$, and it is denoted by $a^{-1}$.

**Proposition 1.3.4** *In a composition algebra an element $a$ has an inverse if and only if $N(a) \neq 0$, and then $a^{-1} = N(a)^{-1}\bar{a}$.*

Proof. Use (i) and (ii) of Lemma 1.3.3.                                    $\square$

## 1.4 Moufang Identities. Alternative Laws

The aim of this section is the proof of three important identities for composition algebras, the *Moufang identities*, and some consequences thereof.

**Proposition 1.4.1** *In any composition algebra, the following identities hold.*

$$(ax)(ya) = a((xy)a); \tag{1.13}$$
$$a(x(ay)) = (a(xa))y; \tag{1.14}$$
$$x(a(ya)) = ((xa)y)a. \tag{1.15}$$

*(Moufang identities)*

Proof. To prove the first identity we form the inner product of the left hand side of (1.13) with an arbitrary $z \in C$ :

$$
\begin{aligned}
\langle (ax)(ya), z \rangle &= \langle ya, (\bar{x}\bar{a})z \rangle \qquad \text{(by (1.10) and Lemma 1.3.1)} \\
&= \langle y, \bar{x}\bar{a} \rangle \langle a, z \rangle - \langle yz, (\bar{x}\bar{a})a \rangle \qquad \text{(by (1.5))} \\
&= \langle xy, \bar{a} \rangle \langle z, a \rangle - N(a)\langle yz, \bar{x} \rangle \qquad \text{(by (1.10) and Lemma 1.3.3).}
\end{aligned}
$$

The inner product of the right hand side of (1.13) with any $z$ becomes

$$
\begin{aligned}
\langle a((xy)a), z \rangle &= \langle (xy)a, \bar{a}z \rangle \\
&= \langle xy, \bar{a} \rangle \langle a, z \rangle - \langle (xy)z, \bar{a}a \rangle \\
&= \langle xy, \bar{a} \rangle \langle z, a \rangle - N(a)\langle (xy)z, e \rangle \\
&= \langle xy, \bar{a} \rangle \langle z, a \rangle - N(a)\langle xy, \bar{z} \rangle \\
&= \langle xy, \bar{a} \rangle \langle z, a \rangle - N(a)\langle yz, \bar{x} \rangle.
\end{aligned}
$$

So the left hand side and the right hand side of (1.13) have the same inner product with every $z$, which implies that they are equal. The proof of the second identity runs along similar lines:

$$\langle (a(xa))y, z \rangle = \langle a(xa), z\bar{y} \rangle$$
$$= \langle xa, \bar{a}(z\bar{y}) \rangle$$
$$= \langle x, (\bar{a}(z\bar{y}))\bar{a} \rangle$$
$$= \langle \bar{x}, a((y\bar{z})a) \rangle \qquad \text{(by Lemma 1.3.1 (vi))}$$
$$= \langle \bar{x}, (ay)(\bar{z}a) \rangle \qquad \text{(by the first Moufang identity)}$$
$$= \langle x, (\bar{a}z)(\bar{y}\bar{a}) \rangle \qquad \text{(by Lemma 1.3.1 (vi) again)}$$
$$= \langle x(ay), \bar{a}z \rangle$$
$$= \langle a(x(ay)), z \rangle,$$

which proves the second Moufang identity. To prove the third identity, conjugate the second one and use Lemma 1.3.1 (ii) and the identity $(ax)a = a(xa)$, which follows from (1.13). □

From the Moufang identities we derive three more associativity relations.

**Lemma 1.4.2** *In a composition algebra the following identities hold.*

$$(xy)x = x(yx); \tag{1.16}$$
$$x(xy) = x^2 y; \tag{1.17}$$
$$(xy)y = xy^2. \tag{1.18}$$

*(Alternative laws).*
*This means that for all $x \in C$, the left multiplication $l_x$ and the right multiplication $r_x$ commute: $l_x r_x = r_x l_x$, and that $l_x^2 = l_{x^2}$, $r_x^2 = r_{x^2}$.*

Proof. The first relation is immediate from the first Moufang identity (1.13). The second relation is a consequence of the following set of equalities:

$$x(\bar{x}y) = (x\bar{x})y \qquad \text{(by Lemma 1.3.3)},$$
$$x(((\langle x, e \rangle e - x)y) = (x(\langle x, e \rangle e - x))y,$$
$$x(\langle x, e \rangle y - xy) = (\langle x, e \rangle x - x^2)y,$$
$$\langle x, e \rangle xy - x(xy) = \langle x, e \rangle xy - x^2 y.$$

The third relation is proved in a similar way. □

The three relations of this lemma permit us to rewrite the Moufang identities in the following form:

$$(ax)(ya) = (a(xy))a; \tag{1.19}$$
$$a(x(ay)) = ((ax)a)y; \tag{1.20}$$
$$x((ay)a) = ((xa)y)a. \tag{1.21}$$

The expression

$$\{x, y, z\} = (xy)z - x(yz)$$

is called the *associator* of $x$, $y$ and $z$. It is trilinear, i.e., linear in each of its three variables, because of the distributive law. Lemma 1.4.2 says that the associator vanishes whenever two of its variables are equal, i.e., it is an alternating function. Any not necessarily associative algebra with this property is called an *alternative algebra*. By linearizing the alternative laws one shows that the associator changes sign whenever two of its variables are interchanged, so it is skew symmetric: $\{x_{\pi(1)}, x_{\pi(2)}, x_{\pi(3)}\} = \text{sg}(\pi)\{x_1, x_2, x_3\}$ for any permutation $\pi$. In characteristic $\neq 2$ the latter property is equivalent to the alternative laws, but in characteristic 2 it is weaker.

In an alternative algebra associativity holds for any product of three elements at most two of which are distinct, by definition. A beautiful result by Artin says that this implies a stronger property, viz., that every subalgebra generated by two elements is associative:

**Theorem 1.4.3** *(E. Artin) The subalgebra generated by any two elements of an alternative algebra is associative.*

For a proof, we refer to [Schaf, Th. 3.1], or to [Ja 74, § 7.6, ex. 9].    □

This implies in particular that any alternative algebra is power associative; for composition algebras we saw this already in Cor. 1.2.5.

## 1.5 Subalgebras. Doubling

In this section we study finite-dimensional subalgebras of composition algebras; these will be the key to the description of the structure of composition algebras. (In the next section we will see that all composition algebras have finite dimension, so then the restriction "finite-dimensional" becomes superfluous.) It is clear that a subalgebra of a composition algebra is closed under conjugation.

Let $C$ be a composition algebra and $D$ a finite-dimensional composition subalgebra. Since $D$ is a nonsingular subspace of finite dimension, $C = D \oplus D^{\perp}$ and $D^{\perp}$ is also nonsingular. If $D \neq C$, there must exist $a \in D^{\perp}$ which is anisotropic, i.e., $N(a) \neq 0$.

**Proposition 1.5.1** *Let $C$ be a composition algebra and $D$ a finite-dimensional composition subalgebra, $D \neq C$. If $a$ is chosen in $D^{\perp}$ with $N(a) \neq 0$, then*

$$D_1 = D \oplus Da$$

*is a composition subalgebra. Product, norm and conjugation on $D_1$ are given by the formulas*

$$(x + ya)(u + va) = (xu + \lambda \bar{v}y) + (vx + y\bar{u})a \qquad (x, y, u, v \in D), \quad (1.22)$$
$$N(x + ya) = N(x) - \lambda N(y) \qquad (x, y \in D), \quad (1.23)$$
$$\overline{x + ya} = \bar{x} - ya \qquad (x, y \in D), \quad (1.24)$$

*respectively, where* $\lambda = -N(a)$, *and* $\dim D_1 = 2 \dim D$.

Proof. We begin by showing that $Da \subseteq D^\perp$, from which it follows that the sum of $D$ and $Da$ is indeed direct. If $x \in D$, then

$$\langle xa, y \rangle = \langle a, \bar{x}y \rangle = 0 \qquad (y \in D),$$

so $xa \in D^\perp$. It is evident that $D_1$ is closed under addition. To prove this for multiplication, it suffices to derive (1.22). This formula is a consequence of the following three formulas:

$$x(va) = (vx)a \qquad (x, v \in D); \tag{1.25}$$
$$(ya)u = (y\bar{u})a \qquad (y, u \in D); \tag{1.26}$$
$$(ya)(va) = -N(a)\bar{v}y \qquad (y, v \in D). \tag{1.27}$$

To prove the first of these three formulas, we form the inner product of the left hand side with an arbitrary $z \in C$:

$$\begin{aligned}
\langle x(va), z \rangle &= \langle va, \bar{x}z \rangle \\
&= -\langle va, \bar{z}x \rangle \qquad \big(\text{by (1.9), since } \langle va, e \rangle = 0\big) \\
&= \langle vx, \bar{z}a \rangle \qquad \big(\text{by (1.5) and } \langle a, x \rangle = 0 \big) \\
&= -\langle vx, za \rangle + \langle z, e \rangle\langle vx, a \rangle \qquad \big(\text{by (1.9)}\big) \\
&= -\langle vx, za \rangle \qquad \big(\text{since } \langle vx, a \rangle = 0\big) \\
&= \langle (vx)a, z \rangle \qquad \big(\text{since } \bar{a} = -a\big).
\end{aligned}$$

Since this holds for all $z \in C$, formula (1.25) follows. The proof of (1.26) is similar, but easier: use (1.5) and $a \in D^\perp$. For the proof of (1.27), we first notice that

$$ya = -\bar{a}\bar{y} + \langle ya, e \rangle e = a\bar{y} \qquad (y \in D), \tag{1.28}$$

since $ya \in D^\perp$ and $\bar{a} = -a$. Now

$$\begin{aligned}
(ya)(va) &= (a\bar{y})(va) \qquad \big(\text{by (1.28)}\big) \\
&= a((\bar{y}v)a) \qquad \big(\text{by the first Moufang identity, (1.13)}\big) \\
&= a(a(\bar{v}y)) \qquad \big(\text{by (1.28) again}\big) \\
&= -N(a)(\bar{v}y) \qquad \big(\text{by Lemma 1.3.3 (i), since } a = -\bar{a}\big).
\end{aligned}$$

This completes the proof of (1.27), whence also that of formula (1.22).
    Since $\langle x, ya \rangle = 0$, we have

$$N(x + ya) = N(x) + N(y)N(a),$$

which proves formula (1.23). From this it easily follows that the norm $N$ is nondegenerate on $D_1$, which completes the proof that $D_1$ is a composition subalgebra.

Formula (1.24) is proved as follows:

$$\overline{x + ya} = \bar{x} + \bar{a}\bar{y} = \bar{x} - a\bar{y} = \bar{x} - ya$$

using (1.28).

Since $r_a : D \to Da$ is bijective, $Da$ has the same dimension as $D$, so $D_1 = D \oplus Da$ has double dimension. $\qquad \square$

$D_1$ is said to be constructed from $D$ by *doubling*.

**Proposition 1.5.2** *Let $C$ be a composition algebra and $D$ a finite-dimensional proper subalgebra. Then $D$ is associative. A subalgebra $D \oplus Da$, where $a \in D^{\perp}$ with $N(a) \neq 0$, is associative if and only if $D$ is commutative and associative.*

Proof. Choose any $a \in D^{\perp}$ with $N(a) \neq 0$; let again $\lambda = -N(a)$. By writing out

$$N((x + ya)(u + va)) = N(x + ya)N(u + va) \qquad (x, y, u, v \in D),$$

we find

$$\langle\, xu, \bar{v}y \,\rangle - \langle\, vx, y\bar{u} \,\rangle = 0,$$

hence

$$\langle\, (xu)\bar{y}, \bar{v} \,\rangle = \langle\, v, (y\bar{u})\bar{x} \,\rangle = \langle\, x(u\bar{y}), \bar{v} \,\rangle.$$

Thus $\langle\, (xy)z, w \,\rangle = \langle\, x(yz), w \,\rangle$ for all $w \in D$, which implies that $(xy)z = x(yz)$ for all $x, y, z \in D$, i.e., $D$ is associative.

Now assume that $D \oplus Da$ is associative, where $a \in D^{\perp}$ with $N(a) \neq 0$. By (1.25), $x(va) = (vx)a$. Using associativity we get $(xv)a = (vx)a$. Multiplying this on the right by $a^{-1}$ yields $xv = vx$, which proves commutativity of $D$. Conversely, a straightforward computation shows that $D \oplus Da$ is associative if $D$ is commutative and associative. $\qquad \square$

As a counterpart of Prop. 1.5.1 we have the following result, which permits us to start from any associative composition algebra $D$ and construct a composition algebra $C$ from it by doubling.

**Proposition 1.5.3** *Let $D$ be a composition algebra and $\lambda \in k^{*}$. Define on $C = D \oplus D$ (direct sum) a product by*

$$(x, y)(u, v) = (xu + \lambda \bar{v}y, vx + y\bar{u}) \qquad (x, y, u, v \in D)$$

*and a quadratic form $N$ by*

$$N((x, y)) = N(x) - \lambda N(y) \qquad (x, y \in D).$$

*If $D$ is associative, then $C$ is a composition algebra. $C$ is associative if and only if $D$ is commutative and associative.*

Proof. The verification of these facts is left to the reader. $\qquad \square$

## 1.6 Structure and Dimension of a Composition Algebra

The results of the previous section enable us to prove a key result on the structure of a composition algebra $C$. Assume $C$ has dimension $> 1$. If $\operatorname{char}(k) \neq 2$, the subspace $D = ke$ is a subalgebra of $C$, and with the aid of Prop. 1.5.1 we find a two-dimensional subalgebra $D_1$. In case $\operatorname{char}(k) = 2$ we have to proceed differently, for then the subspace $ke$ is no longer nonsingular since $\langle e, e \rangle = 0$. In this case, pick any $a$ with $\langle a, e \rangle \neq 0$.

**Lemma 1.6.1** *Let $C$ be a composition algebra over a field $k$ of characteristic 2. If $a \in C$ with $\langle a, e \rangle \neq 0$, the linear subspace $ke \oplus ka$ is a two-dimensional composition subalgebra of $C$.*

Proof. The subspace $ke \oplus ka$ is nonsingular, for $\lambda e + \mu a \perp e, a$ yields $\mu = 0$ and $\lambda = 0$. By Prop. 1.2.3, $D_1 = ke \oplus ka$ is a composition subalgebra.    □

Thus we have found a two-dimensional subalgebra $D_1$ of $C$ in any characteristic. If $\dim C > 2$, we apply Prop. 1.5.1 again to find a four-dimensional subalgebra $D_2 \supset D_1$, and if $\dim C > 4$, we repeat this to obtain an eight-dimensional subalgebra $D_3 \supset D_2$. In this way we get a sequence of composition subalgebras $D_1 \subset D_2 \subset D_3$ of dimensions 2, 4 and 8, respectively.

At this point the doubling process has to stop, as the following argument shows. $D_1$ is evidently a commutative, associative subalgebra. By Prop. 1.5.2, $D_2$ must be associative; however, it is not commutative. For pick $a \in D_1$ with $a \neq \bar{a}$; from the argument at the beginning of this section it follows that $D_2 \cap D_1^\perp$ is nonsingular, so there exists $x \in D_2 \cap D_1^\perp$ with $N(x) \neq 0$. Then

$$xa + \bar{a}\bar{x} = 0,$$

whence

$$xa = \bar{a}x \neq ax.$$

This implies by Prop. 1.5.2 that $D_3$ is not associative, so by the same proposition $D_3$ is not a proper subalgebra of $C$. Thus we have proved the following structure theorem.

**Theorem 1.6.2** *Every composition algebra is obtained by repeated doubling, starting from $ke$ in characteristic $\neq 2$ and from a 2-dimensional composition subalgebra in characteristic 2. The possible dimensions of a composition algebra are 1 (in characteristic $\neq 2$ only), 2, 4 and 8. Composition algebras of dimension 1 or 2 are commutative and associative, those of dimension 4 are associative but not commutative, and those of dimension 8 are neither commutative nor associative.*

A composition algebra of dimension 2 over $k$ is either a quadratic field extension of $k$ or is isomorphic to $k \oplus k$. A composition algebra $C$ of dimension 4 is called a *quaternion algebra* and its elements are called *quaternions*. If $C$ has dimension 8, it is called an *octonion algebra* and its elements are called *octonions*.

**Corollary 1.6.3** *Any octonion algebra $C$ over a field $k$ of characteristic $\neq 2$ has an orthogonal basis of the form $e, a, b, ab, c, ac, bc, (ab)c$ with $N(a)N(b)N(c) \neq 0$.*

*If $\operatorname{char}(k) = 2$, then $C$ has a basis $e, a, b, ab, c, ac, bc, (ab)c$ with*

$$\langle e, a \rangle = 1, \quad \langle b, ab \rangle = N(b), \quad \langle c, ac \rangle = N(c), \quad \langle bc, (ab)c \rangle = N(b)N(c),$$

*all other inner products between distinct basis vectors are zero and*

$$N(a)N(b)N(c) \neq 0.$$

*Similarly for quaternion algebras and for two-dimensional composition algebras, except in one case, viz., $k = \mathbb{F}_2$, $\dim C = 2$, $N$ isotropic.*

Proof. In the characteristic $\neq 2$ case one uses the proof of Th. 1.6.2 and Prop. 1.5.1. If $\operatorname{char}(k) = 2$ one uses in addition Lemma 1.6.1 to construct a basis satisfying the inner product relations between distinct basis vectors and such that $N(b)N(c) \neq 0$. If then $N(a) = 0$ and $\dim C > 2$, one starts again with $a$ replaced by $a' = a + b$ and then proceeds to find $b'$ and $c'$ for the desired basis. If $\operatorname{char}(k) = 2$, $k \neq \mathbb{F}_2$, $\dim C = 2$ and $N(a) = 0$ for some $a$ with $\langle e, a \rangle = 1$, then replace $a$ by $a' = \lambda e + a$ with $\lambda \neq 0, 1$. (In the exceptional case $C \simeq \mathbb{F}_2 \oplus \mathbb{F}_2$.) □

We call a triple $a, b, c$ in an octonion algebra that provides a basis as in the above corollary a *basic triple*.

If $\operatorname{char}(k) \neq 2$, we understand by a *standard orthogonal basis* of an octonion algebra $C$ an orthogonal basis as in the above corollary, but normalized so as to make the product of the norms of all basis vectors equal to 1:

$$e_1 = e, \; e_2 = a, \; e_3 = b, \; e_4 = ab, \; e_5 = c, \; e_6 = ac, \; e_7 = bc,$$

$$e_8 = N(a)^{-2} N(b)^{-2} N(c)^{-2} (ab)c. \tag{1.29}$$

In an octonion algebra $C$ over a field $k$ of characteristic 2 we take a basis as in the above corollary. Then $e_1 = e$, $e_2 = b$, $e_3 = c$, $e_4 = bc$, $e_5 = a$, $e_6 = N(b)^{-1}ab$, $e_7 = N(c)^{-1}ac$, $e_8 = N(b)^{-1}N(c)^{-1}(ab)c$ form a basis of $C$ such that

$$\langle e_i, e_{i+4} \rangle = 1 \quad (1 \le i \le 4) \quad \text{and} \quad \langle e_i, e_j \rangle = 0 \quad (1 \le i < j \le 8, \; j \neq i + 4), \tag{1.30}$$

which we call a *symplectic basis*. After at most three separable quadratic extensions of $k$ we may require in addition that $N(a) = N(b) = N(c) = 1$. Then we get a symplectic basis with $N(e_i) = 1$ for all $i$; we call that a *standard symplectic basis* of $C$.

Let $a$ be any element of a composition algebra that is not a multiple of the identity. If $ke \oplus ka$ is nonsingular, it is a subalgebra by Prop. 1.2.3. But if the restriction of the norm to $ke \oplus ka$ is degenerate, $a$ is not contained in a two-dimensional subalgebra. However, we can prove that $a$ is contained in a four-dimensional subalgebra.

**Proposition 1.6.4** *Every element a of an octonion algebra C over k is contained in a quaternion subalgebra of C.*

Proof. For $a \in ke$ this is obvious, so assume $a \notin ke$. If char$(k) \neq 2$, we may assume that $\langle a, e \rangle = 0$, for if not, we can replace $a$ by $a - \frac{1}{2}\langle a, e \rangle e$. If either char$(k) \neq 2$, $\langle a, e \rangle = 0$ and $N(a) \neq 0$ or char$(k) = 2$ and $\langle a, e \rangle \neq 0$, then the subspace $D = ke \oplus ka$ is nonsingular, so it is a subalgebra. By doubling $D$ we get a quaternion subalgebra containing $a$.

Now assume char$(k) \neq 2$, $\langle a, e \rangle = 0$ and $N(a) = 0$. Pick $b \in e^\perp$ with $\langle a, b \rangle \neq 0$ and $N(b) \neq 0$. The subspace $D = ke \oplus kb$ is a subalgebra. Take $c = a + \lambda b$ with $\langle c, b \rangle = 0$. One easily sees that $N(c) = -\lambda^2 N(b) \neq 0$. So $c \in D^\perp$ with $N(c) \neq 0$ and $D \oplus Dc$ is a quaternion subalgebra containing $a$ (cf. Prop. 1.5.1).

Finally, we have to consider the case where char$(k) = 2$ and $\langle a, e \rangle = 0$. We may assume that $N(a) \neq 0$, for otherwise replace $a$ by $e + a$. Since $e^\perp$ and $a^\perp$ are distinct subspaces of the same dimension, we can find $b$ with $\langle a, b \rangle = 0$ and $\langle e, b \rangle \neq 0$. Then $D = ke \oplus kb$ is a subalgebra. Now $a \in D^\perp$ and $N(a) \neq 0$, so $D \oplus Da$ is a quaternion subalgebra that contains $a$.     $\square$

## 1.7 A Composition Algebra is Determined by its Norm

The metric structure of a composition algebra $C$ already determines its algebraic structure, i.e., a scalar multiple of the norm on $C$ determines $C$ up to isomorphism:

**Theorem 1.7.1** *Let C and C′ be composition algebras over fields k and k′, respectively, and let $\sigma : k \to k'$ be a field isomorphism. If there exists a $\sigma$-similarity of C onto C′, then the two algebras are $\sigma$-isomorphic.*

Proof. The identity elements of $C$ and $C'$ are denoted by $e$ and $e'$, respectively; the two norms are both denoted by $N$. Let $t : C \to C'$ be a $\sigma$-similarity with multiplier $n(t)$. Since $N(t(e)) = n(t)N(e) = n(t) \neq 0$, $t(e)$ is invertible. The map

$$C \to C', \quad x \mapsto t(e)^{-1}t(x),$$

is a $\sigma$-isometry which maps $e$ to $e'$, so we may assume that $t$ is a $\sigma$-isometry with $t(e) = e'$.

We give the rest of the proof for octonion algebras; in lower dimensions one just has to stop earlier. For simplicity of notation, we write down the proof in the case that $t$ is an isometry, i.e., $\sigma = \text{id}$; the general case is a straightforward extension. We are going to construct a linear isomorphism $\varphi : C \to C'$.

Pick $a \in C$ with $\langle a, e \rangle = 0$ and $N(a) \neq 0$ if char $k \neq 2$, and with $\langle a, e \rangle = 1$ if char $k = 2$, and take $a' = t(a)$. Then $N(a') = N(a)$ and $\langle a', e' \rangle = \langle a, e \rangle$.

The two-dimensional subalgebras $D = ke \oplus ka$ and $D' = ke' \oplus ka'$ of $C$ and $C'$, respectively, are isomorphic and

$$\varphi : D \to D', \quad \lambda e + \mu a \mapsto \lambda e' + \mu a' \quad (\lambda, \mu \in k),$$

is an isomorphism; if char $k \neq 2$, this follows from Prop. 1.5.1, and in the characteristic 2 case one has to use Lemma 1.6.1. Notice that $\varphi$ coincides with the restriction of $t$ to $D$.

Pick $b \in D^\perp$ with $N(b) \neq 0$ and take $b' = t(b)$; then $b' \in D'^\perp$ and $N(b') = N(b)$. We extend $\varphi$ to $D \oplus Db$ by defining

$$\varphi(x + yb) = \varphi(x) + \varphi(y)b' \quad (x, y \in D).$$

This is an isomorphism from the subalgebra $E = D \oplus Db$ onto the subalgebra $E' = D' \oplus D'b'$ by Prop. 1.5.1, using that $\varphi(\bar{x}) = \overline{\varphi(x)}$ for $x \in D$.

By Witt's Theorem we can extend $\varphi$ to an isometry $u : C \to C'$. Now we repeat the last step with $c \in E^\perp, N(c) \neq 0$ and $c' = u(c)$, extending $\varphi$ to an isomorphism from $C$ onto $C'$ by

$$\varphi(x + yc) = \varphi(x) + \varphi(y)c' \quad (x, y \in E).$$

$\square$

As a special case of the above theorem we have the following corollary.

**Corollary 1.7.2** *Let $C$ be a composition algebra over $k$ and $\sigma$ a field automorphism of $k$. Then there exists a $\sigma$-automorphism of $C$ if and only if there exists a $\sigma$-isometry of $C$ onto itself.*

From the proof of the Theorem we infer another important corollary.

**Corollary 1.7.3** *Let $C$ be a composition algebra and let $D$ and $D'$ be subalgebras of the same dimension. If $C$ admits a $\sigma$-isometry onto itself, then every $\sigma$-isomorphism from $D$ onto $D'$ can be extended to a $\sigma$-automorphism of $C$. In particular, every linear isomorphism from $D$ onto $D'$ can be extended to an automorphism of $C$.*

Also from the proof of Th. 1.7.1 we can derive a transitivity result, which will be used in the next chapter. First a definition.

**Definition 1.7.4** For $\lambda, \mu \in k$ with $\lambda \neq 0$ and $\mu \neq 0$ if char$(k) \neq 2$, and with $\mu \neq 0$ (and $\lambda$ arbitrary) if char$(k) = 2$, we define a *special $(\lambda, \mu)$-pair* to be an ordered pair $a, b$ with $a, b \in C$ satisfying:

- if char$(k) \neq 2$, then $\langle a, e \rangle = \langle b, e \rangle = \langle a, b \rangle = 0$, $N(a) = \lambda$, $N(b) = \mu$;
- if char$(k) = 2$, then $\langle a, e \rangle = 1$, $\langle b, e \rangle = \langle a, b \rangle = 0$, $N(a) = \lambda$, $N(b) = \mu$.

By a *special pair* we understand a special $(1,1)$-pair if char$(k) \neq 2$ and a special $(0,1)$-pair if char$(k) = 2$.

Notice that every special $(\lambda, \mu)$-pair $a, b$ is contained in a unique quaternion subalgebra, which has $e, a, b, ab$ as a basis, and that every quaternion subalgebra contains special $(\lambda, \mu)$-pairs for suitable $\lambda$ and $\mu$. Now the transitivity result is:

**Corollary 1.7.5** *For any two special $(\lambda, \mu)$-pairs $a, b$ and $a', b'$ in a composition algebra $C$, there exists a linear automorphism $\varphi$ of $C$ with $\varphi(a) = a'$ and $\varphi(b) = b'$.*

## 1.8 Split Composition Algebras

With respect to the metric structure of a composition algebra there is an important dichotomy: either $N$ is isotropic, i.e., there exist nonzero $x \in C$ with $N(x) = 0$, or $N$ is anisotropic, i.e., $N(x) \neq 0$ for all $x \in C$, $x \neq 0$. Since $x\bar{x} = \bar{x}x = N(x)e$ by Lemma 1.3.1, $C$ has zero divisors in the first case, viz., all $x \neq 0$ with $N(x) = 0$. In the second case, every nonzero $x \in C$ has an inverse, viz., $N(x)^{-1}\bar{x}$ (see Prop. 1.3.4); such a $C$ is called a *composition division algebra*.

We further examine the case that $N$ is isotropic. First assume that $\mathrm{char}(k) \neq 2$. We want to find $a \in e^{\perp}$ with $N(a) = -1$. Pick any nonzero $x$ with $N(x) = 0$. If $x \in e^{\perp}$, we pick $a' \in e^{\perp}$ with $\langle a', x \rangle = 1$, then $a = a' - (1 + N(a'))x$ does the job. If $x \notin e^{\perp}$, then we can write $x = \alpha e + y$ with $\alpha \neq 0$ and $y \in e^{\perp}$. Since

$$0 = N(x) = \alpha^2 + N(y),$$

we find $N(y) = -\alpha^2$. We can take $a = \alpha^{-1}y$.

The subspace $ke \oplus ka$ is nonsingular and hence it is a composition subalgebra. Further, it is isotropic, since $N(e + a) = 0$. By repeatedly applying the doubling procedure (cf. Prop. 1.5.1 and Th. 1.6.2) we find that the Witt index, i.e. the dimension of any maximal totally isotropic subspace, must be $\frac{1}{2} \dim C$.

In case $k$ has characteristic 2, we can reach the same conclusion by the following argument. First, we must find $a \in C$ with $N(a) = 0$ and $\langle a, e \rangle \neq 0$. To this end we pick any nonzero $x \in C$ with $N(x) = 0$. If $\langle x, e \rangle \neq 0$, we take $a = x$. If $\langle x, e \rangle = 0$, then we further pick $y \in C$ with $\langle y, e \rangle \neq 0$ and $\langle y, x \rangle \neq 0$; if $N(y) = 0$, we take $a = y$, and if $N(y) \neq 0$, then

$$a = s_y x = x - N(y)^{-1}\langle x, y \rangle y$$

has the required properties. Now the subspace $ke \oplus ka$ is a subalgebra by Lemma 1.6.1, and it is isotropic. As above we find that $C$ has maximal Witt index, i.e., equal to $\frac{1}{2} \dim C$.

Two vector spaces of the same dimension over the same field with a nondegenerate quadratic form of maximal Witt index are necessarily isometric.

It follows from Th. 1.7.1 that any two composition algebras of the same dimension over the same field $k$ which both have isotropic norm are isomorphic. We call these *split composition algebras*. Summing up our results, we have proved:

**Theorem 1.8.1** *In each of the dimensions 2, 4 and 8 there is, up to isomorphism, exactly one split composition algebra, i.e., composition algebra with isotropic norm, over any given field $k$. These are the only composition algebras containing zero divisors.*

The two-dimensional split composition algebra over $k$ is $k \oplus k$ with $N((\xi, \eta)) = \xi\eta$ $(\xi, \eta \in k)$. The four-dimensional one is the algebra of $2 \times 2$ matrices over $k$ with the determinant as norm, the *split quaternion algebra*. An easy computation shows that for a matrix

$$x = \begin{pmatrix} \xi_{11} & \xi_{12} \\ \xi_{21} & \xi_{22} \end{pmatrix}$$

we have $\langle x, e \rangle = \operatorname{tr}(x)$, where $e$ denotes the identity matrix, $\operatorname{tr}(x)$ is the trace of the matrix $x$, and

$$\bar{x} = \begin{pmatrix} \xi_{22} & -\xi_{12} \\ -\xi_{21} & \xi_{11} \end{pmatrix}.$$

The *split octonion algebra* can be constructed from the split quaternion algebra $D$ by the doubling process as in Prop. 1.5.3. We may take an arbitrary $\lambda \neq 0$ in this proposition, since all the resulting algebras are isomorphic. Choosing $\lambda = 1$, we get for the product

$$(x, y)(u, v) = (xu + \bar{v}y, vx + y\bar{u}) \qquad (x, y, u, v \in D)$$

and for the norm

$$N((x, y)) = \det(x) - \det(y) \qquad (x, y \in D).$$

There is another description of split octonions over a field $k$, which uses *vector matrices*

$$\begin{pmatrix} \xi & x \\ y & \eta \end{pmatrix} \qquad (\xi, \eta \in k, \ x, y \in k^3),$$

where $k^3$ is the three-dimensional vector space over $k$. On $k^3$ we have the nondegenerate bilinear form $\langle \ , \ \rangle$ defined by

$$\langle x, y \rangle = \sum_{i=1}^{3} \xi_i \eta_i \qquad (x = (\xi_1, \xi_2, \xi_3) \in k^3, \ y = (\eta_1, \eta_2, \eta_3) \in k^3),$$

and the exterior product $\wedge$ defined by

$$\langle x \wedge y, z \rangle = \det(x, y, z) \qquad (x, y, z \in k^3).$$

Addition of vector matrices is defined entrywise, and multiplication by

$$\begin{pmatrix} \xi & x \\ y & \eta \end{pmatrix} \begin{pmatrix} \xi' & x' \\ y' & \eta' \end{pmatrix} = \begin{pmatrix} \xi\xi' + \langle x, y' \rangle & \xi x' + \eta' x + y \wedge y' \\ \eta y' + \xi' y + x \wedge x' & \eta\eta' + \langle y, x' \rangle \end{pmatrix}.$$

The quadratic form $N$ is defined by

$$N\left( \begin{pmatrix} \xi & x \\ y & \eta \end{pmatrix} \right) = \xi\eta - \langle x, y \rangle.$$

The bilinear form associated with $N$ is obviously nondegenerate, and one easily verifies that $N$ permits composition, using the identity

$$\langle x \wedge x', y \wedge y' \rangle = \langle x, y \rangle \langle x', y' \rangle - \langle x, y' \rangle \langle x', y \rangle$$

for the exterior product. Thus, the vector matrices form an eight-dimensional composition algebra, and since $N$ is isotropic this must be the split octonion algebra.

## 1.9 Center and Associating Elements

By the *center* of a composition algebra $C$ we understand $\{ c \in C \,|\, cx = xc \text{ for all } x \in C \}$. We determine the center of a quaternion or octonion algebra, and also the elements in an octonion algebra that "associate" with all other elements.

**Proposition 1.9.1** *The center of a quaternion or octonion algebra $C$ over $k$ is $ke$.*

Proof. That the center of a quaternion algebra $C$ is $ke$, is seen by extending $k$ to an algebraic closure $K$. The norm $N$ on $C_K = K \otimes_k C$ is isotropic, so $C_K \cong \mathrm{M}(2, K)$, which has center $Ke$.

   Let now $C$ be an octonion algebra and $z$ an element of its center. By Prop. 1.6.4, $z$ is contained in a quaternion subalgebra $D$. Since $z$ is central in $D$, it must lie in $ke$.                                           $\square$

   Thus, a quaternion algebra over $k$ is a central simple associative algebra over $k$ (see [Ja 80, § 4.6]).

**Proposition 1.9.2** *Let $C$ be an octonion algebra. If $a \in C$ satisfies*

$$(xy)a = x(ya) \qquad (x, y \in C), \tag{1.31}$$

*then $a \in ke$. Condition (1.31) is equivalent to the condition $(xa)y = x(ay)$ for all $x, y \in C$, and also to $(ax)y = a(xy)$ for all $x, y \in C$.*

Proof. The latter statement is a consequence of the fact that the associator $\{x, y, z\} = (xy)z - x(yz)$ is an alternating function, which we saw in § 1.4 before Artin's Theorem 1.4.3. Let $a$ satisfy (1.31). By Prop. 1.6.4 we can pick a quaternion subalgebra $D$ which contains $a$. Write $C = D \oplus Db$ with some $b \in D^\perp$, $N(b) = -\lambda \neq 0$. Taking $x = ub$ $(u \in D)$ and $y = b$ in (1.31), we get

$$((ub)b)a = (ub)(ba).$$

Write this out, repeatedly using formula (1.22) in which $a$ is replaced by $b$:

$$\lambda ua = (ub)(\bar{a}b) = \lambda au.$$

This implies that $ua = au$ for $u \in D$, so $a$ lies in the center of $D$, which is $ke$. □

## 1.10 Classification over Special Fields

Over some special fields $k$ one can give a classification of quaternion and octonion algebras; we will do this for algebraically closed fields, the reals, finite fields, complete, discretely valued fields with finite residue fields, and for algebraic number fields.

In a four- or eight-dimensional composition algebra $C$ we choose a basis as in Cor. 1.6.3. If $\mathrm{char}(k) \neq 2$, the norm $N$ gets the form (in coordinates $\xi_0, \ldots, \xi_3$ or $\xi_0, \ldots, \xi_7$, respectively):

$$\xi_0^2 + \alpha \xi_1^2 + \beta \xi_2^2 + \alpha\beta \xi_3^2 \tag{1.32}$$

$$\xi_0^2 + \alpha \xi_1^2 + \beta \xi_2^2 + \alpha\beta \xi_3^2 + \gamma \xi_4^2 + \alpha\gamma \xi_5^2 + \beta\gamma \xi_6^2 + \alpha\beta\gamma \xi_7^2 \tag{1.33}$$

with nonzero constants $\alpha, \beta$ and $\gamma$. If $\mathrm{char}(k) = 2$, we get for $N$:

$$\xi_0^2 + \xi_0\xi_1 + \alpha \xi_1^2 + \beta \xi_2^2 + \beta\xi_2\xi_3 + \alpha\beta \xi_3^2 \tag{1.34}$$

$$\xi_0^2 + \xi_0\xi_1 + \alpha \xi_1^2 + \beta \xi_2^2 + \beta\xi_2\xi_3 + \alpha\beta \xi_3^2 + \\ \gamma \xi_4^2 + \gamma\xi_4\xi_5 + \alpha\gamma \xi_5^2 + \beta\gamma \xi_6^2 + \beta\gamma\xi_6\xi_7 + \alpha\beta\gamma \xi_7^2 \tag{1.35}$$

with nonzero constants $\alpha, \beta$ and $\gamma$. (Quadratic forms of these kinds are called twofold and threefold Pfister forms, respectively.)

Since by Th. 1.7.1 the norm determines a composition algebra up to isomorphism, the classification of quaternion and octonion algebras boils down to the description of the equivalence classes of the above quadratic forms. We do this for some special fields.

*(i) k algebraically closed.*
Here all quadratic forms are isotropic, so there are only split composition algebras.

*(ii) $k = \mathbb{R}$, the field of the reals.*
It is easily seen that there are two classes of quadratic forms (1.32) and
(1.33), respectively, viz., the isotropic ones and the positive definite ones. To
the latter correspond the Hamiltonian quaternions and the Cayley numbers
(real octonion division algebra), respectively.

*(iii) $k$ a finite field.*
If $k$ has characteristic $\neq 2$, every quadratic form in dimension $> 2$ is isotropic
(see, e.g., [O'M, § 62]). In the characteristic 2 case, every element of $k$ is a
square, so a quadratic form as in (1.34) or (1.35) has a nontrivial zero, viz.,
with $\xi_0$ such that $\xi_0^2 = \beta$, $\xi_2 = 1$ and all other $\xi_i = 0$. Hence there are only
the split quaternion and octonion algebras.

*(iv) $k$ a complete, discretely valued field with finite residue class field.*
In this case all quadratic forms in dimension $> 4$ are isotropic (see [O'M,
§ 63]), so there is only the split octonion algebra. In dimension 4 there is
exactly one class of anisotropic quadratic forms, so there exists one isomorphism class of quaternion division algebras.

*(v) $k$ an algebraic number field.*
For the classification of quadratic forms in this case one has to use Hasse's
Theorem, which says that two quadratic forms over $k$ are equivalent if and
only if they are so over all local fields $k_v$, where $v$ runs over all places of $k$
(see [O'M, § 66]).
     If $\dim C = 8$, a form like (1.33) has maximal Witt index at all finite
places, and also at all complex infinite places. So we have to look at the
real places. Let $r$ be the number of these places. At each of them we have
two possibilities as we saw above in (ii), so according to Hasse's Theorem
there are at most $2^r$ octonion algebras over $k$. These can all be realized by a
suitable choice of the signs of $\alpha, \beta$ and $\gamma$ at the real places. For $k = \mathbb{Q}$, the
field of the rationals, we thus find exactly two octonion algebras, the split
one and the division algebra.
     For the case $\dim C = 4$, one needs, moreover, Hilbert's Quadratic Reci-
procity Law (see [O'M, § 71]). The result is as follows. If $S$ is any finite
set of an even number of finite or real places of $k$, then there is, up to isomorphism, exactly one quaternion algebra over $k$ whose norm is anisotropic
precisely at the places $v \in S$. All quaternion algebras are obtained in this way.

*(vi)* Let $k$ be a field with the following property: for any quaternion algebra $D$ over $k$ its norm $N_D$ takes all values in $k$. Then any octonion algebra
$C$ over $k$ is split. For, constructing $C$ by doubling as in § 1.5, one sees from
(1.23) that the norm of $C$ is isotropic, hence $C$ is split (see § 1.8).
Fields which have this property are the perfect fields with cohomological di-

mension $\leq 2$, see [Se 64, 5-e éd., p.98]. Examples of these are: finite fields (which have dimension 1), $p$-adic fields and totally imaginary algebraic number fields, see [loc.cit., p.90, p.96, p.97].

## 1.11 Historical Notes

The four-dimensional division algebra of real *quaternions* was discovered by W.R. Hamilton in 1843, as a result of a vain search for a three-dimensional division algebra over the reals. Hamilton was interested in the latter for geometric reasons: he wanted to use these "triplets" to describe motions in three-space, just as complex numbers are used for this purpose in the plane. J.T. Graves (1843) and, independently, A. Cayley (1845) followed with the eight-dimensional nonassociative real division algebra of the *octaves*, or *Cayley numbers*. We prefer the name *octonions*, of more recent date, in view of its analogy with "quaternions", and since "octave" already refers to other notions, in particular in music.

Quaternions and octonions over the complex numbers followed soon, and in due course one became interested in quaternion and octonion algebras over arbitrary fields, possibly with zero divisors. In [Zo 31], M. Zorn studied them in the framework of *alternative rings*; from this paper stems the representation of the split octonions by *vector matrices*. In [Zo 33], Zorn gave the classification of octonion algebras over algebraic number fields, as in (v) of § 1.10 above.

A typical feature of quaternion and octonion algebras is the presence of a quadratic *norm* $N$ (the square of the absolute value) which permits composition: $N(xy) = N(x)N(y)$. The study of such forms goes back to Gauß and culminated in Hurwitz's Theorem, which says that they can only exist in dimensions 1, 2, 4 and 8 (cf. Th. 1.6.2). N. Jacobson [Ja 58] placed all this in the framework of what he named *composition algebras*. He determined the structure of these algebras over fields of characteristic not two, a restriction removed by T.A. Springer in [Sp 63].

For more historical information about this subject, see [Bl] and [Eb].

# 2. The Automorphism Group of an Octonion Algebra

In this chapter we study the group $G = \mathrm{Aut}(C)$ of automorphisms of an octonion algebra $C$ over a field $k$. By "automorphism" we will in this chapter always understand a *linear automorphism*. Since automorphisms leave the norm invariant, $\mathrm{Aut}(C)$ is a subgroup of the orthogonal group $O(N)$ of the norm of $C$.

Let $K$ be an algebraic closure of $k$ and put $C_K = K \otimes_k C$. The automorphism group $\mathbf{G} = \mathrm{Aut}(C_K)$ is a linear algebraic group in the sense of [Bor], [Hu] and [Sp 81]; we refer to these books for the necessary background from the theory of algebraic groups. In the sequel we will, as a rule, denote algebraic groups by boldface letters.

$\mathbf{G}$ is a closed subgroup of the algebraic group $\mathbf{O}(N)$ (the orthogonal group of the quadratic form on $C_K$ defined by $N$). We will prove in Th. 2.3.5 that $\mathbf{G}$ is a connected, simple algebraic group of type $G_2$.

We will prove in Prop. 2.4.6 that the algebraic group $\mathbf{G}$ is defined over $k$; from this it follows that the automorphism group $G$ is the group $\mathbf{G}(k)$ of $k$-rational points of the algebraic group $\mathbf{G}$.

The results in this chapter are valid in any characteristic, though the characteristic two case sometimes requires a separate treatment.

## 2.1 Automorphisms Leaving a Quaternion Subalgebra Invariant

We begin by studying the automorphisms that map a given quaternion subalgebra $D$ of $C$ onto itself. Let $t$ be such an automorphism. From $t(D) = D$ it follows that $t(D^\perp) = D^\perp$. As we did before, we write $D^\perp = Da$ for a fixed $a \in D^\perp$ with $N(a) \neq 0$; every element of $C$ can then be written as $x + ya$ with $x, y \in D$. We have

$$t(x + ya) = t(x) + t(y)t(a)$$

with $t(x), t(y) \in D$ and $t(a) \in D^\perp = Da$. We define mappings $u, v : D \to D$ by

$$t(x) = u(x) \quad \text{and} \quad t(y)t(a) = v(y)a \qquad (x, y \in D).$$

Hence

$$t(x + ya) = u(x) + v(y)a \qquad (x, y \in D).$$

Clearly, $u = t|_D$ is an automorphism of $D$, and $v$ is an orthogonal transformation. Writing out

$$t((xa)w) = t(xa)t(w) \qquad (x, w \in D)$$

with the aid of (1.22), we find

$$v(x\bar{w}) = v(x)\overline{u(w)} \qquad (x, w \in D).$$

Since $u$ is an automorphism, $\overline{u(w)} = u(\bar{w})$. Hence, writing $y$ for $\bar{w}$,

$$v(xy) = v(x)u(y) \qquad (x, y \in D).$$

Substituting $x = e$ and defining $p = v(e)$ we derive from this

$$v(y) = pu(y) \qquad (y \in D).$$

Here $p \in D$, and since $v$ is an isometry, $N(p) = 1$. Thus we find for $t$:

$$t(x + ya) = u(x) + (pu(y))a \qquad (x, y \in D). \tag{2.1}$$

Conversely, for any $p \in D$ with $N(p) = 1$ and any automorphism $u$ of $D$, equation (2.1) defines an automorphism $t$ of $C$ which leaves $D$ invariant; this is easily verified using (1.22) and the associativity of $D$.

Since the associative algebra $D$ is central simple, every automorphism of $D$ is inner by the Skolem-Noether Theorem (see, e.g., [ArNT, Cor. 7.2D] or [Ja 80, § 4.6, Cor. to Th. 4.9]). Hence by (2.1) every automorphism of $C$ that leaves $D$ invariant can be written as

$$t(x + ya) = cxc^{-1} + (pcyc^{-1})a \qquad (x, y \in D) \tag{2.2}$$

with $c \in D$, $N(c) \neq 0$ and $p \in D$, $N(p) = 1$. Here $c$ is determined by $t$ up to a nonzero scalar factor.

## 2.2 Connectedness and Dimension of the Automorphism Group

For a quaternion subalgebra $D$ of $C$, denote by $\mathbf{G}_D$ the algebraic group of $K$-automorphisms of $C_K$ that fix $D_K$ elementwise and by $G_D$ the group of $k$-automorphisms of $C$ that fix $D$ elementwise. $\mathbf{D}_1$ is the algebraic group of elements $p \in D_K$ with $N(p) = 1$, and $D_1 = \mathbf{D}_1 \cap D$. These are the *norm one groups* in $D_K$ and $D$, respectively. $\mathbf{G}_D$ is a closed subgroup of the algebraic group $\mathbf{G}$. The group $\mathbf{D}_1$ is isomorphic to the special linear group $\mathbf{SL}_2 = \mathrm{SL}_2(K)$, since $D_K$ is isomorphic to the algebra of $2 \times 2$ matrices over $K$ with det as norm. So $\mathbf{D}_1$ is a three-dimensional connected algebraic group.

**Proposition 2.2.1** *There is an isomorphism of algebraic groups* $\varphi : \mathbf{G}_D \to \mathbf{D}_1$ *which induces an isomorphism* $G_D \to D_1$, *such that for every* $t \in \mathbf{G}_D$,

$$t(x + ya) = x + (\varphi(t)y)a \qquad (x, y \in D_K).$$

*Consequently,* $\mathbf{G}_D$ *is 3-dimensional and connected.*

Proof. Use equation (2.1) with $u = \mathrm{id}$ and the associativity of $D$. $\qquad\square$

$\mathbf{G}$ is contained in the stabilizer of $e$ in the orthogonal group $\mathbf{O}(N)$, which also leaves the orthogonal complement $e^\perp$ in $C_K$ invariant. We will need that restriction of the stabilizer of $e$ in $\mathbf{SO}(N)$ to $e^\perp$ yields an isomorphism with the rotation group $\mathbf{SO}(N_1)$, where $N_1$ is the restriction of $N$ to $e^\perp$. In characteristic $\neq 2$ this is an almost trivial matter, but in characteristic 2 there are complications due to the fact that then $e \in e^\perp$. For the purposes of the present chapter we could do with less, but in the next chapter we will use the full isomorphism result.

Let $\mathbf{F}$ be the stabilizer of $e$ in $\mathbf{O}(N)$, and $\mathbf{SF} = \mathbf{F} \cap \mathbf{SO}(N)$. These are closed subgroups of $\mathbf{O}(N)$ and $\mathbf{SO}(N)$, respectively. We denote by $F$ and $SF$ the corresponding groups over $k$. If $t \in \mathbf{F}$, it leaves $e^\perp$ invariant and its restriction to $e^\perp$ is an element of the orthogonal group $\mathbf{O}(N_1)$.
If $\mathrm{char}(k) = 2$ the quadratic form $N_1$ is degenerate. Then $\mathrm{O}(N_1)$ and $\mathbf{O}(N_1)$ are defined to be groups of linear isometries of $e^\perp$, as in §1.1. In this case we put $\mathrm{SO}(N_1) = \mathrm{O}(N_1)$, $\mathbf{SO}(N_1) = \mathbf{O}(N_1)$.

**Proposition 2.2.2** *The restriction homomorphism*

$$\varrho : \mathbf{SF} \to \mathbf{SO}(N_1), t \mapsto t|_{e^\perp},$$

*is an isomorphism of algebraic groups which induces an isomorphism* $SF \to SO(N_1)$.

Proof. It is evident that $\varrho$ is a homomorphism of algebraic groups. If $\mathrm{char}(k) \neq 2$, define for $u \in \mathbf{SO}(N_1)$ its extension $\varepsilon(u)$ to $C$ by $\varepsilon(u)(e) = e$. Clearly, $\varepsilon : \mathbf{SO}(N_1) \to \mathbf{SF}$ is a homomorphism of algebraic groups and $\varepsilon = \varrho^{-1}$, so $\varrho$ is an isomorphism of algebraic groups. The proposition follows. In characteristic 2 the extension of $u \in \mathbf{O}(N_1)$ to an element of $\mathbf{SF}$ cannot be defined in this way: then $e \in e^\perp$, whereas we have to define the image of a vector outside $e^\perp$. We will follow a different approach in this case.

So let $\mathrm{char}(k) = 2$. Note that in this case the restriction $N_1$ is a quadratic form having defect 1, i.e. $\dim V^\perp = 1$ (cf. [Che 54, § 1.2] or [Dieu, Ch. I, § 16]). Every element of $\mathbf{O}(N_1)$ is a product of orthogonal transvections or reflections $s_a$, $N(a) \neq 0$ of the form (1.1) (see [Dieu, Ch. II, § 11]). The algebraic group $\mathbf{O}(N_1)$ is connected; this can be seen in the following manner. The restriction of the polynomial function $N - 1$ to $e^\perp$ is irreducible (over the algebraic closure $K$), since otherwise $N_1$ would be a square of a linear polynomial, which would vanish on a 6-dimensional subspace of $e^\perp$. It would follow that there were 6-dimensional totally isotropic subspaces with respect to the nondegenerate quadratic form $N$ in $C$, which is impossible. Hence $S = \{\, x \in e^{\perp} \mid N(x) = 1 \,\}$ (over $K$) is an irreducible variety. The morphism

$$S \to \mathbf{O}(N_1),\ a \mapsto s_a,$$

maps $S$ onto an irreducible set of generators of $\mathbf{O}(N_1)$ containing the identity, so $\mathbf{O}(N_1)$ is a connected algebraic group (see [Hu, § 7.5] or [Sp 81, Prop. 2.2.6]).

The quadratic form $N$ on $C$ is nondegenerate, so here we have a rotation subgroup $\mathbf{SO}(N)$. *Rotations* are now orthogonal transformations with Dickson invariant 0; every rotation is a product of an even number of reflections (see [Dieu, Ch. II, § 10]).

The restriction homomorphism

$$\mathbf{F} \to \mathbf{O}(N_1),\, t \mapsto t|_{e^{\perp}},$$

is surjective, since every orthogonal transformation of $e^{\perp}$ leaves $e$ invariant and can be written as a product of reflections $s_a$ with $a \in e^{\perp}$. Its kernel has order 2, it is the subgroup generated by the reflection $s_e$. This can be seen by using a symplectic basis $e_1 = e, \ldots, e_8$ of $C$ as in (1.30); the $e_i$ with $i \neq 5$ form a basis of $e^{\perp}$. Since $s_e$ is not a rotation, the restriction homomorphism $\mathbf{SF} \to \mathbf{O}(N_1)$ is an isomorphism of abstract groups. Clearly, $\varrho$ is a homomorphism of algebraic groups. To show that it is an isomorphism of algebraic groups, it suffices by [Sp 81, Cor. 5.3.3], to prove that the Lie algebra homomorphism $d\varrho : \mathrm{L}(\mathbf{SF}) \to \mathrm{L}(\mathbf{O}(N_1))$ is an isomorphism. The Lie algebra $\mathrm{L}(\mathbf{SF})$ is contained in the space $S$ of the linear transformations $t : C_K \to C_K$ satisfying

$$\langle x, tx \rangle = 0 \quad (x \in C_K) \text{ and } te = 0.$$

To see this we work over the ring of dual numbers $K[\epsilon]$, $\epsilon^2 = 0$ (see [Bor, Ch. I, 3.20]). The elements of $\mathrm{L}(\mathbf{SF})$ are linear maps $t$ of $C_K$ with $te = 0$ such that $\mathrm{id} + \epsilon t$, acting on $x + \epsilon y \in K[\epsilon] \otimes C$, satisfies

$$N((\mathrm{id} + \epsilon t)(x + \epsilon y)) = N(x + \epsilon y).$$

It follows that $\langle x, tx \rangle = 0$. Writing this out for the matrix $(\alpha_{ij})$ of $t$ with respect to the above symplectic basis, we find 43 independent equations:

$$\begin{aligned}
\alpha_{i,j+4} + \alpha_{j,i+4} &= 0 & &(i \neq j,\ \text{indices mod } 8),\\
\alpha_{i,i+4} &= 0 & &(i \neq 1),\\
\alpha_{1,j} &= 0 & &(\text{all } j).
\end{aligned}$$

So $S$ has dimension 21. As $\mathrm{L}(\mathbf{SO}(N_1))$ also has dimension 21, we have $S = \mathrm{L}(\mathbf{SO}(N_1))$. The Lie algebra homomorphism $d\varrho$ acts on the matrices of $\mathrm{L}(\mathbf{SF})$

by deleting the fifth row and the fifth column. Since $\alpha_{j,5} = \alpha_{1,j+4}$ for $j \neq 1$, $\alpha_{5,j+4} = \alpha_{j,1}$ for $j \neq 1$ and $\alpha_{1,1} = \alpha_{5,5} = 0$, $d\varrho$ is injective and hence is an isomorphism. The last point is clear. $\qquad\square$

**Proposition 2.2.3** *The automorphism group* $\mathbf{G}$ *of* $C_K$ *is a connected algebraic group of dimension* 14.

Proof. Consider the set of special pairs in $C_K$ (see Def. 1.7.4). Since $K$ is algebraically closed, special pairs certainly exist. For any two special pairs there exists a linear isometry $t$ of $C_K$ fixing $e$ and mapping one pair to the other, by Witt's Theorem; clearly, we may assume that $t$ is a rotation. So the group $\mathbf{SF}$ operates transitively on the set of special pairs.

Fix one special pair $a, b$ with $a, b \in D_K$, a quaternion subalgebra of $C_K$. The stabilizer $\mathbf{H}$ of $a, b$ in $\mathbf{SF}$ is isomorphic to the special orthogonal group $\mathbf{SO}(N_2)$ of a five-dimensional quadratic form $N_2$, under the isomorphism $\varrho$ of $\mathbf{SF}$ with $\mathbf{SO}(N_1)$. If $\operatorname{char}(k) \neq 2$, this isomorphism is restriction to $(Ke \oplus Ka \oplus Kb)^{\perp}$; that this is an isomorphism of algebraic groups is seen as in the first paragraph of the proof of Prop. 2.2.2. If $\operatorname{char}(k) = 2$, we restrict in two steps: first to $(Ke \oplus Ka)^{\perp}$ (same argument as in the characteristic $\neq 2$ case), from there to $(Ke \oplus Ka)^{\perp} \cap b^{\perp} = (Ke \oplus Ka \oplus Kb)^{\perp}$ (argument as in the characteristic 2 case in the proof of Prop. 2.2.2).

The homogeneous space $X = \mathbf{SF}/\mathbf{H}$ can be identified as a set with the set of all special pairs. Counting dimensions we find

$$\dim X = \dim \mathbf{SF}/\mathbf{H} = \dim \mathbf{SO}(N_1) - \dim \mathbf{SO}(N_2) = 21 - 10 = 11.$$

From Cor. 1.7.5 we infer that $\mathbf{G}$ operates transitively on the set of special pairs, i.e., on $X$. Since $e$, $a$, $b$ and $ab$ form a basis of $D_K$, the stabilizer in $\mathbf{G}$ of the special pair $a, b$ is $\mathbf{G}_D$. This implies that $X \cong \mathbf{G}/\mathbf{G}_D$. So another dimension count yields

$$\dim \mathbf{G} = \dim X + \dim \mathbf{G}_D = 11 + 3 = 14.$$

As a homogeneous space of a rotation group, $X$ is an irreducible algebraic variety. Since $\mathbf{G}_D$ is connected, it follows that $\mathbf{G}$ is connected (see [Sp 81, Ex. 5.5.9 (1)]). $\qquad\square$

The transitivity of $\mathbf{G}$ on special pairs implies transitivity on the set of quaternion subalgebras, so

**Corollary 2.2.4** *The automorphism group* $\mathbf{G}$ *of* $C_K$ *acts transitively on the set of quaternion subalgebras.*

(This also follows from Cor. 1.7.3, since over an algebraically closed field $K$ all quaternion algebras are isomorphic to $M_2(K)$.)

We remarked at the beginning of this chapter that $\mathbf{G}$ is contained in $\mathbf{O}(N)$. The connectedness of $G$ implies that it must be contained in the connected component of the identity in $\mathbf{O}(N)$, which is $\mathbf{SO}(N)$. This implies the following corollary.

**Corollary 2.2.5** *The automorphism group $G$ is a subgroup of the rotation group $\mathrm{SO}(N)$.*

## 2.3 The Automorphism Group is of Type $\mathbf{G_2}$

In this section we are going to further determine the structure of the algebraic group $\mathbf{G}$. We view the split octonion algebra $C_K$ as an algebra of pairs $(x, y)$ with $x, y \in D_K$, the algebra of $2 \times 2$ matrices over $K$, and with $N((x, y)) = \det(x) - \det(y)$; see § 1.8. We begin by describing a maximal torus in $\mathbf{G}$. To this end we consider (2.2) for the particular case where $c$ and $p$ are diagonal matrices. We denote the $2 \times 2$ diagonal matrix $\mathrm{diag}(\kappa, \kappa^{-1})$ with $\kappa \in K^*$ by $c_\kappa$.

**Lemma 2.3.1** *The group $\mathbf{T}$ consisting of the automorphisms*

$$t_{\lambda,\mu} : x + ya \mapsto c_\lambda x c_\lambda^{-1} + (c_\mu y c_\lambda^{-1})a \quad (x, y \in D_K), \tag{2.3}$$

*with $\lambda, \mu \in K^*$, is a 2-dimensional torus in $\mathbf{G}$. Every element of $\mathbf{G}$ that commutes with all elements of $\mathbf{T}$ lies in $\mathbf{T}$, so $\mathbf{T}$ is a maximal torus. Hence $\mathbf{G}$ has rank 2.*

Proof. For $a \in D^\perp$ we take $(0, e)$ with $e$ the $2 \times 2$ identity matrix. Notice that $(x, 0)a = (0, x)$. As a basis of $C_K$ we take

$$(e_{11}, 0), (e_{12}, 0), (e_{21}, 0), (e_{22}, 0), (0, e_{11}), (0, e_{12}), (0, e_{21}), (0, e_{22}),$$

where $e_{ij}$ denotes the $2 \times 2$ matrix with $(i, j)$-entry equal to 1 and the other entries 0. The matrix of $t_{\lambda,\mu}$ with respect to this basis is

$$t_{\lambda,\mu} = \mathrm{diag}(1, \lambda^2, \lambda^{-2}, 1, \lambda^{-1}\mu, \lambda\mu, \lambda^{-1}\mu^{-1}, \lambda\mu^{-1}). \tag{2.4}$$

Observe that $t_{-\lambda,-\mu} = t_{\lambda,\mu}$, so the obvious parametrization by $\lambda$ and $\mu$ is not one-to-one. We can reparametrize via $\xi = \lambda\mu$ and $\eta = \lambda\mu^{-1}$; this gives an isomorphism of algebraic groups from $(K^*)^2$ onto $\mathbf{T}$, viz.,

$$(\xi, \eta) \mapsto \mathrm{diag}(1, \xi\eta, \xi^{-1}\eta^{-1}, 1, \eta^{-1}, \xi, \xi^{-1}, \eta).$$

This shows that $\mathbf{T}$ is a 2-dimensional torus.

We now prove that if $t \in \mathbf{G}$ commutes with all $t_{\lambda,\mu}$, then it lies in $\mathbf{T}$. Any such $t$ leaves the eigenspaces of every $t_{\lambda,\mu}$ invariant. From (2.4) we see that these eigenspaces are

$$K(e_{11}, 0) + K(e_{22}, 0), K(e_{12}, 0), K(e_{21}, 0) \quad \text{and all } K(0, e_{ij}),$$

if we choose $\lambda, \mu \in K^*$ in such a way that $t_{\lambda,\mu}$ has seven distinct eigenvalues. This implies that $t$ must leave $D_K$ invariant, so it has the form given by (2.2):

$$t : x + ya \mapsto cxc^{-1} + (pcyc^{-1})a \qquad (x, y \in D_K)$$

with $c, p \in D_K$, $\det(c) = \det(p) = 1$. Since $t$ has to leave the eigenspaces of every $t_{\lambda,\mu}$ invariant, its restriction to $D_K$, i.e., the mapping

$$D_K \to D_K, \; x \mapsto cxc^{-1},$$

has to leave $Ke_{12}$ and $Ke_{21}$ invariant. This implies that $ce_{12} \in Ke_{12}c$ and $ce_{21} \in Ke_{21}c$, so $c = c_\iota$ for some $\iota \in K^*$. In a similar way, the invariance of all $Ke_{ij}$ under the mapping

$$D_K \to D_K, y \mapsto pc_\iota yc_\iota^{-1},$$

implies that $p$ is diagonal. Thus we have proved that $t = t_{\iota,\kappa}$ for some $\iota, \kappa \in K^*$. □

It is easy now to determine the center of **G**.

**Lemma 2.3.2** *The center of* **G** *consists of the identity only.*

Proof. Since a central element commutes with all elements of the maximal torus **T**, it must lie in **T**, so it has the form

$$t_{\lambda,\mu} : x + ya \mapsto c_\lambda xc_\lambda^{-1} + (c_\mu yc_\lambda^{-1})a \qquad (x, y \in D_K),$$

using the notation of the previous lemma. It must commute with all automorphisms $t$ that leave $D_K$ invariant, which by (2.2) have the form

$$t : x + ya \mapsto cxc^{-1} + (pcyc^{-1})a \quad (x, y \in D_K)$$

with $c, p \in D_K$, $N(p) = 1$. This implies that $t_{\lambda,\mu}|_{D_K}$ has to commute with all inner automorphisms of $D_K$, the full $2 \times 2$ matrix algebra over $K$. This can only happen if $\lambda^2 = 1$, so $t_{\lambda,\mu}|_{D_K} = \mathrm{id}$. Since this holds for every quaternion subalgebra of $C_K$ and since every element of $C_K$ is contained in a quaternion subalgebra by Prop. 1.6.4, we conclude that $t_{\lambda,\mu} = \mathrm{id}$. □

In the next theorem, $C$ may be an arbitrary composition algebra. In fact, the proof for octonion algebras uses the case of quaternion algebras, which we therefore have to settle first.

**Theorem 2.3.3** *Let $C$ be any composition algebra over $k$. The only nontrivial invariant subspaces of $C_K$ under the action of* $\mathrm{Aut}(C_K)$ *are $Ke$ and $e^\perp$.*

Proof. If $\dim(C_K) = 2$, the result is trivial if $K$ has characteristic $\neq 2$. If $\mathrm{char}(K) = 2$, we have $Ke = e^{\perp}$. Let $a \notin Ke$ with $\langle e, a \rangle = 1$; set $N(a) = \alpha$. The element $e + a$ satisfies the same conditions, so by Prop. 1.2.3 it satisfies the same minimum equation as $a$. It follows that there exists an automorphism which carries $a$ to $e + a$, so $a$ cannot span another invariant subspace.

Next we deal with the case that $C_K$ is of dimension 4, so $C_K = M(2, K)$. Every automorphism of this algebra is inner by the Skolem-Noether Theorem (see § 2.1). Let $V$ be an invariant subspace of $C_K$ and let $t \in M(2, K)$ be a diagonal matrix with distinct nonzero eigenvalues. Then the eigenspaces of the linear map $x \mapsto txt^{-1}$ of $M(2, K)$ are $Ke_{11} + Ke_{22}$, $Ke_{12}$, $Ke_{21}$ (where the $e_{ij}$ are as in the proof of Lemma 2.3.1). So $V$ is spanned by vectors of the form $\alpha e_{11} + \beta e_{22}$ and multiples of $e_{12}$ and $e_{21}$.

If $e_{12} \in V$ then for all $\lambda \in K$

$$\begin{pmatrix} 1 & 0 \\ \lambda & 1 \end{pmatrix} e_{12} \begin{pmatrix} 1 & 0 \\ -\lambda & 1 \end{pmatrix} = \begin{pmatrix} -\lambda & 1 \\ -\lambda^2 & \lambda \end{pmatrix}$$

lies in $V$, whence $e_{21} \in V$, $e_{11} - e_{22} \in V$. It follows that then $V \supset e^{\perp}$. If $\alpha e_{11} + \beta e_{22} \in V$ and $\alpha \neq \beta$ then similar arguments give that $e_{12}$, $e_{21} \in V$. So if $e_{12}$ and $e_{21}$ do not lie in $V$, we must have $V = Ke$. This proves the Theorem for quaternion algebras.

Finally, let $C_K$ be an octonion algebra over $K$. Consider any subspace $V$ of $C_K$ that is invariant under $\mathrm{Aut}(C_K)$, and any quaternion subalgebra $D_1$. Let $V_1 = V \cap D_1$. Every automorphism of $D_1$ can be extended to an automorphism of $C_K$ by Cor. 1.7.3. So $V_1$ is invariant under $\mathrm{Aut}(D_1)$, whence $V_1 = 0$, $Ke$, $e^{\perp} \cap D_1$ or $D_1$. Let $D_2$ be another quaternion subalgebra of $C_K$; call $V \cap D_2 = V_2$. By Cor. 2.2.4 there exists an automorphism $\varphi$ of $C_K$ carrying $D_1$ to $D_2$. We must have $\varphi(V_1) = V_2$. Consequently, $V_2 = 0$, $Ke$, $e^{\perp} \cap D_2$ or $D_2$ according to whether $V_1 = 0$, $Ke$, $e^{\perp} \cap D_1$ or $D_1$. Since by Prop. 1.6.4 every element of $C_K$ is contained in a quaternion subalgebra, it follows that $V = 0$, $Ke$, $e^{\perp}$ or $C_K$.                                    □

Focussing attention on octonion algebras again, we derive the following important corollary to the above theorem.

**Corollary 2.3.4** *The automorphism group* **G** *of the octonion algebra* $C_K$ *over* $K$ *has a faithful irreducible representation.*

Proof. If $\mathrm{char}(K) \neq 2$, the 7-dimensional representation of **G** in $e^{\perp}$ is faithful and irreducible. If $\mathrm{char}(K) = 2$, this is not true anymore, since in that case $Ke$ is an invariant subspace contained in $e^{\perp}$. But then the representation of **G** in the 6-dimensional quotient space $e^{\perp}/Ke$ is irreducible, and it is not hard to verify that it is faithful.                                    □

We can now identify **G**.

**Theorem 2.3.5** *The algebraic group* **G** *defined by an octonion algebra* $C$ *is a connected, simple algebraic group of type* $G_2$.

Proof. We continue to work in the algebra $C_K$ over the algebraically closed field $K$. From Prop. 2.2.3 and Lemma 2.3.1 we know that **G** is a connected linear algebraic group of dimension 14 and rank 2. By the previous Corollary **G** has a faithful irreducible representation. This implies that **G** is reductive (see [Sp 81, Ex. 2.4.15]). As the center of $G$ is trivial by Lemma 2.3.2, $G$ is semisimple. From Prop. 2.2.3, Lemma 2.3.1 and [Sp 81, 8.1.3] we deduce that the two-dimensional root system of **G** has 12 elements. It then must be irreducible (the only reducible two-dimensional root system has 4 elements), and must be of type $G_2$ (see [Sp 81, 9.1] or [Bour, p.276]).                □

In Prop. 2.4.6 we will see that $k$ is a field of definition of the algebraic group $\mathbf{G} = \mathrm{Aut}(C)$ if $C$ is an octonion algebra over $k$.

## 2.4 Derivations and the Lie Algebra of the Automorphism Group

Let $C$ be an algebra over $k$ with identity element $e$. A *derivation* of $C$ is a linear map of $C$ such that

$$d(xy) = x.d(y) + d(x).y \quad (x, y \in C).$$

Taking $x = y = e$ it follows that $d(e) = 0$. If $d, d'$ are derivations an easy check shows that their commutator

$$[d, d'] = d \circ d' - d' \circ d$$

is also one. It follows that the derivations of $C$ form a Lie algebra $\mathrm{Der}(C)$ or $\mathrm{Der}_k(C)$. Our aim is to prove that if $C$ is an octonion algebra, $\mathrm{Der}_K(C)$ is the Lie algebra $L(\mathbf{G})$, where $G$ is as in the previous section. (From this it will follow that **G** is defined over $k$, see Prop. 2.4.6). For more on derivations of octonion algebras see [Ja 71, §2].

Let $C$ be an arbitrary composition algebra. To deal with $\mathrm{Der}(C)$ we use the doubling process of § 1.5. Let $D$ and $a$ be as in [loc.cit.]. If $d \in \mathrm{Der}(C)$ there are linear maps $d_0$ and $d_1$ of $D$ such that for $x \in D$

$$d(x) = d_0(x) + d_1(x)a.$$

**Lemma 2.4.1** *(i)* $d_0 \in \mathrm{Der}(D)$.
*(ii) For $x, y \in D$ we have $d_1(xy) = d_1(x)\bar{y} + d_1(y)x$.*

Proof. This follows by a straightforward computation, using the multiplication rules (1.22) and the definition of a derivation.                □

If $D$ is a two- or four-dimensional composition algebra denote by $S$ the space of linear maps $d_1$ of $D$ with the property of part (ii) of the lemma.

**Lemma 2.4.2** $\dim S \leq 2$ *if* $\dim D = 2$ *and* $\dim S \leq 8$ *if* $\dim D = 4$.

Proof. As in the case of derivations, we have $d_1(e) = 0$ for $d_1 \in S$. If $\dim D = 2$ this implies the stated inequality. If $\dim D = 4$ then $D$ is generated by two elements $a, b$ (see Cor. 1.6.3), and $d_1 \in S$ is completely determined by $d_1(a)$ and $d_1(b)$. The inequality follows.    □

$C$ and $D$ being as before, denote by $\mathrm{Der}_0(C)$ the space of derivations whose restriction to $D$ is zero.

**Lemma 2.4.3** $\dim \mathrm{Der}_0(C) \leq 1$ *(respectively, $\leq 3$) if* $\dim C = 4$ *(respectively, 8)*.

Proof. Let $d \in \mathrm{Der}_0(C)$. Then $d$ is determined by $da$. From the relations $a^2 = \lambda$, $ax = \bar{x}a$ ($x \in D$) we deduce that

$$(da)a + a(da) = 0, \quad (da)x - \bar{x}(da) = 0. \tag{2.5}$$

Write $da = b + ca$. From the second relation we deduce that $bx + \bar{x}b = 0$ for all $x \in D$. If $\dim D = 2$ then $D$ is commutative, and $(x + \bar{x})b = \langle x, e \rangle b = 0$ for all $x \in D$, whence $b = 0$. If $\dim D = 4$ then

$$b(xy) = -(\bar{y}\bar{x})b = -\bar{y}(\bar{x}b) = \bar{y}(bx) = -(by)x = -b(yx).$$

So $b$ annihilates all elements of the form $xy + yx$. We can conclude that $b = 0$ by producing an invertible element of this form, and it suffices to do this over the algebraic closure $K$. But then $D$ is isomorphic to the algebra of $2 \times 2$-matrices over $K$, and we may take $x = e_{12}$, $y = e_{21}$.

We have now shown that $da = ca$ and from the first relation (2.5) we find that $c + \bar{c} = 0$. The lemma follows.    □

**Lemma 2.4.4** $\mathrm{Der}(C) = 0$ *if* $C$ *is two-dimensional and* $\dim \mathrm{Der}(C) \leq 3$ *(respectively, $\leq 14$) if* $C$ *has dimension 4 (respectively, 8)*.

Proof. If $\dim C = 2$, then $C = k[a]$. Using (1.7) we find that if $d$ is a derivation

$$2a(da) - \langle a, e \rangle da = 0.$$

If char $k \neq 2$ we may assume that $\langle a, e \rangle = 0$ and otherwise that $\langle a, e \rangle = 1$. In both cases $da = 0$ and $d = 0$.

Now let $\dim C > 2$. We use the doubling process. $D$ being as before, it follows from Lemma 2.4.1 that we have an exact sequence

$$0 \to \mathrm{Der}_0(C) \to \mathrm{Der}(C) \to \mathrm{Der}(D) \oplus S.$$

Hence

$$\dim \mathrm{Der}(C) \leq \dim \mathrm{Der}_0(C) + \dim \mathrm{Der}(D) + \dim S.$$

If $\dim C = 4$ using Lemmas 2.4.2 and 2.4.3 we find that $\dim \mathrm{Der}(C) \leq 3$. (In fact we have equality since, as is easily seen, the derivation algebra contains the three-dimensional space of inner derivations $x \mapsto ax - xa$. But we are mainly interested in the octonion case.) If $C$ is an octonion algebra the same argument gives the desired inequality. □

We can now prove the results announced in the beginning.

**Proposition 2.4.5** *Let $C$ be an octonion algebra. Then $\dim \mathrm{Der}_k(C) = 14$ and $\mathrm{Der}_K(C)$ is the Lie algebra of $\mathbf{G}$.*

Proof. Since $\mathrm{Der}_K(C) = K \otimes_k \mathrm{Der}_k(C)$ and $\dim \mathrm{L}(\mathbf{G}) = \dim \mathbf{G} = 14$ (see Th. 2.3.5) it suffices to prove the second point. Let $V = \mathrm{End}(C_K \otimes_K C_K, C_K)$ be the space of $K$-linear maps of $C_K \otimes_K C_K$ to $C_K$. Denote by $\mu$ the multiplication map $x \otimes y \mapsto xy$. The algebraic group of automorphisms $\mathbf{G} = \mathrm{Aut}(C_K)$ is the stabilizer of $\mu$ in the algebraic group $\mathrm{GL}(C_K)$, under the obvious action of $\mathrm{GL}(C_K)$ on $V$, and the Lie algebra $\mathrm{L}(\mathbf{G})$ consists of derivations of $C_K$ (see [Hu, p.77]; one can also see this by working with the ring of dual numbers $K[\epsilon]$, as before in the proof of Prop. 2.2.2). Since $\dim L(\mathbf{G}) = \dim \mathbf{G} = 14$ (by Prop. 2.2.3), it follows from Lemma 2.4.4 that we have the desired equality. □

As a consequence of the proposition we have:

**Proposition 2.4.6** *Let $C$ be an octonion algebra over $k$. Then the automorphism group $\mathbf{G}$ is defined over $k$.*

Proof. We use the notations of the previous proof. $V(k) = \mathrm{End}_k(C \otimes_k C)$ is a $k$-structure on $V$ in the sense of [Sp 81, §11], $\mu \in V(k)$ and the algebraic group $\mathrm{GL}(C_K)$ is defined over $k$ (by [loc.cit., 2.1.5]). Let $\varphi : G \to V$ be the morphism $g \mapsto g.\mu$. It is defined over $k$. To prove that the stabilizer $\mathbf{G}$ is defined over $k$ it suffices by [loc.cit, 12.1.2] to prove that the kernel of the differential $d\varphi_e$ has dimension 14. But this kernel is precisely $\mathrm{Der}(C_K)$, as follows from [Hu, p. 77]. An application of the previous Proposition proves the assertion. □

## 2.5 Historical Notes

É. Cartan remarked in [Ca 14] that the compact real exceptional simple Lie group of type $G_2$ can be realized as the automorphism group of the real octonions; see [Fr 51] for an explicit proof. A similar result for Lie algebras of type $G_2$ over arbitrary fields of characteristic zero was proved independently by N. Jacobson [Ja 39] and E. Bannow (Mrs E. Witt) [Ba].

Generalization to groups over arbitrary base fields was a fairly obvious idea. L.E. Dickson defined analogs of the Lie group $G_2$ over arbitrary fields – without referring to octonion algebras – and proved simplicity over finite fields as early as 1901, after he had dealt with the classical groups; see [Di 01a] and [Di 05]. C. Chevalley refers to this work by Dickson in his Tôhoku paper [Che 55]. In [Ja 58], Jacobson made an extensive study of the groups of automorphisms of octonion algebras over fields of characteristic not two, and proved they are simple if the algebra is split, so in particular over algebraically closed fields and over finite fields.

# 3. Triality

In this chapter we deal with algebraic triality in the group of similarities and in the orthogonal group $O(N)$ of the norm $N$ of an octonion algebra $C$, and with the related triality in the Lie algebras of these groups, usually called local triality. Geometric triality on the quadric $N(x) = 0$ in case $N$ is isotropic will be left aside; the reader interested in the subject may consult [BlSp 60] and [Che 54, Ch. IV].

The results are proved here in all characteristics. In the existing literature, local triality has only been dealt with in characteristic not two. It turns out that for local triality in characteristic two one cannot use the Lie algebra $L(\mathbf{SO}(N))$ of the orthogonal group, but has to pass to the Lie algebra of the group of similarities, which is one dimension higher, or to a subalgebra of $L(\mathbf{SO}(N))$ of codimension one. The proof we give for the characteristic two case is different from that in the other cases.

Algebraic triality defines two outer automorphisms of the projective similarity group of $N$, which generate a group of outer automorphisms isomorphic to the symmetric group $S_3$. Further, we derive a characterization of the automorphism group of $C$ within the group of rotations that leave the identity element of $C$ invariant (cf. Cor. 2.2.5). We use triality to give an explicit description of the spin group of the norm of $C$, and we describe the outer automorphisms induced by triality in this spin group. Finally, we prove that the corresponding algebraic group is defined over the base field of $C$.

Bear in mind that $C$ will always be an octonion algebra in this chapter. In the first section we present material related to quadratic forms that we need in the remainder of the chapter.

## 3.1 Similarities. Clifford Algebras, Spin Groups and Spinor Norms

Let $N$ be a nondegenerate quadratic form on a vector space $V$ (of finite dimension) over a field $k$. We recall from § 1.1 that a *similarity* of $V$ with respect to $N$ is a linear transformation $t : V \to V$ such that

$$N(t(x)) = n(t)N(x) \qquad (x \in V)$$

for some $n(t) \in k^*$, called the *multiplier* of $t$. Then $\langle t(x), t(y) \rangle = n(t) \langle x, y \rangle$ $(x, y \in V)$, so $t$ is bijective. The similarities form a group $\mathrm{GO}(N)$ (or $\mathrm{GO}(N, V)$ if there is danger of confusion), called the *similarity group* of $N$, and $n$ is a homomorphism $\mathrm{GO}(N) \to k^*$. The kernel of $n$ is the orthogonal group $\mathrm{O}(N)$.

If $t$ is a similarity, so is $\lambda t$ for $\lambda \in k^*$, and $n(\lambda t) = \lambda^2 n(t)$. Since we will often have to deal with similarities up to a nonzero scalar multiple, it is useful to introduce the homomorphism

$$\nu : \mathrm{GO}(N) \to k^*/k^{*2}, \ t \mapsto n(t)k^{*2},$$

where $k^{*2}$ denotes the subgroup of squares in $k^*$. We call $\nu(t)$ the *square class* of the multiplier of $t$.

Now consider the case that $N$ is the norm of an octonion algebra $C$. The left multiplication $l_a$ by an element $a$ with $N(a) \neq 0$ is a similarity with $n(l_a) = N(a)$, and so is the right multiplication $r_a$. Any $t \in \mathrm{GO}(N)$ can be written as $t = l_a t'$ with $a = t(e)$ and $t' \in \mathrm{O}(N)$. It follows that every similarity is a product of a left multiplication and a number of reflections. Here we recall the convention of § 1.1 that for $\mathrm{char}(k) = 2$ we understand by a reflection an *orthogonal transvection*.

The remainder of this section is devoted to a digression on Clifford algebras, spin groups and the spinor norm. We will, to a large extent, only give an exposition of the definitions and results we need, referring to the literature for proofs: [Ar, Ch. V, § 4 and § 5], [Che 54, Ch. II], [Dieu, Ch. II, § 7 and § 10], [Ja 80, § 4.8], [KMRT, § 8].

Consider again an arbitrary vector space $V$ of finite dimension $n$ over a field $k$ with a nondegenerate quadratic form $N$. In the tensor algebra

$$\mathrm{T}(V) = k \oplus V \oplus (V \otimes_k V) \oplus \cdots$$

consider the two-sided ideal $I_N$ generated by the elements $x \otimes x - N(x)$ $(x \in V)$. The *Clifford algebra* of $N$ is the algebra $\mathrm{Cl}(N) = \mathrm{T}(V)/I_N$. (In the literature Clifford algebras are commonly denoted by a C, followed or not by something within parentheses. We use the notation Cl to avoid confusion with the $C$ denoting a composition algebra.) The canonical map of $V$, considered as a subspace of $\mathrm{T}(V)$, into $\mathrm{Cl}(N)$ is injective; one identifies $V$ with its image in $\mathrm{Cl}(N)$. For $x, y \in \mathrm{Cl}(N)$ we denote their product in $\mathrm{Cl}(N)$ by $x \circ y$, to avoid confusion with the product in $C$ if $\mathrm{Cl}(N)$ is the Clifford algebra of the norm $N$ on an octonion algebra $C$. As an algebra, $\mathrm{Cl}(N)$ is clearly generated by $V$, and one has the relations

$$x^{\circ 2} = N(x) \text{ and } x \circ y + y \circ x = \langle x, y \rangle \qquad (x, y \in V),$$

using the notation $x^{\circ 2} = x \circ x$. It follows that every $x \in V$ with $N(x) \neq 0$ has an inverse in $\mathrm{Cl}(N)$, viz. $x^{\circ -1} = N(x)^{-1}x$.

If $e_1, e_2, \ldots, e_n$ is a basis of $V$, a basis of $\mathrm{Cl}(N)$ is formed by the elements

$$e_{i_1} \circ e_{i_2} \circ \cdots \circ e_{i_h} \quad \text{with} \quad 1 \le i_1 < i_2 < \ldots < i_h \le n, \quad 0 \le h \le n,$$

so $\dim \mathrm{Cl}(N) = 2^n$. The elements $x \circ y$ with $x, y \in V$ generate a subalgebra $\mathrm{Cl}^+(N)$ of $\mathrm{Cl}(N)$, which has dimension $2^{n-1}$ and which is called the *even Clifford algebra* of $N$. This even Clifford algebra can also be described as follows. Take the even tensor algebra

$$\mathrm{T}^+(V) = k \oplus (V \otimes_k V) \oplus (V \otimes_k V \otimes_k V \otimes_k V) \oplus \cdots.$$

The ideal $I_N^+ = I_N \cap \mathrm{T}^+(V)$ is generated by the elements $x \otimes x - N(x)$ and $u \otimes x \otimes x \otimes v - N(x)(u \otimes v)$ $(u, v, x \in V)$. Now $\mathrm{Cl}^+(N) = \mathrm{T}^+(V)/I_N^+$.

The *Clifford group* of $N$ is the group $\Gamma(N)$ of invertible $u \in \mathrm{Cl}(N)$ such that $u \circ V \circ u^{\circ -1} = V$. The *even Clifford group* is $\Gamma^+(N) = \Gamma(N) \cap \mathrm{Cl}^+(N)$. For $u \in \Gamma(N)$, we define $t_u$ as the restriction to $V$ of conjugation by $u$:

$$t_u : V \to V, \ x \mapsto u \circ x \circ u^{\circ -1}.$$

This is an orthogonal transformation of $V$, since

$$N(u \circ x \circ u^{\circ -1}) = u \circ x \circ u^{\circ -1} \circ u \circ x \circ u^{\circ -1} = N(x).$$

Clearly, $t_{u \circ v} = t_u t_v$ for $u, v \in \Gamma(N)$. If $u \in V$ with $N(u) \neq 0$, then we have for $x \in V$,

$$u \circ x \circ u^{\circ -1} = N(u)^{-1} u \circ x \circ u =$$

$$N(u)^{-1}(-x \circ u^{\circ 2} + \langle x, u \rangle u) = -x + N(u)^{-1} \langle x, u \rangle u,$$

so $u \in \Gamma(N)$ and $t_u = -s_u$. Every rotation is an even product of reflections $s_{a_1} s_{a_2} \cdots s_{a_{2r}}$, so it is of the form $t_u$ with $u = a_1 \circ a_2 \circ \cdots \circ a_{2r} \in \Gamma^+(N)$, all $a_i \in V$, $N(a_i) \neq 0$. All elements of $\Gamma^+(N)$ are of this form, and $u \in \Gamma^+(N)$ is determined up to a factor in $k^*$ by $t_u$, since the intersection of the center of $\mathrm{Cl}(N)$ with $\Gamma^+(N)$ is $k^*$. In other words, there is an exact sequence

$$1 \to k^* \to \Gamma^+(N) \xrightarrow{\chi} \mathrm{SO}(N) \to 1, \tag{3.1}$$

where $\chi$ denotes the homomorphism $u \mapsto t_u$.

The *main involution* $\iota$ of $\mathrm{Cl}(N)$ is the anti-automorphism of order 2 defined by

$$\iota(x_1 \circ x_2 \circ \cdots \circ x_s) = x_s \circ \cdots \circ x_2 \circ x_1 \quad (x_1, x_2, \ldots, x_s \in V).$$

For $u = a_1 \circ \cdots \circ a_{2r} \in \Gamma^+(N)$, $(a_i \in V, N(a_i) \neq 0)$, define

$$N(u) = u \circ \iota(u) = N(a_1) \cdots N(a_{2r}) \in k^*.$$

It is easily verified that $N(u \circ u') = N(u)N(u')$ and $N(\lambda u) = \lambda^2 N(u)$. So

$$N : \Gamma^+(N) \to k^*, \ u \mapsto N(u),$$

is a homomorphism. Its kernel is called the *spin group* Spin($N$). Notice that the spin group is contained in the even Clifford algebra $\mathrm{Cl}^+(N)$.

Since $u \in \Gamma^+(N)$ is determined by $t_u$ up to a nonzero scalar factor, it makes sense to define the homomorphism

$$\sigma : \mathrm{SO}(N) \to k^*/k^{*2}, \; t_u \mapsto N(u)k^{*2}.$$

One calls $\sigma(t)$ the *spinor norm* of $t \in \mathrm{SO}(N)$. If $t = s_{a_1} \cdots s_{a_{2r}}$ with all $a_i \in V$, $N(a_i) \neq 0$, then $\sigma(t) = N(a_1) \cdots N(a_{2r})k^{*2}$. The kernel of $\sigma$ is the *reduced orthogonal group* $\mathrm{O}'(N)$. The homomorphism $\chi$ of (3.1) maps Spin($N$) onto $\mathrm{O}'(N)$.

Let us consider all these objects over the algebraic closure $K$ of $k$. The vector space $V$ with the quadratic form $N$ is replaced by $V_K = K \otimes_k V$ with the extension of $N$ to $V_K$, which we sometimes denote by $N_K$. Clearly, $\mathrm{Cl}(N_K) = K \otimes_k \mathrm{Cl}(N)$. The invertible elements of $\mathrm{Cl}(N_K)$ form an algebraic group, of which $\Gamma(N_K)$ is a closed subgroup. (From the fact that $\Gamma(N_K)$ is defined over $k$ – though we will not prove that here – it follows that $\Gamma(N)$ is the group of $k$-rational points of $\Gamma(N_K)$.) Similarly, we have the algebraic group $\Gamma^+(N_K)$, a closed subgroup of $\Gamma(N_K)$, and a closed subgroup **Spin**($N$) of $\Gamma^+(N_K)$. The exact sequence (3.1) (with $k$ replaced by $K$) is an exact sequence of algebraic groups, and $N : \Gamma^+(N_K) \to K^*$ is a homomorphism of algebraic groups. Every rotation has spinor norm 1 over $K$ since $K^* = K^{*2}$, so $\mathrm{O}'(N)$ coincides with $\mathrm{SO}(N)$ over an algebraically closed field. (Notice that, in general, $\mathrm{O}'(N)$ is not the group of rational points of an algebraic group.)

The algebraic group **Spin**($N$) is connected provided $\dim V \geq 2$; we show this in a similar way as in the proof of Prop. 2.2.2 for **SO**($N$). The polynomial function $N - 1$ on $V_K$ is irreducible, so $S = \{ x \in V \mid N(x) = 1 \}$ is an irreducible algebraic variety in $V$. The morphism

$$S \times S \to \mathbf{Spin}(N), \; (x, y) \mapsto x \circ y,$$

maps the irreducible variety $S \times S$ onto an irreducible set of generators of **Spin**($N$) containing the identity, so this must indeed be a connected algebraic group (cf. [Hu, § 7.5] or [Sp 81, Prop. 2.2.6]). (If $\dim V = 1$, then **Spin**($N$) = $\{ \pm 1 \}$.)

We have a homomorphism of algebraic groups $\pi : \mathbf{Spin}(N) \to \mathbf{SO}(N)$, where $\pi$ is the restriction of $\chi$ to **Spin**($N$) (so $\pi(a_1 \circ a_2 \circ \cdots \circ a_{2r}) = s_{a_1} s_{a_2} \cdots s_{a_{2r}}$), where **Spin**($N$) is the simply connected covering group of **SO**($N$) (cf. [Sp 81, 10.1.4]). The homomorphism $\pi$ is a separable isogeny with kernel of order 2 if $\mathrm{char}(K) \neq 2$; if $\mathrm{char}(K) = 2$, $\pi$ is an inseparable isogeny with kernel $\{ 1 \}$. We will discuss this in detail for the case that $N$ is the norm on an octonion algebra $C$, though in fact the argument will work for the general case. We assume that $k = K$ is algebraically closed. Then $N$ is the unique (up to equivalence) form of maximal Witt index in dimension 8. We denote the corresponding spin group by **Spin**(8), etc.

We have to determine explicitly the Lie algebra $L(\mathbf{Spin}(8))$ of $\mathbf{Spin}(8)$. To this end we introduce again the ring of dual numbers $K[\varepsilon]$ (as in the proof of Prop. 2.2.2), extend $N$ to a quadratic form on $C[\varepsilon] = K[\varepsilon] \otimes_K C$ and consider the corresponding Clifford algebra $K[\varepsilon] \otimes_K \mathrm{Cl}(N)$. The Lie algebra of $\Gamma^+(N_K)$ consists of elements $u \in \mathrm{Cl}^+(N_K)$ such that

$$(1 + \varepsilon u) \circ C[\varepsilon] \circ (1 + \varepsilon u)^{\circ -1} \subseteq C[\varepsilon],$$

that is,

$$(1 + \varepsilon u) \circ (x + \varepsilon y) \circ (1 + \varepsilon u)^{\circ -1} \in C[\varepsilon] \qquad (x, y \in C).$$

Since $(1 + \varepsilon u)^{\circ -1} = 1 - \varepsilon u$, this leads to

$$x + \varepsilon y + \varepsilon(u \circ x - x \circ u) \in C[\varepsilon] \qquad (x, y \in C).$$

Since $x + \varepsilon y \in C[\varepsilon]$, we get the condition

$$u \circ x - x \circ u \in C \qquad (x \in C). \tag{3.2}$$

Now first let $\mathrm{char}(K) \neq 2$. Choose an orthonormal basis $e_1, \ldots, e_8$ of $C$, write $x$ as a linear combination of these and $u = \sum_{l=0}^{4} \sum \alpha_{i_1 \ldots i_{2l}} e_{i_1} \circ \cdots \circ e_{i_{2l}}$, where the second sum is taken over all sequences $i_1, \ldots, i_{2l}$ with $0 \leq i_1 < \ldots < i_{2l} \leq 8$. One easily sees that (3.2) holds if and only if $u = \alpha_0 + \sum_{i<j} \alpha_{ij} e_i \circ e_j$. Since $\mathbf{Spin}(8)$ consists of the $z \in \Gamma^+(N_K)$ with $z \circ \iota(z) = 1$, we need for $L(\mathbf{Spin}(8))$ the extra condition $(1 + \varepsilon u) \circ \iota(1 + \varepsilon u) = 1$, which yields $u + \iota(u) = 0$. Since $u + \iota(u) = 2\alpha_0$, we get $\alpha_0 = 0$, so $L(\mathbf{Spin}(8))$ is contained in the space $S$ of the elements $u = \sum_{i<j} \alpha_{ij} e_i \circ e_j$. Since $\dim S = 28 = \dim L(\mathbf{Spin}(8))$, we have $L(\mathbf{Spin}(8)) = S$. The Lie action of $u$ on $C$, $x \mapsto u \circ x - x \circ u$, has matrix $\left( \gamma_{ij} \right)_{1 \leq i,j \leq 8}$ with $\gamma_{ii} = 0$, $\gamma_{ij} = 2\alpha_{ij}$ for $i < j$ and $\gamma_{ji} = -2\alpha_{ij}$ for $i < j$. So we see that $d\pi : L(\mathbf{Spin}(8)) \to L(\mathbf{SO}(8))$ is an isomorphism if the characteristic $\neq 2$, which means that $\pi$ is a separable isogeny (cf. [Sp 81, Th. 4.3.7]).

In the case $\mathrm{char}(K) = 2$ we have to use a symplectic basis $e_1, \ldots, e_8$ for $C$ (cf. (1.30)). Proceeding as above we find again that the elements $u$ that satisfy condition (3.2) are those of the form $u = \alpha_0 + \sum_{i<j} \alpha_{ij} e_i \circ e_j$. In this case, writing out the extra condition $u + \iota(u) = 0$ for the elements $u \in L(\mathbf{Spin}(8))$ yields

$$\sum_{i=1}^{4} \alpha_{i,i+4} = 0. \tag{3.3}$$

The Lie action of $u = \alpha_0 + \sum_{i<j} \alpha_{ij} e_i \circ e_j$ satisfying (3.3) on $C_K$, $x \mapsto u \circ x - x \circ u$, has matrix $\left( \gamma_{ij} \right)_{1 \leq i,j \leq 8}$ with entries as follows:

$$\gamma_{ij} = \gamma_{j+4,i+4} = \alpha_{i,j+4} \qquad (1 \leq i,j \leq 4),$$
$$\gamma_{i,i+4} = \gamma_{i+4,i} = 0 \qquad (1 \leq i \leq 4),$$
$$\gamma_{i,j+4} = \gamma_{j,i+4} = \alpha_{ij} \qquad (1 \leq i < j \leq 4),$$
$$\gamma_{i+4,j} = \gamma_{j+4,i} = \alpha_{i+4,j+4} \qquad (1 \leq i < j \leq 4).$$

It follows that the kernel of $d\pi$ is $K$ (considered as a subspace of $\mathrm{Cl}(N_K)$). Its image is a 27-dimensional subalgebra of the 28-dimensional Lie algebra $L(\mathbf{SO}(8))$, because of condition (3.3). Thus we see that $\pi$ is an inseparable isogeny if $\mathrm{char}(K) = 2$ (cf. [Sp 81, § 9.6]).

## 3.2 The Principle of Triality

The following theorem is the central result of this chapter.

**Theorem 3.2.1 (Principle of Triality)** *Let $C$ be an octonion algebra over a field $k$, with norm $N$.*
*(i) The elements $t_1 \in \mathrm{GO}(N)$ such that there exist $t_2, t_3 \in \mathrm{GO}(N)$ with*

$$t_1(xy) = t_2(x)t_3(y) \qquad (x, y \in C) \tag{3.4}$$

*form a normal subgroup $\mathrm{SGO}(N)$ of index 2, called the special similarity group. If $(t_1, t_2, t_3)$ and $(t_1', t_2', t_3')$ satisfy (3.4), then so do $(t_1 t_1', t_2 t_2', t_3 t_3')$ and $(t_1^{-1}, t_2^{-1}, t_3^{-1})$.*
*(ii) If $t_1 \in \mathrm{GO}(N)$, then there exist $t_2, t_3 \in \mathrm{GO}(N)$ such that*

$$t_1(xy) = t_2(y)t_3(x) \qquad (x, y \in C) \tag{3.5}$$

*if and only if $t_1 \notin \mathrm{SGO}(N)$.*
*(iii) The elements $t_2$ and $t_3$ in (3.4) and in (3.5) are uniquely determined by $t_1$ up to scalar factors $\lambda$ and $\lambda^{-1}$ in $k^*$, respectively. For $t_1 = l_a$, the similarities $t_2 = l_a r_a$ and $t_3 = l_{a^{-1}} = N(a)^{-1} l_{\bar{a}}$ satisfy (3.4), and similarly with $t_1 = r_a$, $t_2 = r_{a^{-1}} = N(a)^{-1} r_{\bar{a}}$, $t_3 = l_a r_a$; for $t_1 = s_a$, (3.5) is satisfied by $t_2 : y \mapsto -N(a)^{-1} a\bar{y}$, and $t_3 : x \mapsto \bar{x}a$; for $t_1 = s_a s_b$ we can take $t_2 = N(a)^{-1} l_a l_{\bar{b}}$ and $t_3 = N(b)^{-1} r_a r_{\bar{b}}$ to satisfy (3.4). If $t_1 \in \mathrm{SGO}(N)$, then so are $t_2$ and $t_3$.*
*(iv) If $t_1, t_2, t_3$ are as in (3.4) or (3.5), then $n(t_1) = n(t_2)n(t_3)$.*
*(v) Let $t \in \mathrm{GO}(N)$; it is a product of a left multiplication by an invertible element of $C$ and an orthogonal transformation $t'$. Then $t \in \mathrm{SGO}(N)$ if and only if $t'$ is a rotation.*
*(vi) If $t_1, t_2, t_3$ are bijective linear transformations of $C$ that satisfy (3.4) or (3.5), then they are necessarily similarities with respect to $N$.*

Proof. Call $t_1 \in \mathrm{GO}(N)$ *even* or *odd* if there exist $t_2, t_3 \in \mathrm{GO}(N)$ such that (3.4) or (3.5) holds, respectively.
(a) If two triples $(t_1, t_2, t_3)$ and $(t_1', t_2', t_3')$ of invertible linear transformations satisfy (3.4), then so does their product $(t_1 t_1', t_2 t_2', t_3 t_3')$, since

$$t_1 t_1'(xy) = t_1(t_2'(x)t_3'(y)) = t_2 t_2'(x)t_3 t_3'(y).$$

If the first triple satisfies (3.4) and the second one (3.5), then their product satisfies (3.5). If the first triple satisfies (3.5) and the second one (3.4), then

$(t_1 t_1', t_2 t_3', t_3 t_2')$ satisfies (3.5), and this last triple satisfies (3.4) if both initial triples satisfy (3.5). Finally, if $(t_1, t_2, t_3)$ satisfies (3.4), then the same is true for $(t_1^{-1}, t_2^{-1}, t_3^{-1})$. For, substituting in (3.4) $t_2^{-1}(x)$ for $x$ and $t_3^{-1}(y)$ for $y$ yields

$$t_1(t_2^{-1}(x)t_3^{-1}(y)) = xy,$$

from which we infer

$$t_1^{-1}(xy) = t_2^{-1}(x)t_3^{-1}(y) \qquad (x, y \in C).$$

So the product of two even or two odd similarities is even and that of an even and an odd similarity is odd, while the inverse of an even similarity is even. (b) We show now that a similarity cannot be both even and odd. For assume that $(t_1, t_2, t_3)$ satisfies (3.4) and at the same time $(t_1, t_2', t_3')$ satisfies (3.5). Then, by the results of (a), $(\mathrm{id}, u_2, u_3)$ with $u_2 = t_2' t_3^{-1}$ and $u_3 = t_3' t_2^{-1}$ would satisfy (3.5). Thus,

$$xy = u_2(y)u_3(x) \qquad (x, y \in C).$$

Taking $y = e$ we get $x = u_2(e)u_3(x)$, so $u_3 = l_a$ with $a = u_2(e)^{-1}$; taking $x = e$ we get $u_2 = r_b$ with $b = u_3(e)^{-1}$. Thus,

$$xy = (yb)(ax) \qquad (x, y \in C).$$

For $x = y = e$ this yields $e = ba$, so $b = a^{-1}$. If we take $y = a$, we get

$$xa = ax \qquad (x \in C).$$

This means that $a$ belongs to the center of $C$, so $a = \lambda e$ for some $\lambda \in k$ by Prop. 1.9.1, and $b = \lambda^{-1}e$. Thus we arrive at the relation

$$xy = yx \qquad (x, y \in C).$$

Since $C$ is not commutative, we have a contradiction.
(c) Every similarity is a product of a left multiplication by an invertible element and a number of reflections, so to show that we have either (3.4) or (3.5), it suffices by (a) to consider the cases $t_1 = l_a$ and $t_1 = s_a$ for $a \in C$ with $N(a) \neq 0$.

For $t_1 = l_a$ we deduce from the second Moufang identity (1.14) that

$$a(xy) = (axa)(a^{-1}y) \qquad (x, y \in C).$$

Hence if we take $t_2 = l_a r_a$ and $t_3 = l_{a^{-1}}$, then $(t_1, t_2, t_3)$ satisfies (3.4). If we conjugate the above relation and then replace $\bar{a}$ by $a$, $\bar{y}$ by $x$ and $\bar{x}$ by $y$, we get

$$(xy)a = (xa^{-1})(aya) \qquad (x, y \in C),$$

so for $t_1 = r_a$ we find $t_2 = r_{a^{-1}}$ and $t_3 = l_a r_a$ so that $(t_1, t_2, t_3)$ satisfies (3.4).

Next consider a reflection $s_a : x \mapsto x - N(a)^{-1}\langle x, a \rangle a$. Since $\langle x, a \rangle e = x\bar{a} + a\bar{x}$,

$$s_a(x) = x - x - N(a)^{-1}a\bar{x}a = -N(a)^{-1}a\bar{x}a.$$

Using the first Moufang identity (1.13) we get:

$$\begin{aligned} s_a(xy) &= -N(a)^{-1}a(\bar{y}\bar{x})a \\ &= -N(a)^{-1}(a\bar{y})(\bar{x}a) \qquad (x, y \in C), \end{aligned}$$

so if we take $t_1 = s_a$, $t_2 : y \mapsto -N(a)^{-1}a\bar{y}$, and $t_3 : x \mapsto \bar{x}a$, we have a triple of similarities which satisfies (3.5).

Combining this with the results of (a) and (b), we find the statements (i), (ii) and (v) of the Theorem and the statements in the second and third sentence of (iii). The normality of $SGO(N)$ in $GO(N)$ follows from the fact that it has index 2. The relation in (iv) is easily proved by taking norms on both sides of (3.4) and (3.5).

(d) We now prove the uniqueness of $t_2, t_3$ up to a scalar factor. Assume $(t_1, t_2, t_3)$ and $(t_1, t_2', t_3')$ both satisfy (3.4), then with $u_i = t_i^{-1}t_i'$ $(i = 1, 2)$ we have by (a),

$$xy = u_2(x)u_3(y) \qquad (x, y \in C).$$

Taking $y = e$ we find $x = u_2(x)u_3(e)$, so $u_2 = r_a$ with $a = u_3(e)^{-1}$, and similarly $u_3 = l_b$ with $b = u_2(e)^{-1}$. Hence

$$xy = (xa)(by) \qquad (x, y \in C).$$

Substitution of $x = y = e$ yields $ab = e$, so $b = a^{-1}$. Replacing $y$ by $ay$ we find

$$x(ay) = (xa)y \qquad (x, y \in C).$$

By Prop. 1.9.2, $a = \lambda e$ for some $\lambda \in k$, so $b = \lambda^{-1}e$. This implies that $t_2' = \lambda t_2$ and $t_3' = \lambda^{-1}t_3$.

Now let $(t_1, t_2, t_3)$ satisfy (3.5). Define similarities $u_i$ by $u_i(x) = t_i(\bar{x})$. Then $(u_1, u_2, u_3)$ satisfies (3.4). The uniqueness of $u_2$ and $u_3$ up to factors $\lambda$ and $\lambda^{-1}$, respectively, implies the same for $t_2$ and $t_3$.

(e) To prove (vi), finally, consider arbitrary bijective linear transformations $t_1, t_2, t_3$ of $C$ that satisfy (3.4). Taking $y = e$ in (3.4), we get

$$t_1(x) = t_2(x)t_3(e) \qquad (x \in C).$$

As $t_1$ is bijective, there exists $x \in C$ such that $N(t_1(x)) \neq 0$; since $N(t_1(x)) = N(t_2(x))N(t_3(e))$, we see that $N(t_3(e)) \neq 0$, so $t_3(e)$ is invertible. Hence $t_2 = r_a t_1$ with $a = t_3(e)^{-1}$. By substituting $x = e$ in (3.4), we find in a similar way that $t_3 = l_b t_1$ with $b = t_2(e)^{-1}$. We can now rewrite (3.4) in the following form:

$$t_1(xy) = (t_1(x)a)(bt_1(y)) \qquad (x, y \in C). \tag{3.6}$$

From $t_1(e) = t_2(e)t_3(e)$ it follows that $t_1(e)$ is invertible. Set $c = t_1(e)^{-1}$. The similarity $t_1' = l_c$ can be completed to a triple of similarities $(t_1', t_2', t_3')$ satisfying (3.4) as we saw in (c) above. Then the product triple $(t_1't_1, t_2't_2, t_3't_3)$ also satisfies (3.4) as we saw in (a). Now $t_1't_1(e) = l_ct_1(e) = e$, so for the proof of (vi) we may as well assume that $t_1(e) = e$. Taking $x = y = e$ in (3.6), we find $ab = e$. Taking norms of both sides of (3.6), we get, since $N(a)N(b) = N(e) = 1$,

$$N(t_1(xy)) = N(t_1(x))N(t_1(y)).$$

This means that the nondegenerate quadratic form $\tilde{N}$ on $C$ defined by $\tilde{N}(x) = N(t_1(x))$ $(x \in C)$ permits composition. This implies by Cor. 1.2.4 that $\tilde{N} = N$, so $t_1$ is orthogonal. Then $t_2$ and $t_3$ must be similarities, since each of them is a product of $t_1$ and a similarity.

A similar proof shows that bijective linear transformations $t_1, t_2, t_3$ that satisfy (3.5) must be similarities. □

The elements of $\mathrm{SGO}(N)$ are called *proper similarities*, the other similarities *improper*. For $t_1 \in \mathrm{GO}(N)$ we fix the notation $t_2$ and $t_3$ by the rule that the triple $(t_1, t_2, t_3)$ satisfies either (3.4) or (3.5); such a triple $(t_1, t_2, t_3)$ is said to be *related*. Bear in mind that $t_2$ and $t_3$ are determined by $t_1$ up to factors $\lambda$ and $\lambda^{-1}$, respectively. Equation (3.4) is called the *first form of triality* and (3.5) the *second form of triality*. Consider a related triple $(t_1, t_2, t_3)$; if $t_1 \in \mathrm{SO}(N)$, it satisfies the first form of triality, and hence $t_2, t_3 \in \mathrm{SGO}(N)$. With the aid of the spinor norm we can determine when we can take $t_2$ and $t_3$ to be orthogonal.

**Proposition 3.2.2** *For $t_1 \in \mathrm{SO}(N)$, the similarities $t_2$ and $t_3$ such that $(t_1, t_2, t_3)$ satisfies (3.4) have square class of multiplier $\nu(t_i) = \sigma(t_1)$. Hence $t_2$ and $t_3$ can be taken to be orthogonal if and only if $\sigma(t_1) = 1$; in that case they are both rotations.*

Proof. The equality $\nu(t_i) = \sigma(t_1)$ follows from the case $t_1 = s_a s_b$ in (iii) of Th. 3.2.1. We have also seen there that $t_2, t_3 \in \mathrm{SGO}(N)$, so if we take them to be orthogonal, they are rotations by (v) of the same theorem. □

**Remark 3.2.3** If $k^* = k^{*2}$, which is for instance the case if $k$ is algebraically closed, every rotation has spinor norm 1. Then for every $t_1 \in \mathrm{SO}(N)$ one can find $t_2, t_3 \in \mathrm{SO}(N)$, unique up to a common factor $\pm 1$, such that $(t_1, t_2, t_3)$ satisfies the first form of triality (3.4), so the Principle of Triality holds in this form for $\mathrm{SO}(N)$ over algebraically closed fields, more generally over fields in which every element is a square.

## 3.3 Outer Automorphisms Defined by Triality

For $t \in \mathrm{GO}(N)$ we define $\hat{t} \in \mathrm{GO}(N)$ by

$$\hat{t}(x) = n(t)^{-1}\overline{t(\bar{x})} \qquad (x \in C).$$

**Lemma 3.3.1** *For $t \in GO(N)$ we have $n(\hat{t}) = n(t)^{-1}$ and*

$$t(x^{-1}) = (\hat{t}(x))^{-1} \qquad (x \in C, \ N(x) \neq 0).$$

*Further,*

$$\widehat{tu} = \hat{t}\hat{u} \qquad and \qquad \hat{\hat{t}} = t,$$

*so $t \mapsto \hat{t}$ is an involutory automorphism of $GO(N)$.*

Proof. Easy verifications.                                    □

**Lemma 3.3.2** *Let $t_1, t_2, t_3 \in SGO(N)$. If $(t_1, t_2, t_3)$ satisfies the first form of triality, then so do $(t_2, t_1, \hat{t}_3)$, $(t_3, \hat{t}_2, t_1)$, $(\hat{t}_1, \hat{t}_3, \hat{t}_2)$, $(\hat{t}_2, t_3, \hat{t}_1)$ and $(\hat{t}_3, \hat{t}_1, t_2)$.*

Proof. From

$$t_1((xy)\bar{y}) = t_2(xy)t_3(\bar{y})$$

we infer for $N(y) \neq 0$,

$$N(y)t_1(x) = t_2(xy)N(y)t_3(y^{-1}) = N(y)t_2(xy)(\hat{t}_3(y))^{-1}.$$

Hence

$$t_2(xy) = t_1(x)\hat{t}_3(y) \qquad (x, y \in C, \ N(y) \neq 0).$$

Working over $K$, we obtain a polynomial equality which is valid on a Zariski open subset of $C_K \times C_K$. It must hold for all $x, y \in C_K$ and in particular for $x, y \in C$. So $(t_2, t_1, \hat{t}_3)$ is a related triple. In a similar way one proves that $(t_3, \hat{t}_2, t_1)$ is related.

Thus, if we interchange in a related triple satisfying the first form of triality the first and the second (or the third) component and replace the remaining component $t$ by $\hat{t}$, we get another related triple satisfying the first form of triality. Applying this operation a few times in a suitable way, we get the other three related triples.                                    □

**Corollary 3.3.3** *If $t$ is a rotation, then $\sigma(\hat{t}) = \sigma(t)$. If $t_1$ is a rotation with $\sigma(t_1) = 1$ and $(t_1, t_2, t_3)$ is a related triple of rotations, then $\sigma(t_2) = \sigma(t_3) = 1$.*

Proof. If $t$ is a rotation, then $\hat{t}(x) = \overline{t(\bar{x})}$; since conjugation is the negative of the reflection $s_e$, $\hat{t}$ is a rotation with the same spinor norm as $t$. The second statement follows by combining the above lemma with Prop. 3.2.2.                                    □

In dealing with similarities of the octonion algebra $C$ up to a scalar factor, it is natural to interpret them as transformations of the set $\mathbb{P}(C)(k)$ of $k$-rational points of the seven-dimensional projective space $\mathbb{P}(C) = (C \backslash \{0\})/K^*$. These transformations form the *projective similarity group* $PGO(N) =$

$GO(N)/k^*$. The image of $SGO(N)$ in $PGO(N)$ under the natural projection is the *projective special similarity group* $PSGO(N) = SGO(N)/k^*$. For $t \in GO(N)$, we identify its image $tk^*$ in $PGO(N)$ with the projective transformation it induces in $\mathbb{P}(C)$, and we denote both by $[t]$. Every $[t_1] \in PSGO(N)$ uniquely determines $[t_2], [t_3] \in PSGO(N)$ such that $(t_1, t_2, t_3)$ is a related triple.

**Proposition 3.3.4** *The mappings* $\alpha, \beta, \varepsilon : PSGO(N) \to PSGO(N)$,

$$\alpha : [t_1] \mapsto [t_2],$$
$$\beta : [t_1] \mapsto [\hat{t}_3],$$
$$\varepsilon : [t] \mapsto [\hat{t}],$$

*are automorphisms of* $PSGO(N)$. *The automorphisms* $\alpha$ *and* $\beta$ *generate a group* $S$ *of outer automorphisms of* $PSGO(N)$, *which contains* $\varepsilon$, *and which is isomorphic to the symmetric group* $S_3$.

Proof. That $\alpha$ and $\beta$ are homomorphisms follows from part (i) of Th. 3.2.1 and Lemma 3.3.1, and their bijectivity follows from Lemma 3.3.2.

From Lemma 3.3.2 it also follows that

$$\alpha : \quad [t_1] \leftrightarrow [t_2], \quad [t_3] \leftrightarrow [\hat{t}_2], \quad [\hat{t}_1] \leftrightarrow [\hat{t}_3],$$

and

$$\beta : \quad [t_1] \mapsto [\hat{t}_3] \mapsto [\hat{t}_2] \mapsto [t_1], \quad [t_2] \mapsto [t_3] \mapsto [\hat{t}_1] \mapsto [t_2].$$

This implies $\varepsilon = \alpha\beta$. Further one derives the relations

$$\alpha^2 = \beta^3 = (\alpha\beta)^2 = 1,$$

denoting by 1 the identity automorphism. These are the defining relations for the symmetric group $S_3$, if one takes as generators (1 2) and (1 2 3). So there exists a homomorphism

$$\tau : S_3 \to S, \ (1\,2) \mapsto \alpha \text{ and } (1\,2\,3) \mapsto \beta.$$

The only nontrivial normal subgroup of $S_3$ is the alternating group $A_3$. Since $(123) \in A_3$ is not mapped onto 1 by $\tau$, we see that $\tau$ must be an isomorphism.

There remains the proof that all elements $\neq 1$ of $S$ are outer automorphisms. Let $A$ be the quotient of the group of automorphisms of $PSGO(N)$ by the group of inner automorphisms. We have to show that the homomorphism $S_3 \to A$ induced by $\tau$ is injective. If not, its kernel would contain the alternating group $A_3$, and $\beta$ would be an inner automorphism. If this were the case, there would exist $u \in SGO(N)$ such that

$$[ut_1] = \beta([t_1])[u] \qquad (t_1 \in SGO(N)).$$

If $t_1 = l_a$, then $\beta([t_1]) = [\hat{t}_3]$ with $t_3 = N(a)^{-1}l_{\bar{a}}$ by Th. 3.2.1 (iii). Writing this out we see that for each $a \in C$ with $N(a) \neq 0$ there is $\lambda_a \in k$ such that

$$N(a)u(ax) = \lambda_a u(x)a \qquad (x \in C). \qquad (3.7)$$

Taking $x = u^{-1}(a^{-1})$ we find

$$\lambda_a = u(a(u^{-1}(\bar{a}))) \qquad (a \in C, \; N(a) \neq 0). \qquad (3.8)$$

Taking norms of both sides of (3.7) we get

$$N(a)^2 n(u)N(a) = \lambda_a^2 n(u)N(a),$$

from which we infer $\lambda_a^2 = N(a)^2$, so $\lambda_a = \pm N(a)$ for $a \in C, N(a) \neq 0$. We may assume that $k$ is algebraically closed. Since $a \mapsto \lambda_a$ is a polynomial function on $C$ by (3.8) and since $\lambda_e = 1$, we must have $\lambda_a = N(a)$ if $N(a) \neq 0$. By Zariski continuity this must hold for all $a \in C$. Using Prop. 1.9.2 we conclude that $c \in k^*e$. Consequently, $\beta = \mathrm{Inn}(u) = \mathrm{id}$, which is a contradiction.  □

**Remark 3.3.5** Since $\varepsilon \in \mathcal{S}$, it is an outer automorphism of $\mathrm{PSGO}(N)$. One can extend $\varepsilon$ to an automorphism $[t] \mapsto [\hat{t}]$ of $\mathrm{PGO}(N)$, which is an inner automorphism since in $\mathrm{GO}(N)$ we have

$$\hat{t} = n(t)^{-1}ctc^{-1},$$

where $c$ is the similarity $x \mapsto \bar{x}$. Notice that $c$ satisfies (3.5), the second form of triality, so $c \notin \mathrm{SGO}(N)$. Since $\varepsilon$ is an outer automorphism of $\mathrm{PSGO}(N)$, we have $[c] \notin \mathrm{PSGO}(N)$, so $\mathrm{PSGO}(N) \neq \mathrm{PGO}(N)$.

## 3.4 Automorphism Group and Rotation Group of an Octonion Algebra

Let $C$ be an octonion algebra over $k$. We saw in § 2.2 that the algebraic group $\mathbf{G}$ is a closed subgroup of the stabilizer $\mathbf{SF}$ of $e$ in $\mathbf{SO}(N)$, which also leaves $e^\perp$ invariant, and that restriction of elements of $\mathbf{SF}$ to $e^\perp$ defines an isomorphism $\varrho$ of $\mathbf{SF}$ onto $\mathbf{SO}(N_1)$; see Prop. 2.2.2. We will henceforth identify the elements of $\mathbf{SF}$ with their $\varrho$-images in $\mathbf{SO}(N_1)$, and conversely identify every $t \in \mathbf{SO}(N_1)$ with its extension $\varrho^{-1}(t)$ to a rotation fixing $e$. Thus, we have identified $\mathbf{G}$ with a closed subgroup of $\mathbf{SO}(N_1)$. Similarly, we identify $G = \mathrm{Aut}(C)$ with a subgroup of $\mathrm{SO}(N_1)$.

For $x, y \in C$, both $\neq 0$, we denote $k^*x \in \mathbb{P}(C)(k)$ by $[x]$, and sometimes write $[x].[y]$ for $[xy]$. For $t_1 \in \mathrm{SO}(N_1)$ there exist similarities $t_2, t_3$, unique up to scalar factors $\lambda$ and $\lambda^{-1}$, respectively, such that $(t_1, t_2, t_3)$ satisfies the first form of triality (3.4). We define a mapping $\Delta$ from $\mathrm{SO}(N_1)$ to $\mathbb{P}(C)(k)$ by

$$\Delta : \mathrm{SO}(N_1) \to \mathbb{P}(C)(k), \quad t_1 \mapsto [t_2(e)].$$

By Lemma 3.3.2, $(t_2, t_1, \hat{t}_3)$ also satisfies the first form of triality. This implies $\hat{t}_3(e) = t_2(e)$ since $t_1(e) = e$. Using all this, we find for the map $\Delta$:

$$\Delta(t_1 u_1) = [t_2 u_2(e)] = [t_2(u_2(e)e)] = [t_1(u_2(e))\hat{t}_3(e)] = [t_1(u_2(e))].[t_2(e)].$$

Thus,

$$\Delta(tu) = [t](\Delta(u)).\Delta(t) \qquad (t, u \in \mathrm{SO}(N_1)).$$

If $t_1 \in \mathrm{Aut}(C)$, then $t_1 = t_2 = t_3$, so $\Delta(t_1) = [e]$. This necessary condition is also sufficient as we see in the following proposition.

**Proposition 3.4.1** *For $t \in \mathrm{SO}(N_1)$ we have $t \in \mathrm{Aut}(C)$ if and only if $\Delta(t) = [e]$.*

Proof. It remains only to prove the "if" part. If $t_1 \in \mathrm{SO}(N_1)$ and $\Delta(t_1) = [e]$, we may assume $t_2(e) = e$. From (3.4) with $x = e$ it follows that $t_1(y) = t_3(y)$ for all $y$. Then $t_3(e) = t_1(e) = e$, hence taking $y = e$ in (3.4) yields that $t_1(x) = t_2(x)$ for all $x$. Thus we have found that $t_1 = t_2 = t_3$, which means that $t_1 \in \mathrm{Aut}(C)$.                                                       □

If $t_1 \in \mathrm{SO}(N_1)$, $t_2(e)$ has nonzero norm. We claim that every element of $C$ with nonzero norm can be obtained in this way.

**Lemma 3.4.2** *For every $c \in C$ with $N(c) \neq 0$ there exists $t_1 \in \mathrm{SO}(N_1)$ such that $t_2(e) = c$.*

Proof. We try $t_1 = s_a s_b$ with $a, b \in e^\perp$. Then $t_1(e) = e$, so $t_1 \in \mathrm{SO}(N_1)$. Pick any $a \in C$ with $\langle a, e \rangle = \langle a, c \rangle = 0$ and $N(a) \neq 0$. Take $b = \lambda ac$ with $\lambda \in k^*$ still to be chosen. Then $N(b) \neq 0$ and

$$\langle b, e \rangle = \lambda \langle ac, e \rangle = \lambda \langle c, \bar{a} \rangle = -\lambda \langle c, a \rangle = 0.$$

By Th. 3.2.1 (iii), we can take $t_2 = l_a l_{\bar{b}}$. Then

$$t_2(e) = a\bar{b} = -ab = -\lambda aac = \lambda a\bar{a}c = \lambda N(a)c.$$

So if we take $\lambda = N(a)^{-1}$, we have $t_2(e) = c$ as desired.                        □

We define an action $T$ of $\mathrm{SO}(N_1)$ in $\mathbb{P}(C)(k)$ by $T(t) = \alpha([t])$ with $\alpha$ as in Prop. 3.3.4, so $T(t_1) = [t_2]$. If $T(t) = [\mathrm{id}]$, then $[t] = [\mathrm{id}]$ since $\alpha$ is an automorphism; this implies that $t = \mathrm{id}$ since $t \in \mathrm{SO}(N_1)$. Thus, $T$ is faithful. For $k = K$ we have obtained an action of $\mathbf{SO}(N_1)$ on $\mathbb{P}(C)$. Consider in $\mathbb{P}(C)$ the Zariski open subset

$$\mathcal{O} = \{\, [c] \mid N(c) \neq 0 \,\}.$$

It is the complement of a projective quadric which is defined over $k$. Lemma 3.4.2 shows that $T(\mathrm{SO}(N_1))$ operates transitively on the set $\mathcal{O}(k)$ of $k$-rational points of $\mathcal{O}$. According to Prop. 3.4.1, $t \in \mathrm{SO}(N_1)$ is an automorphism of $C$ if and only if $T(t)$ fixes $[e]$, so $G = \mathrm{Aut}(C)$ is the stabilizer of $[e]$. Working over $K$ we see that $T$ induces a bijective morphism $\tau$ of algebraic varieties from the homogeneous space $\mathbf{SO}(N_1)/\mathbf{G}$ onto $\mathcal{O}$. Counting dimensions yields

$$\dim \mathbf{SO}(N_1)/\mathbf{G} = 21 - \dim \mathbf{G} = \dim \mathcal{O} = \dim \mathbb{P}(C) = 7$$

(use [Sp 81, Th. 5.1.6]), from which we derive again that $\dim \mathbf{G} = 14$. By Prop. 2.4.6 we know that $\mathbf{G}$ is defined over $k$.

The group $\mathbf{SO}(N_1)$ is the rotation group of a 7-dimensional quadratic form, it is a quasisimple algebraic group of type $B_3$ (see [Sp 81, 17.2.1 and 17.2.2]) and $\mathbf{G}$ is a closed algebraic subgroup which is an exceptional simple group of type $G_2$. We have found a bijective morphism $\tau$ of $\mathbf{SO}(N_1)/\mathbf{G}$ onto a Zariski open subset $\mathcal{O}$ of the 7-dimensional projective space $\mathbb{P}(C)$, which induces a bijection between $\mathrm{SO}(N_1)/\mathbf{G}(k)$ and $\mathcal{O}(k) \subseteq \mathbb{P}(C)(k)$. If the norm $N$ is anisotropic, i.e., if $C$ is an octonion division algebra, then $\mathcal{O}(k) = \mathbb{P}(C)(k)$.

The bijective morphism $\tau : \mathbf{SO}(N_1)/\mathbf{G} \to \mathcal{O}$ is even an isomorphism of algebraic varieties. According to [Sp 81, 5.3.2] it suffices to show that the tangent map $(d\tau)_e$ is surjective. To prove this, we consider the map

$$\Delta : \mathbf{SO}(N_1) \to \mathcal{O}, \; t \mapsto T(t)[e].$$

The surjectivity of $(d\tau)_e$ will follow if we can show that $(d\Delta)_e$ is surjective. Pick $a \in e^{\perp}$, $N(a) = 1$. The map

$$((e^{\perp}\backslash\{0\})/K^*) \cap \mathcal{O} \to ((ae^{\perp}\backslash\{0\})/K^*) \cap \mathcal{O}, \; [x] \mapsto [a\bar{x}] = [ax],$$

is an isomorphism of algebraic varieties which factors through $\Delta$, for

$$\Delta(s_a s_x) = T(s_a s_x)[e] = [l_a l_{\bar{x}}][e] = [ax] \qquad \left(x \in e^{\perp}\backslash\{0\}, \; N(x) \neq 0\right).$$

The image of $(d\Delta)_e$ must therefore contain $ae^{\perp}$, which we can identify with the tangent space to $\mathcal{O}$ at $[a]$. This shows that $d\tau_e$ is surjective.    □

## 3.5 Local Triality

If $k = K$ is an algebraically closed field, the Principle of Triality holds already for the algebraic group $\mathbf{SO}(N)$, as we noticed in Rem. 3.2.3. It has an analog in the Lie algebra $L(\mathbf{SO}(N))$ of this algebraic group, the Principle of Local Triality, at least in characteristic not two.

In characteristic two the situation is more complicated. In the first instance we have to work in the Lie algebra $L(\mathbf{SGO}(N))$ of the special similarity group, or in the quotient by its one-dimensional center. For local triality in the Lie algebra $L(\mathbf{SO}(N))$ in characteristic two we have to restrict to a subalgebra $\mathbf{M}$ thereof which has codimension 1; in the next section we will see that $\mathbf{M}$ is the commutator subalgebra of $L(\mathbf{SO}(N))$.

Before we formulate and prove local triality, we have to discuss the Lie algebras involved here and, in particular, to find suitable generators for them.

The similarity group $\mathbf{GO}(N)$ is the algebraic group consisting of the invertible linear transformations $t$ which satisfy the equations

$$N(t(x)) = N(t(e))N(x) \qquad (x \in C_K).$$

Notice that for a similarity $t$ with respect to the norm $N$ on $C$ the multiplier $n(t)$ equals $N(t(e))$ (cf. § 3.1). The special similarity group $\mathbf{SGO}(N)$ is the identity component of $\mathbf{GO}(N)$. We denote its Lie algebra by $\mathbf{L}_0$. To find it one can use the ring of dual numbers $K[\varepsilon]$. The method of the proof of Prop. 2.2.2 gives that $\mathbf{L}_0$ consists of the linear maps $t$ of $C_K$ satisfying the condition

$$\langle x, t(x) \rangle = \langle e, t(e) \rangle N(x) \qquad (x \in C_K).$$

We denote by $L_0$ the Lie algebra of linear maps $t$ of $C$ satisfying the same condition. Such a $t$ is called a *local similarity*, and the factor $\langle e, t(e) \rangle$ its *local multiplier*. Examples of local similarities are the left and right multiplications $l_c$ and $r_c$ for $c \in C$, because by (1.3)

$$\langle x, cx \rangle = \langle e, c \rangle N(x),$$

and similarly for the right multiplications.

The subalgebra $\mathbf{L}_1 = \mathrm{L}(\mathbf{SO}(N))$ is given by the equations

$$\langle x, t(x) \rangle = 0 \qquad (x \in C_K).$$

The similarly defined subalgebra of $L_0$ is $L_1$. It consists of the *alternating* linear transformations. These are also *skew symmetric*:

$$\langle t(x), y \rangle + \langle x, t(y) \rangle = 0 \qquad (x, y \in C).$$

Examples of alternating transformations are

$$t_{a,b} : C \to C, \ x \mapsto \langle x, a \rangle b - \langle x, b \rangle a,$$

with $a, b \in C$; further, the $l_a$ with $\langle a, e \rangle = 0$, the $r_a$ with $\langle a, e \rangle = 0$, the $m_c = l_c - r_c$ for all $c \in C$, and finally the $l_a l_{\bar{b}} - l_b l_{\bar{a}}$ and $r_a r_{\bar{b}} - r_b r_{\bar{a}}$ for all $a, b \in C$, as one easily verifies.

**Lemma 3.5.1** *If $t_1$ and $t_2$ are local similarities of $C$, then their Lie commutator $[t_1, t_2] = t_1 t_2 - t_2 t_1$ is alternating.*

Proof. Let $t_1$ and $t_2$ have local multipliers $\lambda_1$ and $\lambda_2$, respectively. The relation

$$\langle x + t_2(x), t_1(x + t_2(x)) \rangle = \lambda_1 N(x + t_2(x)) \qquad (x \in C)$$

leads to

$$\langle x, t_1(x) \rangle + \langle x, t_1 t_2(x) \rangle + \langle t_2(x), t_1(x) \rangle + \langle t_2(x), t_1 t_2(x) \rangle =$$

$$\lambda_1 N(x) + \lambda_1 N(t_2(x)) + \lambda_1 \langle x, t_2(x) \rangle \qquad (x \in C).$$

After cancelling terms on both sides of the equality sign we find

$$\langle x, t_1 t_2(x) \rangle + \langle t_2(x), t_1(x) \rangle = \lambda_1 \lambda_2 N(x) \qquad (x \in C).$$

Now write down this relation with $t_1$ and $t_2$ interchanged, and subtract that from the above relation. Then we find $\langle x, [t_1, t_2]x \rangle = 0$ $(x \in C)$.    □

**Lemma 3.5.2** *(i) $L_1$ is generated as a vector space by the $t_{a,b}$ with $a, b \in C$. (ii) If $\mathrm{char}(k) = 2$, then $L_1$ is generated as a Lie algebra by $\{\, l_a, r_b, m_c \,|\, a, b \in e^\perp, c \in C \,\}$, and $L_0$ is generated as a Lie algebra by $\{\, l_a, r_b \,|\, a, b \in C \,\}$.*

Proof. For $t \in L_1$, define $N_t = \{\, x \in C \,|\, t(x) = 0 \,\}$. For $a = t(e)$ we have $\langle e, a \rangle = 0$. Pick $b \in C$ with $\langle e, b \rangle = -1$. Then $t_{a,b}(e) = a = t(e)$, so $t' = t - t_{a,b}$ maps $e$ to 0, i.e., $e \in N_{t'}$.

Next assume we have $t \in L_1$ with $e \in N_t \neq C$. Pick $u \in C$, $u \notin N_t$. For $a = t(u)$ we have $\langle u, a \rangle = 0$, and for $x \in N_t$,

$$\langle x, a \rangle = \langle x, t(u) \rangle + \langle t(x), u \rangle = 0.$$

Choose $b \in N_t^\perp$ with $\langle u, b \rangle = -1$. Then

$$t_{a,b}(x) = \langle x, a \rangle b - \langle x, b \rangle a = 0 \qquad (x \in N_t),$$
$$t_{a,b}(u) = \langle u, a \rangle b - \langle u, b \rangle a = t(u).$$

Hence for $t' = t - t_{a,b}$ we have $N_{t'} \supseteq N_t$ and $u \in N_{t'}$, so $N_{t'} \neq N_t$. By induction we arrive at statement (i). Notice that after the first step we have only used transformations $t_{a,b}$ with $a, b \in e^\perp$.

For the proof of (ii), consider any $t \in L_1$. Take $a = t(e)$, then $\langle a, e \rangle = 0$. For $t' = t - l_a$ we have $t'(e) = 0$, so $e \in N_{t'}$. Continuing as in the previous paragraph we find that $L_1$ is generated as a vector space by the transformations $l_a$ and $t_{a,b}$ with $a, b \in e^\perp$. In order to prove (ii) it therefore suffices to express these $t_{a,b}$ in left and right multiplications by elements of $e^\perp$ and transformations $m_c$ with $c \in C$.

Let $a, b \in e^\perp$. From $\bar{a} = -a$ we infer that $-ax + \bar{x}a = \langle x, a \rangle e$, so $ax = \bar{x}a - \langle x, a \rangle e$, and similarly $xb = b\bar{x} - \langle x, b \rangle e$. Using this, we find:

$$\begin{aligned}
[l_a, r_b](x) &= a(xb) - (ax)b \\
&= a(b\bar{x}) - (\bar{x}a)b + \langle x, a \rangle b - \langle x, b \rangle a \\
&= a(b\bar{x}) - (\bar{x}a)b + t_{a,b}(x) \qquad (x \in C),
\end{aligned}$$

so

$$t_{a,b}(x) = [l_a, r_b](x) - a(b\bar{x}) + (\bar{x}a)b \qquad (x \in C). \tag{3.9}$$

Since the associator in $C$ is alternating,

$$(\bar{x}a)b - \bar{x}(ab) = (ab)\bar{x} - a(b\bar{x}) \qquad (x \in C),$$

so

$$a(b\bar{x}) + (\bar{x}a)b = (ab)\bar{x} + \bar{x}(ab) \qquad (x \in C).$$

Now assuming that $\mathrm{char}(k) = 2$, we substitute this into (3.9), thus getting

$$\begin{aligned}
t_{a,b}(x) &= [l_a, r_b](x) + (ab)\bar{x} + \bar{x}(ab) \\
&= [l_a, r_b](x) + (ab)x - (ab)\langle x, e \rangle + x(ab) - \langle x, e \rangle(ab) \\
&= [l_a, r_b](x) + (ab)x + x(ab).
\end{aligned}$$

Hence we find

$$t_{a,b} = [l_a, r_b] + m_{ab} \qquad (a, b \in e^\perp),$$

which shows that $L_1$ is generated as a Lie algebra by the $l_a$, $r_b$ and $m_c$ as stated in (ii).

For $L_0$, finally, fix any $d \in C$ with $\langle e, d \rangle = 1$; then by (1.4),

$$\langle x, dx \rangle = \langle e, d \rangle N(x) = N(x) \qquad (x \in C).$$

If $t \in L_0$ has local multiplier $\lambda$, then $t - \lambda l_d \in L_1$. Since $m_c = l_c - r_c$, we see that $L_0$ is generated by all $l_a$ and $r_b$. $\qquad\qquad \square$

As in the last paragraph of the above proof we see that in any characteristic $L_0$ is spanned by $L_1$ and $l_d$ for some $d \notin e^\perp$. If $\mathrm{char}(k) \neq 2$, we can take $d = \frac{1}{2}e$, but in characteristic 2 we have $l_e \in L_1$. Thus,

**Lemma 3.5.3** *(i)* $\dim L_0 = \dim L_1 + 1$.
*(ii) If* $\mathrm{char}(k) \neq 2$, *then* $L_0 = L_1 \oplus k\,\mathrm{id}$, *a direct sum of Lie algebras. If* $\mathrm{char}(k) = 2$, *then* $k\,\mathrm{id} \subset L_1$.

For later use, we formulate another consequence of Lemma 3.5.2, which is of a technical nature and which easily follows from the proof of part (ii) of that lemma.

**Lemma 3.5.4** *Assume* $\mathrm{char}(k) = 2$. *Let $M$ be the subalgebra of $L_1$ generated as a Lie algebra by the $l_a$ and $r_b$ with $a, b \in e^\perp$, and let $d$ be some element of $C$ with $\langle d, e \rangle = 1$. Then $L_1 = M + km_d$.*

We will see later that $m_d \notin M$.

We are now ready to formulate and prove the result on local triality.

**Theorem 3.5.5 (Principle of Local Triality)** *Let $C$ be an octonion algebra with norm $N$ over a field $k$.*
*(i) If* $\mathrm{char}(k) \neq 2$, *then for every $t_1 \in L_1$ there exist unique $t_2, t_3 \in L_1$ such that*

$$t_1(xy) = t_2(x)y + xt_3(y) \qquad (x, y \in C). \tag{3.10}$$

*The mappings $\vartheta_2 : t_1 \mapsto t_2$, and $\vartheta_3 : t_1 \mapsto t_3$, respectively, with $t_1, t_2, t_3$ satisfying (3.10), are Lie algebra automorphisms of $L_1$.*

*For $t_1 = t_{a,b}$, (3.10) is satisfied by $t_2 = \frac{1}{2}(l_b l_{\bar{a}} - l_a l_{\bar{b}})$ and $t_3 = \frac{1}{2}(r_b r_{\bar{a}} - r_a r_{\bar{b}})$.*

*(ii) If* char$(k) = 2$, *then for every $t_1 \in L_0$ there exist $t_2, t_3 \in L_0$ such that (3.10) holds. These $t_2, t_3$ are unique up to a change $t_2 \mapsto t_2 + \lambda\,\mathrm{id}$, $t_3 \mapsto t_3 + \lambda\,\mathrm{id}$, for some $\lambda \in k$. The mappings*

$$\vartheta_2 : t_1 + k\,\mathrm{id} \mapsto t_2 + k\,\mathrm{id}, \quad and \quad \vartheta_3 : t_1 + k\,\mathrm{id} \mapsto t_3 + k\,\mathrm{id},$$

*with $t_1, t_2, t_3$ satisfying (3.10), are Lie algebra automorphisms of $L_0/\,k\,\mathrm{id}$.*

*(3.10) is satisfied by the triples $t_1 = l_a$, $t_2 = l_a + r_a$, $t_3 = l_a$ and $t_1 = r_b$, $t_2 = r_b$, $t_3 = l_b + r_b$.*

Proof. Let $N$ be the set of $t_1 \in L_1$ (or $\in L_0$ in characteristic 2) for which there exist $t_2, t_3 \in L_1$ ($\in L_0$, respectively) such that (3.10) is satisfied. If $(t_1', t_2', t_3')$ also satifies (3.10), then clearly so does $(\lambda t_1 + \mu t_1', \lambda t_2 + \mu t_2', \lambda t_3 + \mu t_3')$ for $\lambda, \mu \in k$. Further,

$$\begin{aligned}
t_1 t_1'(xy) &= t_1(t_2'(x)y + x t_3'(y)) \\
&= t_1(t_2'(x)y) + t_1(x t_3'(y)) \\
&= t_2 t_2'(x)y + t_2'(x)t_3(y) + t_2(x)t_3'(y) + x t_3 t_3'(y) \quad (x, y \in C).
\end{aligned}$$

Interchanging the $t_i$ and the $t_i'$ and then subtracting from the above relation yields

$$[t_1, t_1'](xy) = [t_2, t_2'](x)y + x[t_3, t_3'](y) \quad (x, y \in C).$$

This shows that $N$ is a Lie subalgebra of $L_1$ (or $L_0$ in characteristic 2). Hence, to prove the Theorem it suffices to prove that $N$ contains a set of generators of $L_1$ or $L_0$, respectively.

First, we consider the char$(k) \neq 2$ case for $L_1$. We use the $t_{a,b}$ as generators.

$$\begin{aligned}
t_{a,b}(xy) &= \langle\, xy, a\,\rangle b - \langle\, xy, b\,\rangle a \\
&= \langle\, y, \bar{x}a\,\rangle b - \langle\, y, \bar{x}b\,\rangle a \\
&= (b\bar{y})(\bar{x}a) + (b(\bar{a}x))y - (a\bar{y})(\bar{x}b) - (a(\bar{b}x))y \\
&= (b(\bar{a}x) - a(\bar{b}x))y + (b\bar{y})(\bar{x}a) - (a\bar{y})(\bar{x}b). \tag{3.11}
\end{aligned}$$

From the definition of $t_{a,b}$ it is immediate that $t_{a,b}(z) = \overline{t_{\bar{a},\bar{b}}(\bar{z})}$, so using the above expression we get

$$\begin{aligned}
t_{a,b}(xy) &= \overline{t_{\bar{a},\bar{b}}(\bar{y}\bar{x})} \\
&= x((y\bar{a})b - (y\bar{b})a) + (a\bar{y})(\bar{x}b) - (b\bar{y})(\bar{x}a). \tag{3.12}
\end{aligned}$$

Adding (3.11) and (3.12) and dividing by 2 yields

$$t_{a,b}(xy) = \frac{1}{2}(b(\bar{a}x) - a(\bar{b}x))y + \frac{1}{2}x((y\bar{a})b - (y\bar{b})a).$$

Just as we noticed before Lemma 3.5.1, $\frac{1}{2}(l_b l_{\bar{a}} - l_a l_{\bar{b}})$ and $\frac{1}{2}(r_b r_{\bar{a}} - r_a r_{\bar{b}})$ are elements of $L_1$. This proves existence in case (i).

For existence in case (ii), we take as generators of $L_0$ the $l_a$ and $r_b$. Since the associator is symmetric in characteristic 2,

$$a(xy) + (ax)y = x(ay) + (xa)y,$$

so

$$a(xy) = (ax + xa)y + x(ay).$$

Hence (3.10) is satisfied by $t_1 = l_a$, $t_2 = l_a + r_a$ and $t_3 = l_a$. Similarly with $t_1 = r_b$, $t_2 = r_b$ and $t_3 = l_b + r_b$.

Now about uniqueness. Assume that the triples $(t_1, t_2, t_3)$ and $(t_1, t'_2, t'_3)$ both satisfy (3.10). Then $u_2 = t_2 - t'_2$ and $u_3 = t_3 - t'_3$ satisfy

$$u_2(x)y + xu_3(y) = 0 \qquad (x, y \in C).$$

Taking $y = e$ we find $u_2(x) = xa$ with $a = -u_3(e)$, while with $x = e$ we find $u_3(y) = -u_2(e)y = -ay$. Then

$$(xa)y - x(ay) = 0 \qquad (x, y \in C).$$

By Prop. 1.9.2 this implies that $a = \lambda e$ for some $\lambda \in k$, so $u_2 = -u_3 = \lambda\,\mathrm{id}$. If $\mathrm{char}(k) \neq 2$, this transformation does not belong to $L_1$ if $\lambda \neq 0$, so in that case $t_2$ and $t_3$ are uniquely determined by $t_1$.

The statements about the automorphisms $\vartheta_2$ and $\vartheta_3$ are consequences of the first paragraph of this proof.    □

A triple $(t_1, t_2, t_3)$ of local similarities which satisfies (3.10) is said to be *related*. The following result is immediate from the first paragraph of the proof of the Theorem combined with Lemma 3.5.1.

**Corollary 3.5.6** *If $(t_1, t_2, t_3)$ and $(t'_1, t'_2, t'_3)$ are related triples of local similarities, then the triple of alternating transformations $([t_1, t'_1], [t_2, t'_2], [t_3, t'_3])$ is related.*

**Remark 3.5.7** In characteristic $\neq 2$, the Principle of Local Triality holds in $L_0$ as well. Then $t_1$ determines $t_2$ and $t_3$ up to adding opposite multiples of the identity. Taking $t_1 = \mathrm{id}$ as an extra generator besides the $t_{a,b}$, the triple $t_1 = \mathrm{id}$, $t_2 = \mathrm{id}$, $t_3 = 0$ satisfies (3.10).

Assume $\mathrm{char}(k) = 2$. Then the principle of local triality does not hold in $L_1$. For consider $t_1 = m_d = l_d + r_d \in L_1$ for some $d \in C$ with $\langle d, e \rangle = 1$. Since the associator is invariant under cyclic permutation, we have

$$d(xy) + (dx)y = x(yd) + (xy)d,$$

so

$$d(xy) + (xy)d = (dx)y + x(yd).$$

The Principle of Local Triality in $L_0$ says that $t_2 = l_d + \lambda\,\mathrm{id}$ and $t_3 = r_d + \lambda\,\mathrm{id}$ are the only elements in $L_0$ such that the triple $(t_1, t_2, t_3)$ satisfies (3.10). But none of these $t_2$ and $t_3$ are in $L_1$ if $d \notin e^\perp$.

One way out is to formulate a Principle of Local Triality for $L_1$ in combination with two other subalgebras $L_2$ and $L_3$ of $L_0$ which we define as follows: $L_i$ is the subalgebra containing $k\,\mathrm{id}$ such that $L_i/k\,\mathrm{id} = \vartheta_i(L_1/k\,\mathrm{id})$ for $i = 2, 3$. For $t_1 \in L_1$ one can then find $t_2 \in L_2$ and $t_3 \in L_3$ such that (3.10) holds, and these are unique up to adding to both of them one and the same multiple of the identity.

Another approach is to restrict to a subalgebra of $L_1$. Let $M$ be the Lie subalgebra of $L_1$ generated by all $l_a$ and $r_b$ with $a, b \in e^\perp$. We saw in Lemma 3.5.4 that $L_1 = M + km_d$, so $L_2 = M + kl_d$ and $L_3 = M + kr_d$. Since $l_d$ and $r_d$ have local multiplier $\langle e, d \rangle = 1$, they do not belong to $L_1$. On the other hand, if $t_1 \in M$, then $t_2, t_3 \in M$ by Th. 3.5.5. Hence $m_d \notin M$, and $L_1$, $L_2$ and $L_3$ are three distinct Lie algebras which have $M$ in common, while the sum of any two of them is $L_0$. Recall that $\dim L_1 = 28$, hence $\dim M = 27$. Thus we have proved the following theorem.

**Theorem 3.5.8** *Assume* $\mathrm{char}(k) = 2$. *Let $M$ be the Lie subalgebra of $L_1$ generated as a Lie algebra by all $l_a$ and $r_a$ with $a \in e^\perp$, and let $d$ be a fixed element of $C$ with $d \notin e^\perp$.*
*(i)* $\dim M = 27$, $L_1 = M \oplus k(l_d + r_d)$, $L_2 = M \oplus l_d$ *and* $L_3 = M \oplus r_d$. *Further, $L_i \cap L_j = M$ and $L_i + L_j = L_0$ for $1 \le i \ne j \le 3$.*
*(ii) For $t_1 \in L_1$ there exist $t_2 \in L_2$ and $t_3 \in L_3$ such that (3.10) holds, and these are unique up to adding a common multiple of the identity to them.*
*(iii) For $t_1 \in M$ there exist $t_2, t_3 \in M$, unique up to adding a common multiple of the identity, such that (3.10) holds. For $t_1 \in L_1$, $T_1 \notin M$ such $t_2$ and $t_3$ in $L_1$ do not exist.*

Lemma 3.3.2 about permutations of related triples of proper similarities has a local analog. First some notation. For a local similarity $t$, the local similarity $\hat{t}$ is defined by

$$\hat{t}(x) = \overline{t(\bar{x})} \qquad (x \in C).$$

$\hat{t}$ and $t$ have the same local multiplier. The mapping $t \mapsto \hat{t}$ is an involutory automorphism of the Lie algebra $L_0$, and it maps $L_1$ onto itself.

**Lemma 3.5.9** *Assume $t_1, t_2, t_3 \in L_1$ if $\mathrm{char}(k) \ne 2$, and $t_1, t_2, t_3 \in M$ if $\mathrm{char}(k) = 2$. If the triple $(t_1, t_2, t_3)$ is related, then so are the triples $(t_2, t_1, \hat{t}_3)$, $(t_3, \hat{t}_2, t_1)$, $(\hat{t}_1, \hat{t}_3, \hat{t}_2)$, $(\hat{t}_2, t_3, \hat{t}_1)$ and $(\hat{t}_3, \hat{t}_1, t_2)$. In the characteristic 2 case, $t \in M$ implies $\hat{t} \in M$.*

Proof. Substitution in (3.10) of $xy$ for $x$ and of $\bar{y}$ for $y$ yields

$$t_1((xy)\bar{y}) = t_2(xy)\bar{y} + (xy)t_3(\bar{y}) \qquad (x, y \in C),$$

hence

$$N(y)t_1(x) = t_2(xy)\bar{y} + (xy)t_3(\bar{y}) \qquad (x, y \in C).$$

Multiplying all the terms in the latter equation on the right by $y$, we get

$$N(y)t_1(x)y = N(y)t_2(xy) + ((xy)t_3(\bar{y}))y \qquad (x, y \in C).$$

Now

$$
\begin{aligned}
((xy)t_3(\bar{y}))y &= ((xy)\overline{\hat{t}_3(y)})y \\
&= -((xy)\bar{y})\hat{t}_3(y) + \langle y, \hat{t}_3(y)\rangle xy \qquad \text{(by Lemma 1.3.3 (iv))} \\
&= -N(y)x\hat{t}_3(y) \qquad \text{(since } \hat{t}_3 \in L_1\text{)}.
\end{aligned}
$$

Thus we get

$$t_1(x)y = t_2(xy) - x\hat{t}_3(y) \qquad (x, y \in C,\ N(y) \neq 0),$$

and by Zariski continuity this holds for all $x, y \in C$, so $(t_2, t_1, \hat{t}_3)$ is related. Further,

$$
\begin{aligned}
\hat{t}_1(xy) &= \overline{t_1(\bar{y}\bar{x})} \\
&= \overline{t_2(\bar{y})\bar{x} + \bar{y}t_3(\bar{x})} \\
&= x\hat{t}_2(y) + \hat{t}_3(x)y \qquad (x, y \in C),
\end{aligned}
$$

from which it follows that $(\hat{t}_1, \hat{t}_3, \hat{t}_2)$ is related. In particular, we see that $\hat{t}_1 \in M$ if $t_1 \in M$.

The other three cases follow by applying the above results.    $\square$

In § 3.1 we considered the projection $\pi : \mathbf{Spin}(N) \to \mathbf{SO}(N)$, which sends $a_1 \circ a_2 \circ \cdots \circ a_{2r}$ to $s_{a_1} s_{a_2} \cdots s_{a_{2r}}$. We saw that in the characteristic 2 case $d\pi$ maps $L(\mathbf{Spin}(N))$ onto a 27-dimensional subalgebra of $\mathbf{L}_1$. We will see in Cor. 3.6.7 that this image is $\mathbf{M}$ (defined as in Th. 3.5.8). Using this result we will then derive that $M$ is the commutator subalgebra $[L_1, L_1]$, which provides a characterization of $M$ that does not depend on a special set of generators.

The fact that, in characteristic 2, $M$ is the commutator subalgebra of $L_1$ could also be derived in a different way. From Cor. 3.5.6 and Th. 3.5.8 one infers that $[L_0, L_0] \subseteq M$, so $M$ is a 27-dimensional ideal in $L_1$. The latter is a Lie algebra of type $D_4$ of intermediate type, i.e., belonging to the orthogonal group and not to the simply connected nor to the adjoint group. From general results about Lie algebras of classical type in nonzero characteristic it follows that $M$ must be the commutator subalgebra $[L_1, L_1]$ (see [Ho 78, Th. (8.20)] or [Ho 82, Th. (2.1)]). These general results also tell us that $L_1$ has a 1-dimensional center in this case; in the following lemma we prove this directly, and also that $L_1$ has trivial center in the other characteristics.

**Lemma 3.5.10** *(i) If* char$(k) \neq 2$, *the center of* $L_1$ *is trivial, and the center of* $L_0$ *as well as the centralizer of* $L_1$ *in* $L_0$ *is* $k$ id.
*(ii) If* char$(k) = 2$, *then* $L_0$, $L_1$ *and* $M$ *have center* $k$ id, *and this is also the centralizer in* $L_0$ *of each of these Lie algebras.*

Proof. Let $t \in L_0$ commute with all $l_a$ and $r_a$ for $a \in e^\perp$. From $tl_a = l_a t$ it follows that

$$t(ax) = at(x) \qquad (a \in e^\perp,\ x \in C).$$

Taking $x = e$ we get that $t(a) = au$ for $a \in e^\perp$, where $u = t(e)$. Replacing $l_a$ by $r_a$ we find that $t(a) = ua$ for $a \in e^\perp$. Thus,

$$t(x) = xu = ux \qquad (x \in e^\perp). \tag{3.13}$$

If char$(k) \neq 2$, we can drop the restriction $x \in e^\perp$ in (3.13), since this equation also holds for $x = e$. This implies that $u$ is central in $C$, so by Prop. 1.9.1 $u = \lambda e$ for some $\lambda \in k$. Hence $t = \lambda$ id; this is an element of $L_0$, but not of $L_1$.

Now assume char$(k) = 2$. Then (3.13) and Prop. 1.2.3 together imply

$$\langle x, u \rangle e + \langle u, e \rangle x = 0 \qquad (x \in e^\perp).$$

Since $x$ need not be a multiple of $e$, we see that $\langle u, e \rangle = 0$, and hence $\langle x, u \rangle = 0$ for all $x \in e^\perp$. This proves that $u = \lambda e$ for some $\lambda \in k$. Finally, consider $x \notin e^\perp$. Choose $a \in e^\perp \cap x^\perp$ with $N(a) \neq 0$. Then $ax \in e^\perp$, since $\langle ax, e \rangle = \langle x, \bar{a} \rangle = \langle x, a \rangle = 0$. Hence

$$t(x) = t(a^{-1}(ax)) = N(a)^{-1}t(a(ax)) = N(a)^{-1}at(ax) = N(a)^{-1}a(\lambda ax) = \lambda x,$$

so $t = \lambda$ id. This is contained in $M$ if char$(k) = 2$ by Th. 3.5.8 (iii).    □

## 3.6 The Spin Group of an Octonion Algebra

We consider again the rotation group $\mathbf{SO}(N)$ of the norm $N$ of an octonion algebra $C$ over $k$. The phenomenon of triality makes it possible to give another description of the simply connected covering of $\mathbf{SO}(N)$. In fact, we will construct an algebraic group $\mathbf{RT}(C)$ which we show to be isomorphic to the spin group $\mathbf{Spin}(N)$ of § 3.1.

We let the group $\mathrm{SO}(N)^3 = \mathrm{SO}(N) \times \mathrm{SO}(N) \times \mathrm{SO}(N)$ act componentwise on the vector space

$$V = C^3 = \{ (x_1, x_2, x_3) \mid x_i \in C,\ i = 1, 2, 3 \}.$$

In $\mathrm{SO}(N)^3$ we consider the subgroup $\mathrm{RT}(C)$ of related triples, i.e., $\mathrm{RT}(C)$ consists of the triples that satisfy the first form of triality:

$$RT(C) = \{\,(t_1, t_2, t_3)\,|\,t_1, t_2, t_3 \in SO(N),\ (t_1, t_2, t_3)\ \text{satisfies (3.4)}\,\}.$$

Recall that for related triples $(t_1, t_2, t_3)$ of rotations we must have all $t_i \in O'(N)$, the reduced orthogonal group. By (i) of Th. 3.2.1, $RT(C)$ is closed under componentwise multiplication and taking inverses.

A rotation $t_1$ with $\sigma(t_1) = 1$ can be written as

$$t_1 = s_{a_1} s_{b_1} \cdots s_{a_r} s_{b_r},$$

with $a_i, b_i \in C$, $\prod_i N(a_i) N(b_i) = 1$. According to Th. 3.2.1 (iii) the corresponding rotations $t_2, t_3$ such that $(t_1, t_2, t_3)$ is related are given by

$$(t_2, t_3) = \pm(l_{a_1} l_{\bar{b}_1} \cdots l_{a_r} l_{\bar{b}_r}, r_{a_1} r_{\bar{b}_1} \cdots r_{a_r} r_{\bar{b}_r}).$$

We can get the plus sign here by replacing $a_1$ by $-a_1$, if necessary. Hence, $RT(C)$ consists of the elements

$$(s_{a_1} s_{b_1} \cdots s_{a_r} s_{b_r}, l_{a_1} l_{\bar{b}_1} \cdots l_{a_r} l_{\bar{b}_r}, r_{a_1} r_{\bar{b}_1} \cdots r_{a_r} r_{\bar{b}_r}),$$

$$\left(a_i, b_i \in C,\ \prod_i N(a_i) N(b_i) = 1\right).$$

where $a_i, b_i \in C$, $\prod_i N(a_i) N(b_i) = 1$.

We denote $RT(C_K)$ by $\mathbf{RT}(C)$. Over the algebraic closure $K$ of $k$ the reduced orthogonal group $O'(N)$ coincides with $\mathbf{SO}(N)$, so $\mathbf{RT}(C)$ is a closed subgroup of the algebraic group $\mathbf{SO}(N)^3$. We will see in Prop. 3.7.1 below that the algebraic group $\mathbf{RT}(C)$ is defined over $k$. It then will follow that $RT(C)$ is the group of rational points $\mathbf{RT}(C)(k)$.

**Proposition 3.6.1** $\mathbf{RT}(C)$ *is a connected algebraic group, and*

$$\varrho_1 : \mathbf{RT}(C) \to \mathbf{SO}(N),\ (t_1, t_2, t_3) \mapsto t_1,$$

*is a surjective homomorphism of algebraic groups which has a kernel of order* 2 *if* $\mathrm{char}(k) \neq 2$ *and which is bijective if* $\mathrm{char}(k) = 2$. *The group* $RT(C)$ *is mapped onto* $SO(N)$ *by* $\varrho_1$.

Proof. It is obvious that $\varrho_1$ is a surjective homomorphism of algebraic groups. Since $t_2$ and $t_3$ are determined by $t_1$ up to a common factor $\pm 1$, $\ker(\varrho_1) = \{\,1, 1^-\,\}$, with 1 and $1^-$ denoting $(1, 1, 1)$ and $(1, -1, -1)$, respectively. Note that $1^- = 1$ if $\mathrm{char}(k) = 2$.

To prove the connectedness of $\mathbf{RT}(C)$, consider in $C_K$ the set $S = \{\,x \in C_K\,|\,N(x) = 1\,\}$; this is an irreducible variety as we remarked already in the proof of the connectedness of $\mathbf{Spin}(N)$ in § 3.1, hence so is $S \times S$. The elements $(s_a s_b, l_a l_{\bar{b}}, r_a r_{\bar{b}})$ with $N(a) = N(b) = 1$ generate $\mathbf{RT}(C)$. There is a morphism of algebraic varieties

$$\gamma : S \times S \to \mathbf{RT}(C),\ (a, b) \mapsto (s_a s_b, l_a l_{\bar{b}}, r_a r_{\bar{b}}).$$

Its image under $\gamma$ is an irreducible subvariety of $\mathbf{RT}(C)$ which contains the identity and generates $\mathbf{RT}(C)$, hence $\mathbf{RT}(C)$ is connected (see [Hu, § 7.5] or [Sp 81, Prop. 2.2.6]). The last point is clear.                                           □

To find the Lie algebra $\mathrm{L}(\mathbf{RT}(C))$, we work again over the ring $K[\varepsilon]$ of dual numbers. We must have the triples $(t_1, t_2, t_3)$ such that $(1+\varepsilon t_1, 1+\varepsilon t_2, 1+\varepsilon t_3)$ is a related triple of rotations in $C_K[\varepsilon]$. This leads to triples of $t_i \in \mathbf{L}_1$ which satisfy

$$t_1(xy) = t_2(x)y + xt_3(y) \qquad (x, y \in C).$$

Thus we find in characteristic $\neq 2$:

$$\mathrm{L}(\mathbf{RT}(C)) \subset \{ (t_1, t_2, t_3) \in (\mathbf{L}_1)^3 \mid (t_1, t_2, t_3) \text{ is related} \}.$$

Recall that, in this case, $t_1 \in L_1$ uniquely determines $t_2$ and $t_3$. So the right-hand side has dimension $\dim \mathbf{L}_1 = 28 = \dim \mathbf{RT}(C)$, and it follows that the inclusion must be an equality. If $\mathrm{char}(k) = 2$, the components $t_i$ of a related triple of alternating transformations must lie in $\mathbf{M}$. We obtain that

$$\mathrm{L}(\mathbf{RT}(C)) \subset \{ (t_1, t_2, t_3) \in (\mathbf{M})^3 \mid (t_1, t_2, t_3) \text{ is related} \}.$$

Now $t_1$ determines the couple $(t_2, t_3)$ modulo $k(1, 1)$. Again, the inclusion is an equality.

**Corollary 3.6.2** $\varrho_1$ *is a separable isogeny if* $\mathrm{char}(k) \neq 2$, *and an inseparable isogeny with* $\ker d\varrho_1$ *of dimension 1 if* $\mathrm{char}(k) = 2$.

Proof. If $\mathrm{char}(k) \neq 2$, then

$$d\varrho_1 : \mathrm{L}(\mathbf{RT}(C)) \to \mathbf{L}_1, \ (t_1, t_2, t_3) \mapsto t_1,$$

is bijective, so $\varrho_1$ is a separable isogeny (see [Sp 81, Th. 4.3.7]). If $\mathrm{char}(k) = 2$, then $d\varrho_1$ has a 1-dimensional kernel, viz. $k(0, 1, 1)$, so then $\varrho_1$ is an inseparable isogeny (see [Sp 81, § 9.6]).                                           □

We see that the situation of $\mathbf{RT}(C)$ with its projection $\varrho_1$ onto $\mathbf{SO}(N)$ is the same as that of $\mathbf{Spin}(N)$ and its projection $\pi$ onto $\mathbf{SO}(N)$ (treated at the end of § 3.1). Hence $\mathbf{RT}(C)$ must be the simply connected covering of $\mathbf{SO}(N)$ and is therefore isomorphic to $\mathbf{Spin}(N)$, at least in characteristic $\neq 2$. Rather than exactly figuring out the situation in characteristic 2 (with root systems and all that), we prefer to exhibit directly an isomorphism between $\mathbf{RT}(C)$ and $\mathbf{Spin}(8)$ in all characteristics.

**Proposition 3.6.3** *There is an isomorphism of algebraic groups*

$$\varphi : \mathbf{Spin}(N) \to \mathbf{RT}(C)$$

*defined by*

$$\varphi(a_1 \circ b_1 \circ \cdots \circ a_r \circ b_r) = (s_{a_1} s_{b_1} \cdots s_{a_r} s_{b_r}, l_{a_1} l_{\bar{b}_1} \cdots l_{a_r} l_{\bar{b}_r}, r_{a_1} r_{\bar{b}_1} \cdots r_{a_r} r_{\bar{b}_r}),$$

*where* $a_i, b_i \in C_K$, $\prod_i N(a_i) N(b_i) = 1$. *It commutes with the projections of* $\mathbf{Spin}(N)$ *and* $\mathbf{RT}(C)$ *on* $\mathbf{SO}(N)$ *and induces an isomorphism of* $\mathrm{Spin}(N)$ *onto* $\mathrm{RT}(C)$.

Proof. We follow a detour via the even tensor algebra of $C$,

$$\mathrm{T}^+(C) = k \oplus (C \otimes_k C) \oplus (C \otimes_k C \otimes_k C \otimes_k C) \oplus \cdots,$$

and the algebra $\mathrm{End}(C)$ of $k$-linear transformations of $C$ into itself. The bilinear transformation

$$C \times C \to \mathrm{End}(C), \ (a, b) \mapsto l_a l_{\bar{b}},$$

defines a linear transformation

$$C \otimes_k C \to \mathrm{End}(C), \ a \otimes b \mapsto l_a l_{\bar{b}}.$$

This can be extended to an algebra homomorphism

$$\varphi_{\mathrm{T}} : \mathrm{T}^+(C) \to \mathrm{End}(C) \quad \text{with} \quad \varphi_{\mathrm{T}}(a \otimes b) = l_a l_{\bar{b}} \quad \text{for} \quad a, b \in C.$$

$\varphi_{\mathrm{T}}$ maps the ideal $I_N^+$ in $\mathrm{T}^+(C)$ generated by the elements $x \otimes x - N(x)$ and $u \otimes x \otimes x \otimes v - N(x)(u \otimes v)$ $(u, v, x \in C)$ onto 0, so it can be factored through the even Clifford algebra $\mathrm{Cl}^+(N) = \mathrm{T}^+(C)/I_N^+$ of $C$, that is to say, we find an algebra homomorphism

$$\varphi_1 : \mathrm{Cl}^+(N) \to \mathrm{End}(C) \quad \text{with} \quad \varphi_1(a \circ b) = l_a l_{\bar{b}} \quad \text{for} \quad a, b \in C.$$

In a similar way we define an algebra homomorphism

$$\varphi_2 : \mathrm{Cl}^+(N) \to \mathrm{End}(C) \quad \text{with} \quad \varphi_2(a \circ b) = r_a r_{\bar{b}} \quad \text{for} \quad a, b \in C.$$

We determine the kernels of $\varphi_1$ and $\varphi_2$. First assume $\mathrm{char}(k) \neq 2$. The center of $\mathrm{Cl}^+(N)$ has dimension 2 and is spanned by 1 and $z = e_1 \circ e_2 \circ \cdots \circ e_8$ where $e_1 = e, e_2, \ldots, e_8$ is a standard orthogonal basis of $C$ (see (1.29)). Since $z^{\circ 2} = 1$, the elements $u_1 = \frac{1}{2}(1 + z)$ and $u_2 = \frac{1}{2}(1 - z)$ are orthogonal central idempotents. The algebra $\mathrm{Cl}^+(N)$ is the direct sum of two simple two-sided ideals, viz., $I_1 = u_1 \circ \mathrm{Cl}^+(N)$ and $I_2 = u_2 \circ \mathrm{Cl}^+(N)$, both of dimension 64 (see [Che 54, Th. II.2.3 and Th. II.2.4]). Since $\dim \mathrm{Cl}^+(N) = 128$ and $\dim \mathrm{End}(C) = 64$, the kernel of $\varphi_1$ must have dimension at least 64. Since $\varphi_1 \neq 0$, $\ker(\varphi_1)$ is either $I_1$ or $I_2$. It follows that $\varphi_1(u_1) = 0$ or $\varphi_1(u_2) = 0$, so $\varphi_1(z) = 1$ or $-1$. In the former case we replace $e_8$ by $-e_8$, so we may assume $\varphi_1(z) = -1$. This implies $\ker(\varphi_1) = I_1$. We have

$$l_{e_1} l_{e_2} \cdots l_{e_7} l_{e_8} = l_{e_1} l_{\bar{e}_2} \cdots l_{e_7} l_{\bar{e}_8} = \varphi_1(z) = -1.$$

By conjugating we obtain $r_{\bar{e}_1} r_{\bar{e}_2} \cdots r_{\bar{e}_7} r_{\bar{e}_8} = -1$. Since $\bar{e}_1 = e_1$ and $\bar{e}_i = -e_i$ for $i > 1$, this implies

$$\varphi_2(z) = r_{e_1} r_{\bar{e}_2} \cdots r_{e_7} r_{\bar{e}_8} = 1.$$

Therefore, $\ker(\varphi_2) = I_2$.

If $\mathrm{char}(k) = 2$, we proceed similarly, but there is some difference in the details. We choose a standard symplectic basis $e_1 = e, e_2, \ldots, e_8$, i.e., a basis as in (1.30) with all $N(e_i) = 1$. The center of $\mathrm{Cl}^+(N)$ is again two-dimensional and is spanned by 1 and $z = e_1 \circ e_5 + e_2 \circ e_6 + e_3 \circ e_7 + e_4 \circ e_8$ (see [Dieu, Ch. II, § 10]). The element $z$ satifies the equation $z^{\circ 2} + z = 0$. It follows that $u_1 = z$ and $u_2 = z + 1$ are orthogonal central idempotents. As above we see that either $\varphi_1(z) = 0$ or $\varphi_1(z) = 1$. In the latter case we replace $e_5$ by $e + e_5$, which interchanges $z$ and $z + 1$ and hence $u_1$ and $u_2$. So we may assume that $\varphi_1(z) = 0$. One easily computes that $\varphi_2(z) = \varphi_1(z) + 1 = 1$, since $\bar{e}_5 = e + e_5$ and $\bar{e}_i = e_i$ for $i \neq 5$. As in the characteristic $\neq 2$ case above one finds that $\ker(\varphi_1) = I_1 = u_1 \circ \mathrm{Cl}^+(N)$ and $\ker(\varphi_2) = I_2 = u_2 \circ \mathrm{Cl}^+(N)$.

Thus we find that in every characteristic the algebra homomorphism

$$(\varphi_1, \varphi_2) : \mathrm{Cl}^+(N) \to \mathrm{End}(C) \times \mathrm{End}(C), \; x \mapsto (\varphi_1(x), \varphi_2(x)) \; (x \in \mathrm{Cl}^+(N)),$$

is injective and hence an isomorphism since both algebras have dimension 128.

Working over $K$ we obtain a homomorphism of algebraic groups

$$\psi : \mathbf{RT}(C) \to \mathbf{SO}(N) \times \mathbf{SO}(N), \; (t_1, t_2, t_3) \mapsto (t_2, t_3).$$

$\mathbf{RT}'(C) = \psi(\mathbf{RT}(C))$ is a closed subgroup of $\mathbf{SO}(N) \times \mathbf{SO}(N)$. For $(t_2, t_3) \in \mathbf{RT}'(C)$, the element $t_1 \in \mathbf{SO}(N)$ such that $(t_1, t_2, t_3)$ satisfies the first form of triality is unique, since (3.4) implies

$$t_1(x) = t_2(x) t_3(e) \qquad (x \in C_K).$$

Hence $\psi : \mathbf{RT}(C) \to \mathbf{RT}'(C)$, has an inverse $\psi^{-1}$, which is a morphism. Thus, $\psi$ induces an isomorphism of algebraic groups between $\mathbf{RT}(C)$ and $\mathbf{RT}'(C)$.

From Th. 3.2.1 (iii) it follows that $\mathbf{RT}'(C)$ consists of the elements of the form

$$(l_{a_1} l_{\bar{b}_1} \cdots l_{a_r} l_{\bar{b}_r}, r_{a_1} r_{\bar{b}_1} \cdots r_{a_r} r_{\bar{b}_r}) \qquad (a_i, b_i \in C, \; \prod_i N(a_i) N(b_i) = 1).$$

Since $\mathbf{Spin}(N)$ consists of the elements

$$a_1 \circ b_1 \circ \cdots \circ a_r \circ b_r \qquad (a_i, b_i \in C, \; \prod_i N(a_i) N(b_i) = 1),$$

the algebra isomorphism $(\varphi_1, \varphi_2)$ induces an isomorphism of algebraic groups from $\mathbf{Spin}(N)$ onto $\mathbf{RT}'(C)$. Combining this with $\psi^{-1}$ we obtain an isomorphism $\varphi = \psi^{-1}(\varphi_1, \varphi_2)$ of algebraic groups from $\mathbf{Spin}(N)$ onto $\mathbf{RT}(C)$. A similar argument with generators shows that $\varphi$ maps $\mathrm{Spin}(N)$ onto $\mathrm{RT}(C)$. $\square$

We saw in § 3.3 how triality induces in the projective special similarity group PSGO($N$) a group $\mathcal{S}$ of outer automorphisms which is isomorphic to the symmetric group $S_3$. We have similar automorphisms of the simply connected covering group $\mathbf{RT}(C) \cong \mathbf{Spin}(N)$, coming from permutations of the three "components" of related triples (cf. Lemma 3.3.2). More precisely, consider $\tau_i : \mathbf{RT}(C) \to \mathbf{RT}(C)$, for $i = 1, 2, 3$, defined by

$$\tau_1 : (t_1, t_2, t_3) \leftrightarrow (\hat{t}_1, \hat{t}_3, \hat{t}_2),$$
$$\tau_2 : (t_1, t_2, t_3) \leftrightarrow (t_3, \hat{t}_2, t_1),$$
$$\tau_3 : (t_1, t_2, t_3) \leftrightarrow (t_2, t_1, \hat{t}_3).$$

By Th. 3.2.1 (i) and Lemma 3.3.2 these are automorphisms of $\mathbf{RT}(C)$. We have similar automorphisms of $\mathrm{RT}(C)$.

**Proposition 3.6.4** $\tau_2$ and $\tau_3$ generate a group $\tilde{\mathcal{S}}$ of automorphisms of $\mathrm{RT}(C)$ which is isomorphic to the symmetric group $S_3$. The nontrivial elements of $\tilde{\mathcal{S}}$ are outer automorphisms.

Proof. We have $\tau_2^2 = \tau_3^2 = \mathrm{id}$ and $\tau_2\tau_3\tau_2 = \tau_3\tau_2\tau_3 = \tau_1$. It follows that $\tilde{\mathcal{S}}$ consists of id, $\tau_1$, $\tau_2$, $\tau_3$, $\tau_2\tau_3$, $\tau_3\tau_2$, so has order 6.

Assume $\mathrm{char}(k) \neq 2$. The central elements $(1, -1, -1)$, $(-1, 1, -1)$ and $(-1, -1, 1)$ of $\mathrm{RT}(C)$ are permuted by the elements of $\tilde{\mathcal{S}}$. This defines an isomorphism of $\tilde{\mathcal{S}}$ onto $S_3$. An inner automorphism induces the trivial permutation of the center, so the nontrivial elements of $\tilde{\mathcal{S}}$ must be outer automorphisms.

The above argument breaks down in characteristic 2, since then the above central elements coincide. In that case we work over $K$, and use the center $\mathbf{Z}$ of $\mathrm{L}(\mathbf{RT}(C))$). This center consists of the related triples $(t_1, t_2, t_3)$ with all $t_i$ belonging to the center of $\mathbf{M}$, so by Lemma 3.5.10,

$$\mathbf{Z} = \{ (\lambda, \mu, \lambda + \mu) \mid \lambda, \mu \in K \}.$$

The above automorphisms $\tau_i$ induce in $\mathrm{L}(\mathbf{RT}(C))$ automorphisms $d\tau_i$ which are described by the same formulas as the $\tau_i$, but with $t_1$, $t_2$ and $t_3$ denoting this time elements of $\mathbf{L}_1$ (see also Lemma 3.5.9). So they induce permutations of the elements $(1, 1, 0)$, $(1, 0, 1)$ and $(0, 1, 1)$ of $\mathbf{Z}$. Thus, we get an isomorphism of $\tilde{\mathcal{S}}$ onto $S_3$ again. Since the inner automorphisms of $\mathbf{RT}(C)$ act trivially on $\mathbf{Z}$, the nontrivial elements of $\tilde{\mathcal{S}}$ are outer automorphisms. □

**Remark 3.6.5** $\mathbf{RT}(C) \cong \mathbf{Spin}(8)$ is an algebraic group of type $D_4$. The quotient of $\mathrm{Aut}(\mathbf{RT}(C))$ by the group of inner automorphisms $\mathrm{Inn}(\mathbf{RT}(C))$ is isomorphic to the automorphism group of the Dynkin diagram of $D_4$, which is $S_3$ (see [Hu, § 27.4] or [Stei, § 10]). The above proposition gives an explicit splitting $\mathrm{Aut}(\mathbf{RT}(C)) = \tilde{\mathcal{S}}.\mathrm{Inn}(\mathbf{RT}(C))$, with $\tilde{\mathcal{S}} \cong S_3$.

We have three representations $\varrho_i$ of $\mathbf{RT}(C)$ in $C_K$, given by

$$\varrho_i(t_1, t_2, t_3) = t_i.$$

**Proposition 3.6.6** *The representations $\varrho_i$ are irreducible and pairwise inequivalent.*

Proof. The image of $\varrho_i$ is $\mathbf{SO}(N)$, which acts irreducibly in $C_K$.

If $\mathrm{char}(k) \neq 2$, then the kernels of $\varrho_1$, $\varrho_2$ and $\varrho_3$ are $< (1, -1, -1) >$, $< (-1, 1, -1) >$ and $< (-1, -1, 1) >$, respectively. Since these kernels are distinct, the representations can not be equivalent.

In the case that $\mathrm{char}(k) = 2$, we seek refuge in the Lie algebra $\mathrm{L}(\mathbf{RT}(C))$ again. There, $d\varrho_1$, $d\varrho_2$ and $d\varrho_3$ have kernels $k(0, 1, 1)$, $k(1, 0, 1)$ and $k(1, 1, 0)$, respectively. These being distinct, the representations are not equivalent. $\square$

Composing the $\varrho_i$ with the isomorphism $\varphi : \mathbf{Spin}(8) = \mathbf{Spin}(N) \to \mathbf{RT}(C)$ of Prop. 3.6.3, we get three surjective homomorphisms $\pi_i$ of $\mathbf{Spin}(8)$ onto $\mathbf{SO}(8)$, which yield inequivalent representations in $C$. Prop. 3.6.3 enables us to write these out explicitly. We will see that $\pi_1$ is the same as the homomorphism $\pi$ we considered at the end of § 3.1.

For these explicit computations we need to fix a nontrivial idempotent $z$ in the center of $\mathrm{Cl}^+(N, K)$. If $\mathrm{char}(k) \neq 2$, we choose a standard orthogonal basis $e_1, e_2, \ldots, e_8$ in $C_K$ (see (1.29)), and take $z = e_1 \circ e_2 \circ \cdots \circ e_8$. A straightforward computation shows that $\varphi_1(z) = -1$ and $\varphi_2(z) = 1$. (Hint: in the proof of Prop. 3.6.3 it was shown that $\varphi_1(z) = \pm 1$ and that $\varphi_2(z) = -\varphi_1(z)$, so it suffices, e.g., to check that $\varphi_2(z)(e) = e$, which is easy using (1.8), Moufang identities and Lemma 1.3.3.)

The image of $\mathrm{L}(\mathbf{Spin}(8))$ under $d\pi_i$ is $\mathbf{L}_1 = \mathrm{L}(\mathbf{SO}(N))$ in characteristic $\neq 2$, as follows from Th. 3.5.5.

If $\mathrm{char}(k) = 2$, we use a standard symplectic basis $e_1 = e, e_2, \ldots, e_8$ (as in (1.30)) with all $N(e_i) = 1$ and take $z = \sum_{i=1}^4 e_i \circ e_{i+4}$. Then $\varphi_1(z) = 1$ and $\varphi_2(z) = 0$. (Hint: from the proof of Prop. 3.6.3 we know already that $\varphi_1(z) = 0$ or $1$, and that $\varphi_2(z) = \varphi_1(z) + 1$, and one easily checks that $\varphi_1(z)(e) = e$ in this case.) As we saw at the end of § 3.1, the Lie algebra $\mathrm{L}(\mathbf{Spin}(8))$ consists of the elements $u = \alpha_0 + \sum_{i<j} \alpha_{ij} e_i \circ e_j$ with $\alpha_0, \alpha_{ij} \in K$, $\sum_{i=1}^4 \alpha_{i,i+4} = 0$. Its image under $d\varphi$ is

$$\mathrm{L}(\mathbf{RT}(C)) = \{ (t_1, t_2, t_3) \in \mathbf{M}^3 \mid (t_1, t_2, t_3) \text{ related} \}.$$

From Th. 3.5.8 we infer that $d\varrho_i(\mathrm{L}(\mathbf{RT}(C))) = \mathbf{M}$ for $i = 1, 2, 3$. We thus find:

**Corollary 3.6.7** *The homomorphisms of algebraic groups $\pi_i : \mathbf{Spin}(8) \to \mathbf{SO}(8)$ which are defined by*

$$\pi_1(a_1 \circ b_1 \circ \cdots \circ a_r \circ b_r) = s_{a_1} s_{b_1} \cdots s_{a_r} s_{b_r},$$

$$\pi_2(a_1 \circ b_1 \circ \cdots \circ a_r \circ b_r) = l_{a_1} l_{\bar{b}_1} \cdots l_{a_r} l_{\bar{b}_r},$$

$$\pi_3(a_1 \circ b_1 \circ \cdots \circ a_r \circ b_r) = r_{a_1} r_{\bar{b}_1} \cdots r_{a_r} r_{\bar{b}_r} \ (a_i, b_i \in C, \prod_i N(a_i)N(b_i) = 1),$$

are surjective. If $\mathrm{char}(k) \neq 2$, they are separable isogenies with kernels of order 2, viz., $<-1>$, $<-z>$ and $<z>$, respectively. If $\mathrm{char}(k) = 2$, then the $\pi_i$ are bijective, inseparable isogenies; the kernels of $d\pi_1$, $d\pi_2$ and $d\pi_3$ are the subspaces $k\,1$, $k(1+z)$ and $kz$ of $\mathrm{L}(\mathbf{Spin}(8))$, respectively, and the $d\pi_i$ all have image $\mathbf{M}$.

We can now prove the result announced after Lemma 3.5.9.

**Lemma 3.6.8** (i) If $\mathrm{char}(k) = 2$, then $M = [M, M] = [L_1, L_1] = [L_0, L_0]$.
(ii) $[\mathrm{L}(\mathbf{Spin}(8)), \mathrm{L}(\mathbf{Spin}(8))] = \mathrm{L}(\mathbf{Spin}(8))$.

Proof. We know already that $[L_0, L_0] \subseteq M$ (see Cor. 3.5.6 and Th. 3.5.8), hence

$$[M, M] \subseteq [L_1, L_1] \subseteq [L_0, L_0] \subseteq M.$$

Therefore, to get (i) it suffices to prove that $[M, M] = M$. This will follow from $[\mathbf{M}, \mathbf{M}] = \mathbf{M}$. Since $M = d\pi_1(\mathrm{L}(\mathbf{Spin}(8)))$, it suffices to prove (ii). We do this by picking generators of $\mathrm{L}(\mathbf{Spin}(8))$ and expressing these as sums of commutators. As generators we take 1, the $e_i \circ e_j$ with $1 \leq i < j \leq 8$, $j \neq i + 4$, and the elements $e_i \circ e_{i+4} + e_j \circ e_{j+4}$ with $1 \leq i < j \leq 4$, using the notations introduced just before the above corollary. In the following commutator relations, which are easily verified, the indices $i, j, k$ are distinct and $1 \leq i, j, k \leq 4$.

$$[e_i \circ e_k, e_{i+4} \circ e_{k+4}] = e_i \circ e_{i+4} + e_k \circ e_{k+4} + 1,$$

hence

$$[e_i \circ e_k, e_{i+4} \circ e_{k+4}] + [e_j \circ e_k, e_{j+4} \circ e_{k+4}] = e_i \circ e_{i+4} + e_j \circ e_{j+4}.$$

This also enables us to write 1 as a sum of commutators. Further,

$$[e_i \circ e_k, e_{k+4} \circ e_j] = e_i \circ e_j,$$
$$[e_i \circ e_k, e_{k+4} \circ e_{j+4}] = e_i \circ e_{j+4},$$
$$[e_{i+4} \circ e_k, e_{k+4} \circ e_{j+4}] = e_{i+4} \circ e_{j+4}.$$

Thus we see that all generators of $\mathrm{L}(\mathbf{Spin}(8))$ are in the commutator subalgebra. □

## 3.7 Fields of Definition

The results of the foregoing sections enable us to prove that the algebraic group $\mathbf{RT}(C)$ is defined over $k$.

**Proposition 3.7.1** *Let $C$ be an octonion algebra over $k$. The algebraic group* $\mathbf{RT}(C)$ *of related triples of rotations is defined over $k$.*

Proof. We proceed as in the proof of Th. 2.4.6. Let $\mathbf{A} = \mathrm{GL}(C_K)^3$, the product of three copies of $\mathrm{GL}(C_K)$. This is an algebraic group which is defined over $k$. Further, $V = \mathrm{End}_K(C_K \otimes_K C_K, C_K)$ is a vector space over $K$ with $k$-structure $V(k) = \mathrm{End}(C \otimes_k C, C)$. In a similar way we have the vector space $Q(C)(K)$ over $K$ of quadratic forms on $C_K$, which has $Q(C)(k)$ as a $k$-structure. The multiplication map $\mu : x \otimes y \mapsto xy$ lies in $V(k)$ and in $Q(C)(k)$ we have the norm $N$ of $C$. On the vector space

$$X = V \times Q(C)(K)^3$$

we have an action of $\mathbf{A}$ defined by

$$(t_1, t_2, t_3) : X \to X, \ (\nu, q_1, q_2, q_3) \mapsto (t_1^{-1} \circ \nu \circ (t_2 \otimes t_3), q_1 \circ t_1, q_2 \circ t_2, q_3 \circ t_3).$$

Now notice that $t_1^{-1} \circ \mu \circ (t_2 \otimes t_3) = \mu$ if and only if the triple $(t_1, t_2, t_3)$ satisfies the first form of triality (3.4). Therefore, the stabilizer $G_x$ of $x = (\mu, N, N, N) \in X$ consists of the triples $(t_1, t_2, t_3)$ of orthogonal transformations that satisfy (3.4), i.e., of the related triples of rotations. Hence, this stabilizer is just the group $\mathbf{RT}(C)$.

By [Sp 81, 12.1.2] the stabilizer will be defined if the kernel of the differential $d\varphi_e$ has dimension $28 = \dim \mathbf{RT}(C)$, where $\varphi(a) = a.x$ ($a \in \mathbf{A}$). Arguments like those used after the proof 3.6.1 show that this is indeed the case. $\qquad\square$

By Prop. 3.6.3 we identify $\mathbf{RT}(C)$ with $\mathbf{Spin}(N)$. The previous Proposition then gives a structure of algebraic group over $k$ on $\mathbf{Spin}(N)$.

## 3.8 Historical Notes

The study of *triality* originated with the treatment of geometric triality between points, spaces of one kind and spaces of the other kind on a quadric in complex seven-dimensional projective space, by E. Study [Stu] and E. Cartan [Ca 25]. Octonions appeared in this connection in papers by F. Vaney [Va] and E.A. Weiss [Weiss]. *Local triality* in Lie algebras of type $D_4$ was studied by H. Freudenthal [Fr 51] for the real and complex case. F. van der Blij and T.A. Springer [BlSp 60] treated algebraic triality over arbitrary fields and local triality over fields of characteristic not two. N. Jacobson [Ja 64b], [Ja 71] dealt with local triality in characteristic not two, and with algebraic triality in [Ja 60]. The results on local triality in characteristic two presented here are new, to our knowledge.

In the literature one often finds the Principle of Triality formulated in a different way: for any similarity $t$, there exist similarities $t_1$ and $t_2$ such that either $t_1(xy) = t(x)t_2(y)$ for all $x, y \in C$ or $t_1(xy) = t(y)t_2(x)$ for all $x, y \in C$, and similarly with local triality; this convention goes back to Cartan [Ca 25]. The formulations used here (and also in [Ja 64b] and [Ja 71]) are more convenient in applications and make some proofs simpler. Lemmas 3.3.2 and 3.5.9 ensure that the different formulations are in fact equivalent.

# 4. Twisted Composition Algebras

Twisted composition algebras are somewhat similar to composition algebras, and the two notions are strongly interrelated. On the other hand, twisted composition algebras have close connections with Jordan algebras, which we will deal with in the next chapter. In fact, the motivation to set up this theory comes from exceptional Jordan algebras and twisted algebraic groups of type $D_4$. In particular, the notion of reduced twisted composition algebra comes from reduced Jordan algebras. We refer to Ch. 6 for the details.

In this chapter, $k$ will denote a field and $l$ a separable cubic field extension of $k$. The normal closure of $l$ over $k$ is $l' = l(d)$, where $d$ satisfies a separable quadratic equation over $k$ (see, e.g., [Ja 74, Th. 4.13 and ex. 3 in § 4.8]). If $\operatorname{char}(k) \neq 2$, we can take $d = \sqrt{D}$, i.e., either of the square roots of the discriminant D of $l$ over $k$. We set $k' = k(d)$. So either $l$ is a Galois extension of $k$ with cyclic Galois group of order three (a *cubic cyclic* extension), and then $l' = l$ and $k' = k$, or $l'$ and $k'$ are quadratic extensions of $l$ and $k$, respectively, and $l'$ is a cubic cyclic extension of $k'$. We identify $l'$ with $k' \otimes_k l$. Fix a generator $\sigma$ of $\operatorname{Gal}(l'/k')$. We view it also as an element of $\operatorname{Gal}(l'/k)$ and as a $k$-isomorphism of $l$ into $l'$. We denote by $\tau$ the element of $\operatorname{Gal}(l'/k)$ whose fixed field is $l$ if $l' \neq l$, and $\tau = \operatorname{id}$ if $l' = l$. We denote by $N_{l/k}$ and $\operatorname{Tr}_{l/k}$ norm and trace function of the extension $l$ of $k$.

We first introduce twisted composition algebras for the case that the separable cubic extension $l$ of $k$ is normal, so a cubic cyclic extension; we call algebras of this type "normal twisted composition algebras". The basic theory of these algebras will be developed in § 4.1. In the next section we generalize this to the case that the separable cubic extension $l$ of $k$ is not normal; we then get the general twisted composition algebras. In both cases we have a vector space with an additional operation; this is binary in the normal case (a product), and unary in the nonnormal case (a squaring operation). There will be close relations between the two types.

In the theory of normal twisted composition algebras the characteristic of the field $k$ can be arbitrary, but when dealing with the nonnormal twisted composition algebras, more precisely with their relations to the normal algebras, we have to exclude characteristic two and three. For the applications to Jordan algebras in Ch. 6 this is not a problem, since there a restriction to characteristic not two or three will be in force anyway.

To study the automorphism group of a twisted composition algebra in § 4.4, we also have to consider twisted composition algebras over a split cubic extension of a field $K$, i.e., over $K \oplus K \oplus K$; we develop that theory in § 4.3. But in all other sections we work over a separable cubic field extension $l$ of $k$.

In the last sections we discuss explicit constructions of twisted composition algebras, involving connections with cubic central simple algebras.

## 4.1 Normal Twisted Composition Algebras

We first deal with the case of a cubic cyclic field extension $l$ of $k$, so $l' = l$ and $k' = k$.

**Definition 4.1.1** A vector space $F$ over $l$ provided with a nondegenerate quadratic form $N$, and associated bilinear form $\langle \, , \, \rangle$, is said to be a *normal twisted composition algebra* over $l$ and $\sigma$ if there is a $k$-bilinear product $*$ on $F$ which satisfies the following three conditions.
(i) The product $x * y$ is $\sigma$-linear in $x$ and $\sigma^2$-linear in $y$, that is

$$(\lambda x) * y = \sigma(\lambda)(x * y),$$
$$x * (\lambda y) = \sigma^2(\lambda)(x * y) \qquad (x, y \in F, \lambda \in l);$$

(ii) $N(x * y) = \sigma(N(x))\sigma^2(N(y)) \qquad (x, y \in F);$
(iii) $\langle x * y, z \rangle = \sigma(\langle y * z, x \rangle) = \sigma^2(\langle z * x, y \rangle) \qquad (x, y, z \in F).$
One calls $N$ the *norm* of the algebra. The normal twisted composition algebra is also denoted by $(F, *, N)$, or simply by $F$.

A bijective $l$-linear mapping $t : F_1 \to F_2$ between normal twisted composition algebras $F_1$ and $F_2$ over $l$ satisfying

$$t(x * y) = t(x) * t(y) \qquad (x, y \in F_1)$$

is called an *isomorphism*.

We will see in Cor. 4.1.5 that the norm of a normal twisted composition algebra is already determined by the linear structure and the product $*$, and that an isomorphism preserves the norm.
If $\lambda \in k^*$ we define for $x, y \in F$

$$x *_\lambda y = \lambda(x * y), \quad N_\lambda(x) = \sigma(\lambda)\sigma^2(\lambda)N(x).$$

It is immediate from the definitions that $(F, *_\lambda, N_\lambda)$ is a normal twisted composition algebra, denoted by $F_\lambda$. We say that $F_\lambda$ is an *isotope* of $F$, and that $F$ and $F_\lambda$ are *isotopic.* . It is easy to see that $F_{\lambda\mu} = (F_\lambda)_\mu \ (\lambda, \mu \in l^*)$.

From condition (iii) it follows that

$$T(x) = \langle x * x, x \rangle \qquad (x \in F)$$

is invariant under $\sigma$, so $T(x) \in k$ for all $x \in F$. Thus, $T$ is a cubic form over $k$ in $3 \dim_l F$ variables.

In the following two lemmas we give a number of useful identities.

**Lemma 4.1.2** *In any normal twisted composition algebra $F$ we have for* $x, y, z, w \in F$,

$$\langle x * z, y * z \rangle = \sigma(\langle x, y \rangle)\sigma^2(N(z)), \qquad (4.1)$$

$$\langle x * z, x * w \rangle = \sigma(N(x))\sigma^2(\langle z, w \rangle), \qquad (4.2)$$

$$\langle x * z, y * w \rangle + \langle x * w, y * z \rangle = \sigma(\langle x, y \rangle)\sigma^2(\langle z, w \rangle). \qquad (4.3)$$

Proof. From (ii) in the definition we infer

$$N((x + y) * (z + w)) = \sigma(N(x + y))\sigma^2(N(z + w)).$$

Write this out and, using (ii), cancel all terms which contain only two of the variables; this yields a relation we call ($*$). Taking $w = 0$ in ($*$), we get (4.1) and taking $y = 0$ we get (4.2). Now cancel all terms in ($*$) containing only three variables by using (4.1) and (4.2), then what remains is precisely equation (4.3). □

**Lemma 4.1.3** *In any normal twisted composition algebra $F$ we have for* $x, y, z \in F$,

$$x * (y * x) = \sigma(N(x))y, \qquad (4.4)$$

$$x * (y * z) + z * (y * x) = \sigma(\langle x, z \rangle)y, \qquad (4.5)$$

$$(x * y) * x = \sigma^2(N(x))y, \qquad (4.6)$$

$$(x * y) * z + (z * y) * x = \sigma^2(\langle x, z \rangle)y, \qquad (4.7)$$

$$(x * x) * (x * x) = T(x)x - N(x)(x * x). \qquad (4.8)$$

Proof. Since

$$\langle x * (y * x), u \rangle = \sigma^2(\langle u * x, y * x \rangle) \qquad \text{(by (iii) of Def. 4.1.1)}$$
$$= \sigma(N(x))\langle u, y \rangle \qquad \text{(by (4.1))}$$
$$= \sigma(N(x))\langle y, u \rangle$$

for all $u \in F$, we get (4.4). Linearizing this equation, we find (4.5). The following two equations are proved in the same way. Finally, to prove (4.8) take $y = x$ and $z = x * x$ in (4.5):

$$x * (x * (x * x)) + (x * x) * (x * x) = T(x)x.$$

Since $x * (x * x) = \sigma(N(x))x$ by (4.4), we get

$$(x * x) * (x * x) = T(x)x - x * (\sigma(N(x))x) = T(x)x - N(x)(x * x).$$

$\square$

Consider for $a \in F$ the left multiplication $l_a^* : F \to F$, $x \mapsto a * x$, which is $\sigma^2$-linear, and the right multiplication $r_a^* : F \to F$, $x \mapsto x * a$, which is $\sigma$-linear. From (4.4) and (4.6) we infer:

**Lemma 4.1.4** *If $a \in F$ has $N(a) \neq 0$, then $l_a^*$ and $r_a^*$ are invertible, with inverses $r_{N(a)^{-1}a}^*$ and $l_{N(a)^{-1}a}^*$, respectively.*

The following result is also immediate from (4.4).

**Lemma 4.1.5** *The norm $N$ on a normal twisted composition algebra $F$ over $l$ and $\sigma$ is determined by the linear structure over $l$ and the product $*$ on $F$. Isomorphisms of normal twisted composition algebras are norm preserving.*

There exists a close relationship between normal twisted composition algebras and ordinary composition algebras; in fact, either kind can be related to the other one. First, consider a normal twisted composition algebra $F$ over $l$ and $\sigma$. Pick $a, b \in F$ with $N(a)N(b) \neq 0$. Define

$$xy = (a * x) * (y * b).$$

This product is $l$-bilinear, and the norm $N$ satisfies:

$$\begin{aligned}
N(xy) &= N((a * x) * (y * b)) \\
&= \sigma(N(a * x))\sigma^2(N(y * b)) \\
&= \sigma^2(N(a))N(x)N(y)\sigma(N(b)) \\
&= \lambda N(x)N(y)
\end{aligned}$$

with $\lambda = \sigma^2(N(a))\sigma(N(b)) \neq 0$. Take as new norm $\tilde{N} = \lambda N$. This obviously permits composition (see Def. 1.2.1). An identity element is

$$e = (N(b)^{-1}b) * (N(a)^{-1}a) = \sigma(N(b))^{-1}\sigma^2(N(a))^{-1}(b * a),$$

as a straightforward computation using (4.4) and (4.6) shows. We have thus obtained a structure of composition algebra on $F$. As a consequence one finds using Th. 1.6.2:

**Proposition 4.1.6** *A normal twisted composition algebra over $l$ can only have dimension 1, 2, 4 or 8 over $l$.*

From the considerations below it will follow that in each of the dimensions 1, 2, 4 and 8 there do exist normal twisted composition algebras. We will mainly be interested in such algebras of dimension 8; we also call these *normal twisted octonion algebras*.

We further exploit the connection between ordinary composition algebras and normal twisted composition algebras. Let $F$ be as before. Denote by $N(F)^*$ the set of nonzero values of the norm $N$ and by $M(N)$ the group of multipliers of similarities of $N$ (see 1.1).

**Proposition 4.1.7** $M(N)$ *is a $\sigma$-stable subgroup of $l^*$ and $N(F)^*$ is a coset* $\lambda M(N)$, *with* $N_{l/k}(\lambda) \in M(N)$.

Proof. Let $\tilde{N}$ be as before. Clearly, $M(N)$ coincides with the similar group $M(\tilde{N})$, which is the set of nonzero values $\tilde{N}(x)$ (see the beginning of the proof of Theorem 1.7.1). It follows from what we established above that if $a, b \in F$, $N(a)N(b) \neq 0$ we have

$$M(N) = \sigma^2(N(a))\sigma(N(b))N(F)^*,$$

from which we conclude (using that $M(N)$ contains $(l^*)^2$) that $N(F)^*$ is a coset of $M(N)$ and also that $M(N)$ is the set of nonzero elements of the form $\sigma^2(N(a))\sigma(N(b))N(c)$ $(a, b, c \in F)$. This implies that $M(N)$ is $\sigma$-stable. Taking $a = b = c$ we see that $N_{l/k}(\lambda) \in M(N)$ for all $\lambda \in N(F)^*$.   □

We now give a construction of a normal twisted composition algebra from a composition algebra $C$ over $k$. As before, $l$ is cubic cyclic field extension of $k$ and $\sigma$ a generator of the Galois group. Let $N_C$ be the norm of $C$. Extend the base field to $l$:

$$F = l \otimes_k C,$$

extend $N_C$ to a quadratic form over $l$ on $F$, denoted by $N$, and similarly for conjugation and the product. Define a $\sigma$-automorphism $\varphi$ of $F$ by

$$\varphi(\xi \otimes x) = \sigma(\xi) \otimes x.$$

Notice that $N(\varphi(z)) = \sigma(N(z))$ for $z \in F$. Define

$$x * y = \varphi(\bar{x})\varphi^2(\bar{y}) \qquad (x, y \in F). \tag{4.9}$$

A straightforward computation shows that $(F, *, N)$ is a normal twisted composition algebra. For the verification of point (iii) in Def. 4.1.1, use (1.10). In this verification one sees that it is necessary to take the conjugates of $x$ and $y$ in definition (4.9) of the product $*$ ; without conjugation, things go wrong. We denote this twisted composition algebra by $F(C)$.

**Definition 4.1.8** A normal twisted composition algebra $F$ over $l$ and $\sigma$ is said to be *reduced* if there is a composition algebra $C$ over $k$ and $\lambda \in l^*$ such that $F$ is isomorphic to the isotope $F(C)_\lambda$.

**Proposition 4.1.9** *(i) If the reduced normal twisted composition algebras* $F = F(C)_\lambda$ *and* $F' = F(C')_{\lambda'}$ *over $l$ and $\sigma$ are isomorphic, then the composition algebras $C$ and $C'$ are isomorphic.*
*(ii) The normal twisted composition algebras $F(C)_\lambda$ and $F(C)_{\lambda'}$ over $l$ and $\sigma$ have the same automorphism group.*

Proof. (i) An isomorphism from $F$ onto $F'$ preserves the norm by Lemma 4.1.5, hence the norms $N_C$ of $C$ and $N_{C'}$ of $C'$ are similar over $l$. It follows (see the first paragraph of the proof of Th. 1.7.1) that $N_C$ and $N_{C'}$ are equivalent over $l$. By a result of Springer (see [Sp 52, p. 1519,$b$] or [Lam, p. 198]), they must be similar over $k$, so by Th. 1.7.1 $C$ and $C'$ are isomorphic.

(ii) The condition that a linear bijection is an isomorphism is invariant under isotopy. .                                                                                        □

We will develop several criteria for a normal twisted composition algebra to be reduced. The following theorem is the first result in this line.

**Theorem 4.1.10** *Let $F$ be a normal twisted composition algebra over $l$. The following conditions are equivalent.*
*(i)  $F$ is reduced.*
*(ii) $T$ represents zero nontrivially, i.e., there exists $x \neq 0$ in $F$ such that $T(x) = \langle x * x, x \rangle = 0$.*
*(iii) There exists $x \neq 0$ in $F$ such that $x * x = \lambda x$ for some $\lambda \in l$.*

Proof. (i) $\Rightarrow$ (ii). Assume $F$ is reduced. We may assume that $F = F(C)_\lambda$ as before. We write $\langle \ , \ \rangle$ and $\langle \ , \ \rangle_C$ for the bilinear forms associated with $N$ and $N_C$, respectively. For $x, y \in C$ we have $x * y = \lambda \bar{x} \bar{y}$. Pick $x \in C$, $x \neq 0$, $\langle x, e \rangle_C = 0$, then $\bar{x} = -x$ and $x^2 = -N_C(x)e$ by (1.7). Using (1.10) we find

$$\langle x * x, x \rangle = \lambda \langle \bar{x}^2, x \rangle_C = -\lambda N_C(x) \langle e, x \rangle_C = 0,$$

which proves (ii).

(ii) $\Rightarrow$ (iii). If $T$ represents 0 nontrivially, we pick $y \in F$, $y \neq 0$, such that $T(y) = 0$. By (4.8),

$$(y * y) * (y * y) = -N(y)y * y.$$

We take $x = y$ if $y * y = 0$, and $x = y * y$ if $y * y \neq 0$; in either case, $x \neq 0$ and $x * x = \lambda x$ for some $\lambda \in l$. So (iii) holds.

(iii) $\Rightarrow$ (i). Under the assumption of (iii), we have to construct a composition algebra $C$ over $k$ such that $F \cong F(C)_\lambda$ for some $\lambda \in l^*$. We divide the proof into a number of steps. Pick $x \in F$, $x \neq 0$, such that $x * x = \lambda x$ with $\lambda \in l$. We first introduce $e \in F$ which is going to play the role of identity element in $C$.

(a) If $N(x) = \mu \neq 0$, we put $e = x$. Using (4.6) we see that

$$\sigma^2(\mu)x = \sigma^2(N(x))x = (x * x) * x = \lambda \sigma(\lambda)x.$$

It follows that $\sigma^2(\mu) = \lambda \sigma(\lambda)$, so $\mu = \sigma(\lambda)\sigma^2(\lambda)$ and $\lambda \neq 0$. Hence $e$ has the properties

$$e * e = \lambda e \quad \text{with } \lambda \in l, \ \lambda \neq 0, \tag{4.10}$$
$$N(e) = \mu = \sigma(\lambda)\sigma^2(\lambda) \neq 0. \tag{4.11}$$

(b) Assume now $N(x) = 0$. By (4.8),

$$(x * x) * (x * x) = T(x)x = \langle x * x, x \rangle x = \lambda \langle x, x \rangle x = 0.$$

Now either $x * x = 0$, or $y = x * x \neq 0$ satisfies $y * y = 0$ and

$$N(y) = N(x * x) = \sigma(N(x))\sigma^2(N(x)) = 0.$$

So we may as well assume that we have $x \neq 0$, $N(x) = 0$ and $x * x = 0$. Since $x$ is contained in a hyperbolic plane (see § 1.1), there exists $y \in F$ with $N(y) = 0$ and $\langle x, y \rangle = -1$. Using (4.5) we find:

$$x * (z * y) + y * (z * x) = -z \qquad (z \in F), \tag{4.12}$$

since $\langle x, y \rangle = -1$. Hence

$$F = x * F + y * F.$$

The sets $x * F$ and $y * F$ are subspaces of the vector space $F$. They are totally isotropic for $N$, since

$$N(x * u) = \sigma(N(x))\sigma^2(N(u)) = 0 \qquad (u \in F),$$

and similarly for $y * F$. It follows that both subspaces have dimension $\leq \frac{1}{2} \dim F$ (see § 1.1 again). Together they span $F$, so they must both have dimension equal $\frac{1}{2} \dim F$ and we have a direct sum decomposition

$$F = x * F \oplus y * F.$$

Consider the right multiplication $r_x^*$ in $F$. If $z * x = 0$, then $z = -x * (z * y)$ by (4.12), so $\ker r_x^* \subseteq x * F$. By (4.6), on the other hand, $r_x^*(x * z) = (x * z) * x = \sigma^2(N(x))z = 0$. Hence

$$\ker r_x^* = x * F.$$

In the same way we derive from (4.7) the following analog of (4.12):

$$(x * z) * y + (y * z) * x = -z \qquad (z \in F). \tag{4.13}$$

With the same arguments as above one derives:

$$F = F * x \oplus F * y.$$

For the left multiplication $l_x^*$ one derives using (4.13) and (4.4),

$$\ker l_x^* = F * x.$$

We define $a = x * y$ and $b = y * x$. Using the identities of Lemma 4.1.3 one verifies the following relations, where $\alpha = \langle y, x * y \rangle = \langle y, a \rangle$.

$$x * a = -x, \quad a * x = 0, \tag{4.14}$$
$$x * b = 0, \quad b * x = -x, \tag{4.15}$$
$$a * a = b + \sigma(\alpha)x, \tag{4.16}$$
$$b * b = a + \alpha x, \tag{4.17}$$
$$a * b = 0, \tag{4.18}$$
$$b * a = \sigma^2(\alpha)x. \tag{4.19}$$

From (4.14) and (4.15) we infer that $a \neq 0$ and $b \neq 0$, respectively. Using (4.3) and the relations in Def. 4.1.1 one further sees

$$\langle a, b \rangle = 1, \tag{4.20}$$
$$N(a) = N(b) = 0, \tag{4.21}$$
$$\langle a, x \rangle = \langle b, x \rangle = 0. \tag{4.22}$$

In this case we put $e = \sigma^2(\alpha)x + a + b$. We see that

$$
\begin{aligned}
e * e &= (\sigma^2(\alpha)x + a + b) * (\sigma^2(\alpha)x + a + b) \\
&= -\alpha x + b + \sigma(\alpha)x - \sigma(\alpha)x + \sigma^2(\alpha)x + a + \alpha x \\
&= \sigma^2(\alpha)x + a + b = e
\end{aligned}
$$

and

$$
\begin{aligned}
N(e) &= N(\sigma^2(\alpha)x + a + b) \\
&= \sigma^2(\alpha)^2 N(x) + N(a) + N(b) + \sigma^2(\alpha)\langle x, a \rangle + \sigma^2(\alpha)\langle x, b \rangle + \langle a, b \rangle \\
&= \langle a, b \rangle = 1.
\end{aligned}
$$

Again, $e$ satisfies equations (4.10) and (4.11), this time with $\lambda = \mu = 1$.
(c) The next step towards the construction of a composition algebra $C$ is the definition of conjugation in $F$ and of a $\sigma$-linear mapping $\varphi$ of order 3 (cf. (4.9) and the discussion preceding that equation).
    We define conjugation as in (1.9):

$$\bar{x} = -x + \langle x, e \rangle N(e)^{-1} e \qquad (x \in F). \tag{4.23}$$

Then $^-$ is a linear map, $\bar{e} = e$, $N(\bar{x}) = N(x)$, so $\langle \bar{x}, \bar{y} \rangle = \langle x, y \rangle$, and $\bar{\bar{x}} = x$. Further, we have for $x \in F$,

$$
\begin{aligned}
\overline{x * e} &= -x * e + \langle x * e, e \rangle N(e)^{-1} e \\
&= -x * e + \sigma(\langle e * e, x \rangle) N(e)^{-1} e \\
&= -x * e + \sigma(\lambda)\sigma(\langle e, x \rangle) N(e)^{-1} e \\
&= -x * e + \sigma(\langle e, x \rangle)(\lambda \sigma^2(\lambda))^{-1}(e * e) \\
&= (-x + \langle x, e \rangle N(e)^{-1} e) * e \\
&= \bar{x} * e,
\end{aligned}
$$

and similarly

$$\overline{e * x} = e * \bar{x}.$$

Notice that if $F = F(C)_\lambda$, then $x * e = \lambda\varphi(\bar{x})$ by (4.9). This leads us to define the $\sigma$-linear mapping

$$\varphi : F \to F, \; x \mapsto \lambda^{-1}(\bar{x} * e). \tag{4.24}$$

We show that $\varphi$ has order 3:

$$\begin{aligned}
\varphi^2(x) &= \lambda^{-1}\big((\lambda^{-1}\overline{(\bar{x} * e)}) * e\big) \\
&= \lambda^{-1}\sigma(\lambda)^{-1}((x * e) * e) \\
&= \lambda^{-1}\sigma(\lambda)^{-1}\big(-(e * e) * x + \sigma^2(\langle e, x\rangle)e\big) \\
&= \lambda^{-1}\big(-e * x + \lambda^{-1}\sigma(\lambda)^{-1}\sigma^2(\langle e, x\rangle)e * e\big) \\
&= \lambda^{-1}(e * \bar{x}). \\
\varphi^3(x) &= \varphi(\varphi^2(x)) \\
&= \lambda^{-1}\big(\overline{\lambda^{-1}(e * \bar{x})} * e\big) \\
&= \lambda^{-1}\sigma(\lambda)^{-1}((e * x) * e) \\
&= \lambda^{-1}\sigma(\lambda)^{-1}\sigma^2(N(e))x \; = \; x.
\end{aligned}$$

Further, $\varphi(e) = e$ and

$$\overline{\varphi(x)} = \overline{\lambda^{-1}(\bar{x} * e)} = \lambda^{-1}(x * e) = \varphi(\bar{x}).$$

(d) We now define a new product on $F$ (cf. (4.9)):

$$xy = \lambda^{-1}(\varphi^2(\bar{x}) * \varphi(\bar{y})) \qquad (x, y \in F). \tag{4.25}$$

We show that this product defines a structure of composition algebra over $l$ on the vector space $F$; see Def. 1.2.1. First, for $x \in F$,

$$xe = \lambda^{-1}(\varphi^2(\bar{x}) * e) = \lambda^{-1}\big((\lambda^{-1}(e * x)) * e\big) = \lambda^{-1}\sigma(\lambda)^{-1}\sigma^2(N(e))x = x,$$

and in a similar way we find that $ex = x$.

To prove that the norm $N$ permits composition up to a scalar factor we first show that $\varphi$ is a $\sigma$-similarity:

$$N(\varphi(x)) = N(\lambda^{-1}(\bar{x} * e)) = \lambda^{-2}\sigma(N(x))\sigma^2(N(e)) = \lambda^{-1}\sigma(\lambda)\sigma(N(x)). \tag{4.26}$$

Using this we find

$$\begin{aligned}
N(xy) &= N(\lambda^{-1}(\varphi^2(\bar{x}) * \varphi(\bar{y}))) \\
&= \lambda^{-2}\sigma(N(\varphi^2(\bar{x})))\sigma^2(N(\varphi(\bar{y}))) \\
&= \lambda^{-2}(\sigma(\lambda)^{-1}\lambda N(x)\sigma^2(\lambda)^{-1}\lambda N(y)) \\
&= \mu^{-1}N(x)N(y) \qquad (x, y \in F).
\end{aligned}$$

Hence, if we define

$$\tilde{N}(x) = \mu^{-1} N(x) \qquad (x \in F), \qquad (4.27)$$

$\tilde{N}$ permits composition. Thus we have proved that the vector space $F$ with the new, bilinear, product and the norm $\tilde{N}$ is a composition algebra over $l$; we denote this by $\tilde{C}$.

The conjugation $^-$ we defined in step (c) is the normal conjugation in this composition algebra $\tilde{C}$ with respect to the bilinear form of its norm $\tilde{N}$. So $\overline{xy} = \bar{y}\bar{x}$, $x\bar{x} = \tilde{N}(x)$, etc.

(e) In this step we prove that $\varphi$ is a $\sigma$-automorphism of the composition algebra $\tilde{C}$.

$$\begin{aligned}
\varphi(xy) &= \lambda^{-1}(\overline{xy} * e) = \lambda^{-1}(\bar{y}\bar{x} * e) \\
&= \lambda^{-1}\sigma(\lambda)^{-1}\big((\varphi^2(y) * \varphi(x)) * e\big) \\
&= \lambda^{-1}\sigma(\lambda)^{-1}\big(-(e * \varphi(x)) * \varphi^2(y) + \sigma^2(\langle \varphi^2(y), e\rangle)\varphi(x)\big) \\
&= \lambda^{-1}\sigma(\lambda)^{-1}\big((e * \varphi(x)) * (-\varphi^2(y) + \langle\varphi^2(y), e\rangle N(e)^{-1}e)\big) \\
&= \lambda^{-1}\sigma(\lambda)^{-1}\big((e * \varphi(x)) * \varphi^2(\bar{y})\big) \\
&= \lambda^{-1}\sigma(\lambda)^{-1}\big((\lambda\varphi^2(\overline{\varphi(x)})) * \varphi^2(\bar{y})\big) \\
&= \lambda^{-1}\big(\varphi^2(\overline{\varphi(x)}) * \varphi^2(\bar{y})\big) \\
&= \varphi(x)\varphi(y) \qquad (x, y \in \tilde{C}).
\end{aligned}$$

From (4.26) it follows that

$$\tilde{N}(\varphi(x)) = \sigma(\tilde{N}(x)) \qquad (x \in \tilde{C}). \qquad (4.28)$$

(f) Define

$$C = \{\, x \in \tilde{C} \mid \varphi(x) = x \,\}. \qquad (4.29)$$

Then $C$ is a vector space over $k$ such that $\tilde{C} \cong l \otimes_k C$ (see [Sp 81, 11.1.6]). From the multiplicativity of $\varphi$ it follows that $C$ is closed under multiplication, and (4.28) implies that $\tilde{N}(x) \in k$ for $x \in C$. Since $C$ contains the identity element $e$, it is a composition algebra over $k$. To verify condition (4.9), we compute

$$\begin{aligned}
\lambda(\varphi(\bar{x})\varphi^2(\bar{y})) &= \varphi^2(\overline{\varphi(\bar{x})}) * \varphi(\varphi^2(\bar{y})) \\
&= x * y \qquad (x, y \in F).
\end{aligned}$$

It follows that $F \cong F(C)_\lambda$.          $\square$

The above theorem enables us to prove, for some special fields $k$, that every normal twisted octonion algebra over a cubic cyclic extension field $l$ of $k$ is reduced, namely, if the cubic form $T$ in 24 variables over $k$ represents 0 nontrivially. This is so in the following two cases.

*(i) k a finite field.*
A theorem of Chevalley [Che 35, p. 75] (also in [Gre, Th. (2.3)], [Lang, third ed., 1993, p. 214, ex. 7]), [Se 70, § 2.2, Th. 3] or [Se 73, p. 5]) implies that every cubic form in more than three variables over a finite field represents 0 nontrivially.

*(ii) k a complete, discretely valuated field with finite residue class field.*
Every cubic form in more than nine variables over such a field represents 0 nontrivially; see [Sp 55, remark after Prop. 2], or also [De] or [Le].

If $k$ is an algebraic number field, then also every normal twisted octonion algebra over a cubic cyclic extension $l$ of $k$ is reduced; see the end of 4.8. This will follow from a further study: in §§ 4.5 and 4.6 we will thoroughly explore the structure of normal twisted octonion algebras, especially of those whose norm $N$ is isotropic, leading to another criterion (Th. 4.8.1) for normal twisted octonion algebras to be reduced. But first we introduce general twisted composition algebras in the following section, and in the next section we will identify the automorphism groups of reduced, normal or nonnormal, twisted composition algebras as twisted forms of an algebraic group of type $D_4$.

## 4.2 Nonnormal Twisted Composition Algebras

In the present section the separable cubic extension field $l$ of $k$ is not necessarily normal. Notations are as fixed in the introduction to this chapter.

Suppose $l' \neq l$. Then $\sigma(\lambda) \notin l$ if $\lambda \in l$, $\lambda \notin k$, so we cannot carry over the definition of a normal twisted composition algebra in Def. 4.1.1 to a vector space $F$ over $l$. However, $\sigma(\lambda)\sigma^2(\lambda) \in l$. If $F' = \lambda' \otimes_l F$ is a normal twisted composition algebra over $l'$ and if $x * x$ happens to lie in $F$ for $x \in F$, then also $(\lambda x) * (\lambda x) \in F$. This suggests that instead of the product $*$ we must consider the square $x^{*2} = x * x$. Thus we are led to the the following definition. For $\lambda \in l$ we write $\lambda^{*2} = \sigma(\lambda)\sigma^2(\lambda)$.

**Definition 4.2.1** Let $l$ be a separable cubic extension of $k$ and $\sigma$ a nontrivial $k$-isomorphism of $l$ into its normal closure $l'$. By a *twisted composition algebra* over $l$ and $\sigma$ we understand a vector space $F$ over $l$ provided with a unary operation $^{*2} : F \to F$, called the *squaring operation*, and a nondegenerate quadratic form $N$, called the *norm*, with associated bilinear form $\langle\ ,\ \rangle$, such that the following four axioms are satisfied:
(i) $(\lambda x)^{*2} = \lambda^{*2} x^{*2}$        $(\lambda \in l,\ x \in F)$;
(ii) $f : F \times F \to F$ defined by

$$f(x, y) = (x + y)^{*2} - x^{*2} - y^{*2} \qquad (x, y \in F)$$

is $k$-bilinear;

(iii) $N(x^{*2}) = N(x)^{*2} \qquad (x \in F)$;

(iv) $T(x) = \langle x^{*2}, x \rangle \in k \qquad (x \in F)$.

We use the notation $(F, {}^{*2}, N)$, or simply $F$, for such a twisted composition algebra.

An *isomorphism* of twisted composition algebras $F_1$ and $F_2$ as above is a bijective $l$-linear map $t : F_1 \to F_2$ such that $t(x^{*2}) = t(x)^{*2}$ for $x \in F_1$.

Notice that in this definition it is allowed that the cubic extension $l/k$ be Galois, so $l' = l$ and $k' = k$.

If $F$ is a twisted composition algebra as before, and $\lambda \in l^*$, we define the *isotope* $F_\lambda$ to be $(F, {}^{*'2}, N_\lambda)$, where $x^{*'2} = \lambda(x^{*2})$ and $N_\lambda(x) = \lambda^{*2} N(x)$. It is immediate that $F_\lambda$ verifies the axioms of a twisted composition algebra.

The relations between normal and nonnormal twisted composition algebras are given in the following two propositions.

**Proposition 4.2.2** *(i) If $l$ is cubic cyclic over $k$ and $(F, *, N)$ is a normal twisted composition algebra over $l$, then $F$ with the same norm $N$ and squaring operation defined by $x^{*2} = x * x$ $(x \in F)$ is a twisted composition algebra over $l$ and $\sigma$.*

*(ii) Let $\mathrm{char}(k) \neq 2, 3$. If $l$ is cubic over $k$, but not necessarily Galois, and $(F, {}^{*2}, N)$ is a twisted composition algebra over $l$ and $\sigma$, then $F' = l' \otimes_l F$ with the extension of $N$ to a quadratic form over $l'$ on $F'$ carries a unique structure of normal twisted composition algebra over $l'$ and $\sigma$ such that $x^{*2} = x * x$ for $x \in F$. (For uniqueness it is necessary that the extension of the isomorphism $\sigma$ of $l$ into $l'$ to an automorphism of $l'$ is given.)*

*(iii) If $F_1$ and $F_2$ are twisted composition algebras over $l$ and $\sigma$ with $\mathrm{char}(l) \neq 2, 3$ and $F_1'$ and $F_2'$, respectively, are their extensions to normal twisted composition algebras over $l'$ as in (ii), then an $l$-linear bijection $t : F_1 \to F_2$ is an isomorphism of twisted composition algebras if and only if its $l'$-linear extension is an isomorphism of normal twisted composition algebras between $F_1'$ and $F_2'$.*

Proof. (i) being obvious, we tackle (ii). Extend the $k$-bilinear mapping $f : F \times F \to F$ as in (ii) of Def. 4.2.1 to a $k'$-bilinear mapping $F' \times F' \to F'$. Pick a basis $e_1, \ldots, e_n$ of $F$ over $l$; this is also a basis of $F'$ over $l'$. Write

$$f(x, y) = \sum_{i=1}^{n} \varphi_i(x, y) e_i \qquad (x, y \in F')$$

with symmetric $k'$-bilinear $\varphi_i : F' \times F' \to l'$. By Dedekind's Theorem, the automorphisms id, $\sigma$ and $\sigma^2$ form a basis of $\mathrm{End}_{k'}(l')$ over $l'$. It follows that for each $i$,

$$\varphi_i(\lambda x, y) = \sum_{j=0}^{2} \sigma^j(\lambda) \psi_{i,j}(x,y) \qquad (\lambda \in l', \ x,y \in F')$$

with unique $k'$-bilinear mappings $\psi_{i,j} : F' \times F' \to l'$. Thus,

$$f(\lambda x, y) = \sum_{j=0}^{2} \sigma^j(\lambda) f_j(x,y) \qquad (\lambda \in l', \ x,y \in F')$$

with unique $k'$-bilinear $f_j : F' \times F' \to F'$. Repeating this argument, we find unique $k'$-bilinear mappings $g_{i,j} : F' \times F' \to F'$ such that

$$f(\lambda x, \mu y) = \sum_{i,j=0}^{2} \sigma^i(\lambda)\sigma^j(\mu) g_{i,j}(x,y) \qquad (\lambda,\mu \in l', \ x,y \in F'). \qquad (4.30)$$

From the symmetry of $f$ we infer, using Dedekind's Theorem again, that

$$g_{i,j}(x,y) = g_{j,i}(y,x) \qquad (x,y \in F') \qquad (4.31)$$

for $1 \le i,j \le n$. By considering $f(\lambda \alpha x, \mu \beta y)$ as a function of $\lambda$ and $\mu$, we find with Dedekind from (4.30) that

$$g_{i,j}(\alpha x, \beta y) = \sigma^i(\alpha)\sigma^j(\beta) g_{i,j}(x,y) \qquad (\alpha,\beta \in l', \ x,y \in F') \qquad (4.32)$$

for $1 \le i,j \le n$. Using this relation and the fact that $f(x,x) = 2x^{*2}$, we rewrite condition (i) of Def. 4.2.1 in the form

$$\sigma(\lambda)\sigma^2(\lambda) f(x,x) = f(\lambda x, \lambda x)$$
$$= \sum_{i,j=0}^{2} \sigma^i(\lambda)\sigma^j(\lambda) g_{i,j}(x,x) \qquad (\lambda \in l', \ x \in F').$$

Linearizing in $\lambda$, we get for $\lambda, \mu \in l'$, $x \in F'$

$$(\sigma(\lambda)\sigma^2(\mu) + \sigma^2(\lambda)\sigma(\mu)) f(x,x) = \sum_{i,j=0}^{2} (\sigma^i(\lambda)\sigma^j(\mu) + \sigma^j(\lambda)\sigma^i(\mu)) g_{i,j}(x,x)$$
$$= \sum_{i,j=0}^{2} \sigma^i(\lambda)\sigma^j(\mu)(g_{i,j}(x,x) + g_{j,i}(x,x)).$$

By Dedekind this implies

$$g_{i,j}(x,x) + g_{j,i}(x,x) = 0 \qquad (x \in F', \ (i,j) \ne (1,2),(2,1)).$$

If $\mathrm{char}(k) \ne 2$, it follows by (4.31) that $g_{i,j}(x,x) = 0$, so $g_{i,j}$ is antisymmetric for $(i,j) \ne (1,2),(2,1)$. From this we derive, using the symmetry of $f$, (4.30) and (4.31),

$$f(x,y) = \frac{1}{2}(f(x,y) + f(y,x))$$
$$= \frac{1}{2}(g_{1,2}(x,y) + g_{2,1}(x,y) + g_{1,2}(y,x) + g_{2,1}(y,x))$$
$$= g_{1,2}(x,y) + g_{1,2}(y,x) \qquad (x,y \in F').$$

Hence if we define

$$x * y = g_{1,2}(x,y) \qquad (x,y \in F'),$$

we have a $k'$-bilinear product on $F'$ which by (4.32) satisfies condition (i) of Def. 4.1.1 and such that $x^{*2} = x * x$ for $x \in F$. The uniqueness of the product $*$ is obvious from the proof.

Extend the norm $N$ on $F$ to a quadratic form over $l'$ on $F'$. In condition (iii) of Def. 4.2.1 we replace $x$ by $\lambda x + \mu y + \nu z + \varrho w$, with $x,y,z,w \in F$ and $\lambda, \mu, \nu, \varrho \in k$. Writing this as a polynomial in $\lambda$, $\mu$, $\nu$ and $\varrho$ and equating the terms with $\lambda\mu\nu\varrho$, we find

$$\langle f(x,y), f(z,w) \rangle + \langle f(x,z), f(y,w) \rangle + \langle f(x,w), f(y,z) \rangle =$$
$$\sigma(\langle x,y \rangle)\sigma^2(\langle z,w \rangle) + \sigma(\langle z,w \rangle)\sigma^2(\langle x,y \rangle) +$$
$$\sigma(\langle x,z \rangle)\sigma^2(\langle y,w \rangle) + \sigma(\langle y,w \rangle)\sigma^2(\langle x,z \rangle) +$$
$$\sigma(\langle x,w \rangle)\sigma^2(\langle y,z \rangle) + \sigma(\langle y,z \rangle)\sigma^2(\langle x,w \rangle).$$

Here we use that $k$ has more than four elements. The above relation is four-linear over $k$, so it remains valid if we extend $k$ to $k'$. Hence it also holds for $x,y,z,w \in F'$. Replace $x,y,z,w$ in the above relation by $\lambda x, \mu y, \nu z, \varrho w$, respectively, with $x,y,z,w \in F'$ and $\lambda, \mu, \nu, \varrho \in l'$. In the relation we thus obtain, the terms with $\sigma(\lambda)\sigma^2(\mu)\sigma(\nu)\sigma^2(\varrho)$ on either side must be equal by Dedekind, so

$$\langle x * y, z * w \rangle + \langle x * w, z * y \rangle = \sigma(\langle x,z \rangle)\sigma^2(\langle y,w \rangle).$$

Replacing $z$ by $x$ and $w$ by $y$ yields the validity of condition (ii) of Def. 4.1.1 for $F'$.

Applying a similar argument to (iv) of Def. 4.2.1, viz., trilinearization over $k$ and then extension of this field to $k'$, yields that $T(x) \in k'$ for all $x \in F'$; here we use char$(k) \neq 3$. Replacing $x$ by $\lambda x + \mu y + \nu z$ and using Dedekind then proves that condition (iii) of Def. 4.1.1 holds for $F'$. This completes the proof of part (ii) of the Proposition.

As to (iii), let $t : F_1 \to F_2$ be an isomorphism of twisted composition algebras. Denote its $l'$-linear extension also by $t$. On $F_1'$, $t^{-1}(t(x) * t(y))$ is a product for a normal twisted composition algebra which extends the squaring operation $x \mapsto x^{*2}$ on $F_1$, since $t^{-1}(t(x)^{*2}) = x^{*2}$ for $x \in F_1$. By uniqueness in (ii), $t^{-1}(t(x) * t(y)) = x * y$ for $x,y \in F_1'$, that is, the extension $t$ is an isomorphism of normal twisted composition algebras. The converse is obvious. $\qquad\square$

If $F$ is a twisted composition algebra over $l$ with char$(l) \neq 2, 3$, then the normal twisted composition algebra $F'$ over $l'$ determined by $F$ as in part (ii) of the above proposition will be called the *normal extension of $F$*. If $l$ is a cubic cyclic extension of $k$ of characteristic $\neq 2, 3$, a twisted composition algebra over $l$ may be identified with the normal twisted composition algebra it determines. The restriction to fields of characteristic $\neq 2, 3$ is not too much of a nuisance; the theory of twisted composition algebras is set up in view of applications to Jordan algebras (see Ch. 5 and 6), and there we need the same restriction on the characteristic.

It is clear that if two twisted compositions algebras are isotopic, the same holds for their normal extensions.

**Corollary 4.2.3** *Let* char$(l) \neq 2, 3$. *The norm $N$ of a twisted composition algebra $F$ over $l$ is uniquely determined by the linear structure and the squaring operation* $^{*2}$. *Isomorphisms of twisted composition algebras preserve the norm. A twisted composition algebra can only have dimension 1, 2, 4 or 8 over $l$.*

Proof. Let $F'$ be the normal extension of $F$. In the proof of part (ii) of the above proposition, the product $*$ on $F'$ is determined by the squaring operation $^{*2}$ on $F$ and the linear structure; the norm plays no role there. By Lemma 4.1.5, the norm on $F'$ is determined by the product and the linear structure. This proves the first statement. The second one is proved in a similar way. The last statement follows from Prop. 4.1.6 on the dimensions of twisted composition algebras.                                        □

As in the normal case, we speak in the case of dimension 8 about *twisted octonion algebras*.

If $F'$ is a normal twisted composition algebra over $l'$ and $l' \neq l$, how do we find twisted composition algebras $F$ over $l$ such that $F'$ is the normal extension of $F$ ? The following proposition gives an answer to this question. Recall that $\tau$ is the generator of Gal$(k'/k)$.

**Proposition 4.2.4** *Assume* char$(k) \neq 2, 3$ *and* $l' \neq l$, *so* $[l' : k] = 6$. *Let $F'$ be a normal twisted composition algebra over $l'$.*
*(i) If $F'$ is the normal extension of a twisted composition algebra over $l$ and $\sigma$, then there exists a unique bijective $\tau$-linear endomorphism $u$ of $F'$ satisfying $u^2 = $ id and*

$$u(x * y) = u(y) * u(x) \qquad (x, y \in F') \qquad (4.33)$$

*such that $F = \mathrm{Inv}(u) = \{ x \in F' \mid u(x) = x \}$.*
*(ii) Conversely, for any $u$ as in (i), $\mathrm{Inv}(u)$ is a twisted composition algebra over $l$ which has $F'$ as its normal extension.*
*(iii) Every $u$ as in (i) satisfies*

$$N(u(x)) = \tau(N(x)) \qquad (x \in F').$$

(iv) *If $F = \text{Inv}(u)$ and $F_1 = \text{Inv}(u_1)$ are twisted composition algebras over $l$ and $\sigma$ which both have $F'$ as their normal extension, then $F \cong F_1$ if and only if there exists an automorphism $t$ of $F'$ such that $u_1 = tut^{-1}$, and every isomorphism: $F \overset{\sim}{\to} F_1$ extends to such an automorphism. In particular,*

$$\text{Aut}(F) = \{ t|_F \,|\, t \in \text{Aut}(F'), \ tu = ut \,\}.$$

Proof. (i) Identify $F'$ with $l' \otimes_l F$. The transformation $u = \tau \otimes \text{id}$ is bijective $\tau$-linear with $u^2 = \text{id}$ and $F = \text{Inv}(u)$. To prove (4.33), consider for $\lambda, \mu \in l$ and $x, y \in F$,

$$z = (\lambda x) * (\mu y) + (\mu y) * (\lambda x) \in F,$$

so $z$ is invariant under $u$. By the $\tau$-linearity of $u$,

$$u(z) = \tau(\sigma(\lambda)\sigma^2(\mu))u(x * y) + \tau(\sigma(\mu)\sigma^2(\lambda))u(y * x).$$

Since $\tau\sigma = \sigma^2\tau$ and since $\lambda$ and $\mu$ are $\tau$-invariant, $u(z) = z$ implies

$$\sigma^2(\lambda)\sigma(\mu)u(x * y) + \sigma(\lambda)\sigma^2(\mu)u(y * x) =$$

$$(\lambda x) * (\mu y) + (\mu y) * (\lambda x) =$$

$$\sigma(\lambda)\sigma^2(\mu)(u(x) * u(y)) + \sigma^2(\lambda)\sigma(\mu)(u(y) * u(x)).$$

By Dedekind's Theorem,

$$u(x * y) = u(y) * u(x) \qquad (x, y \in F).$$

Since $l' = k' \otimes_k l$ we have $F' = l' \otimes_l F = k' \otimes_k F$. Using the $\tau$-linearity of $u$ we see that 4.33 holds.

(iii) Apply $u$ to both sides of equation (4.4) and use (4.6).

(ii) $F = \text{Inv}(u)$ is a vector space over $\text{Inv}(\tau) = l$ of the same dimension as the dimension of $F'$ over $l'$ (see, e.g., [Sp 81, 11.1.6]). If $x \in F$, then $u(x * x) = u(x) * u(x) = x * x$, so $x * x \in F$, and further $N(x) = N(u(x)) = \tau(N(x))$, so $N(x) \in l$. It is straightforward now that $F$ with $x^{*2} = x * x$ and the restriction of $N$ as norm is a twisted composition algebra which has $F'$ as its normal extension.

(iv) Let $s : F_1 \to F_2$ be an isomorphism of twisted composition algebras over $l$ and $\sigma$ which both have $F'$ as their normal extension. Let $F_i = \text{Inv}(u_i)$ ($i = 1, 2$). Define $t : F' \to F'$ as the $l'$-linear extension of $s$. From

$$s(x^{*2}) = s(x)^{*2} \qquad (x \in F)$$

we derive as in the proof of part (i) above that

$$t(x * y) = t(x) * t(y) \qquad (x, y \in F').$$

From $t(F_1) = F_2$ and it follows $u_2 = tu_1t^{-1}$. This implies the "only if" part of (iv). The "if" part is immediate.                    $\square$

A $\tau$-linear mapping $u$ as in the above proposition is called an *involution* of $F'$. If $F = \mathrm{Inv}(u)$, then $u$ is said to be the involution *associated with $F$.*

Let $F$ be any twisted composition algebra over $l$, with norm $N$. We have a (partial) analogue of Prop. 4.1.7. Let, as before, $N(F)^*$ to be the set of nonzero values of $N$ on $F$, and $M(N)$ the multiplier group of $N$.

**Proposition 4.2.5** $N(F)^*$ *is a coset* $\lambda M(N)$ *with* $\mathrm{N}_{l/k}(\lambda) \in M(N)$.

Proof. We represent $F$ as in the preceding Proposition, via $F'$ and $u$. We proceed as in the proof of Prop. 4.1.7, with $a \in F$ and $b = a$. We obtain a structure of composition algebra $C'$ on $F'$, with norm $\tilde{N} = N(a)^{*2} N$, and identity element $e = (N(a)^{*2})^{-1} a^{*2} \in F$. Moreover, we have $u(xy) = u(y)u(x)$ for $x, y \in C$. Let $v = s_e \circ u$. Then $v$ is a $\tau$-linear automorphism of $C'$ with fixed point set $F$. Now $C'$ induces on $F$ a structure of composition algebra $C$, with norm $\tilde{N}|_F$. It follows that $N(F)^* = N(a)^{*2} M(N)$, and $\mathrm{N}_{l/k}(N(a)) = N(a)^{*2} N(a) \in (N(a)^{*2})^2 M(N) = M(N)$.    $\square$

**Corollary 4.2.6** *(i) If $\lambda$ is as in the proposition, then $N(F_\lambda)^* = M(N)$. (ii) If $F_\mu \simeq F$, then $\mu \in k^* M(N)$.*

Proof. The first point follows from the last equality of the proof. If $F_\mu \simeq F$, then $N(F_\mu)^* = N(F)^*$. Since the multiplier groups of $N$ and $N_\mu$ are the same, it follows from the proposition that $\mu^{*2} \in M(N)$. But then $\mu = \mathrm{N}_{l/k}(\mu)(\mu^{*2})^{-1} \in k^* M(N)$, proving the second point.    $\square$

It can be shown that if $F$ is a normal twisted octonion algebra over $l$ and $\sigma$ (and char $l \neq 2, 3$) the converse of (ii) is also true, see [KMRT, Th. (36.9)]. The proof is rather delicate.

In view of the close connection between normal and nonnormal twisted composition algebras, one can expect properties of the former to be inherited by the latter. We give one identity, to be used later in Lemma 4.1.3.

**Lemma 4.2.7** *In a twisted composition algebra $F$ over a field of characteristic $\neq 2, 3$, the following identity holds for $x \in F$, $\alpha \in l$:*

$$(\alpha x + x^{*2})^{*2} = (T(x) - \alpha N(x) + \mathrm{Tr}_{l/k}(\alpha N(x)))x + (\alpha^{*2} - N(x))x^{*2}. \quad (4.34)$$

Proof. In the normal case this follows from the formulas of Lemma 4.1.3. If $F$ is nonnormal, work in the normal extension.    $\square$

It would be rather natural to call a nonnormal twisted composition algebra $F$ reduced if its normal extension $F'$ is so. We prefer an apparently stronger definition; we will see in Th. 4.2.10 that these two definitions are in fact equivalent.

If $C$ is a composition algebra over $k$, we have over $l'$ the normal twisted composition algebra $F(k' \otimes_k C)$. Its underlying vector spaces is

$$F' = l' \otimes_{k'} (k' \otimes_k C) = l' \otimes_k C.$$

On $F'$ we have the $\tau$-linear automorphism $u$ with $u(\xi \otimes_k x) = \tau(\xi) \otimes \bar{x}$ ($\xi \in l', x \in C$). Let $F = \text{Inv}(u)$. It is straightforward to check that $F'$ and $u$ are as in Prop. 4.2.4. By that Proposition we obtain a structure of twisted composition algebra on $F$. We denote this twisted composition algebra by $F(C)$.

**Definition 4.2.8** A twisted composition algebra $F$ over a field $l$ of characteristic $\neq 2, 3$ is said to be *reduced* if there exist a composition algebra $C$ over $k$ and $\lambda \in l^*$ such that $F$ is isomorphic to the isotope $F(C)_\lambda$.

If $l$ is cubic cyclic over $k$, then this boils down to the definition of 4.1.8. If $F(C)_\lambda$ is a nonnormal reduced twisted composition algebra over $l$, then its normal extension is $F(C')_\lambda$ with $C' = k' \otimes_k C$.

If $F = F(C)_\lambda$ is a reduced nonnormal composition algebra for $l$ and $\sigma$, then
$$F = (l \otimes_k e) \oplus (l\sqrt{D} \otimes_k C_0),$$
where $D$ is the discriminant of $l$ over $k$ (so $\tau(\sqrt{D}) = -\sqrt{D}$) and $C_0 = e^\perp$ in $C$. We have the following formulas for the squaring operation and the norm in $F$ (where $\xi, \eta \in l$, $x \in C_0$) :

$$(\xi \otimes e + \eta\sqrt{D} \otimes x)^{*2} = \lambda(\sigma(\xi)\sigma^2(\xi) - D\sigma(\eta)\sigma^2(\eta)N_C(x)) \otimes e$$
$$-\lambda\sqrt{D}(\sigma(\xi)\sigma^2(\eta) + \sigma^2(\xi)\sigma(\eta)) \otimes x, \quad (4.35)$$
$$N(\xi \otimes e + \eta\sqrt{D} \otimes x) = \sigma(\lambda)\sigma^2(\lambda)(\xi^2 + \eta^2 D N_C(x)). \quad (4.36)$$

In Prop. 4.1.9 (ii) we saw that the automorphism group of a reduced normal twisted composition algebra $F(C)_\lambda$ is independent of $\lambda$. The same holds for reduced nonnormal twisted composition algebras.

**Proposition 4.2.9** *The automorphism group of the twisted composition algebra $F = F(C)_\lambda$ does not depend on $\lambda$.*

Proof. By Prop. 4.2.4, $\text{Aut}(F) = \{ t|_F \mid t \in \text{Aut}(F'), tu = ut \}$, where $F'$ is the normal twisted composition algebra $F(C')_\lambda$ with $C' = k' \otimes_k C$. Now $\text{Aut}(F')$ is independent of $\lambda$, and the same holds for action on it of the involution $u$. $\square$

Th. 4.1.10 on the characterization of reduced normal twisted composition algebras carries over to the following result for the general case.

**Theorem 4.2.10** *Let $F$ be a twisted composition algebra over $l$, char$(l) \neq 2, 3$, and let $F'$ be its normal extension. The following conditions are equivalent.*
*(i)  $F$ is reduced.*
*(ii)  $F'$ is reduced.*
*(iii) $T$ represents 0 nontrivially on $F$, i.e., there exists $x \neq 0$ in $F$ such that $T(x) = \langle x^{*2}, x \rangle = 0$.*
*(iv) There exists $x \neq 0$ in $F$ such that $x^{*2} = \lambda x$ for some $\lambda \in l$.*

Proof. We may assume that $F$ is nonnormal. If $F$ is reduced, then $F' = F(C')_\lambda$ with $C' = k' \otimes_k C$ and $C$ as in Def. 4.2.8, so (i) implies (ii). If $F'$ is reduced, $T$ represents zero nontrivially on $F'$ by Th. 4.1.10, hence so it does on $F$ by the lemma below. Thus, (ii) implies (iii). If (iii) holds, then (iv) follows by the same argument as in the proof of Th. 4.1.10.

Finally, assume (iv) holds. We follow the lines of the proof of the implication (iii) $\Rightarrow$ (i) in Th. 4.1.10, with some adaptations; we refer to that proof as to "the old proof". Choose $x \in F$ such that $x^{*2} = \lambda x$ with $\lambda \in l$. If $N(x) \neq 0$, we take $e = x$ as in part (a) of the old proof. If $N(x) = 0$, we follow part (b) of that proof. We may assume that $x \neq 0$, $N(x) = 0$ and $x * x = 0$. Pick $y \in F$ with $N(y) = 0$ and $\langle x, y \rangle = -1$. Then $F' = x * F' + y * F'$.

We take again $e = \sigma^2(\alpha)x + x * y + y * x$ with $\alpha = \langle y, x * y \rangle$. Notice that

$$\sigma^2(\alpha) = \sigma^2(\langle y, x * y \rangle) = \langle x, y^{*2} \rangle \in l$$

and that $x * y + y * x = (x + y)^{*2} - x^{*2} - y^{*2} \in F$, so $e \in F$. In either case, whether $N(x)$ equals zero or not, we have found $e \in F$ satisfying equations (4.10) and (4.11). Now proceeding as in step (c) and (d) of the old proof, one arrives at an $l'$-bilinear product and a new norm $\tilde{N}$ which define on $F'$ a structure of composition algebra over $l'$ with identity element $e$; we call this $\tilde{C}$. As in step (e) of the old proof, we show that $\varphi$ (as in the old proof) is a $\sigma$-automorphism of $\tilde{C}$ of order 3 which commutes with conjugation.

Let $u$ be the involution of $F'$ associated with $F$. Straightforward computations show that $u$ commutes with conjugation and that $u\varphi = \varphi^2 u$. From this it follows that $u$ is a $\tau$-linear anti-automorphism of $\tilde{C}$. We define by $u_\sigma = \varphi$ and

$$u_\tau(x) = u(\bar{x}) \qquad (x \in F')$$

an isomorphism $\varrho \mapsto u_\varrho$ of $\mathrm{Gal}(l'/k)$ onto a subgroup of the group of semilinear automorphisms of $\tilde{C}$ such that each $u_\varrho$ is a $\varrho$-automorphism. As in step (f) of the old proof one proves that

$$C = \mathrm{Inv}(\, u_\varrho \mid \varrho \in \mathrm{Gal}(l'/k)\,).$$

with the restriction of $\tilde{N}$ as norm is a composition algebra over $k$ which has the properties required by Def. 4.2.8. Thus, $F$ is reduced. $\qquad\square$

We still owe the reader the lemma we used in the beginning of the above proof.

**Lemma 4.2.11** Let $U$ be any vector space over a field $k$, $T$ a cubic form on $U$, and $k'$ a quadratic extension of $k$. If $T$ represents zero nontrivially on $U' = k' \otimes_k U$, then so it does on $U$ itself.

Proof. Pick a basis $1, \varepsilon$ of $k'$ over $k$ and write the elements of $U'$ as $x + \varepsilon y$ with $x, y \in U$. Let $T(x + \varepsilon y) = 0$ for some $x + \varepsilon y \neq 0$. If $y = 0$ or if $y \neq 0$ and $T(y) = 0$, we are done. So let $T(y) \neq 0$. Consider the cubic polynomial

$T(x + Xy) \in k[X]$. This has a root $\varepsilon$ in the quadratic extension $k'$ of $k$, so it must have a root $\alpha$ in $k$ itself. Hence $T(x + \alpha y) = 0$. If $x + \alpha y \neq 0$, we are done. If $x + \alpha y = 0$, we find

$$T(x + \varepsilon y) = T(-\alpha y + \varepsilon y) = (\varepsilon - \alpha)^3 T(y) \neq 0,$$

a contradiction. □

The following lemma will be used later. Let $F$ be an arbitrary twisted composition algebra. The notations are as in Def. 4.1.1.

**Lemma 4.2.12** *There exists $a \in F$ such that the following conditions are satisfied.*
*(i)  $T(a) = \langle a^{*2}, a \rangle \neq 0$;*
*(ii)  $a$ and $a^{*2}$ are linearly independent over $l$;*
*(iii)  the restriction of $\langle \, , \, \rangle$ to the two-dimensional subspace $la + la^{*2}$ is nondegenerate, or equivalently (provided (ii) holds), $T(a)^2 - 4\,\mathrm{N}_{l/k}(N(a)) \neq 0$.*

*If $N$ is isotropic, there exists isotropic $a$ with $T(a) \neq 0$. Such an $a$ also satisfies the conditions (ii) and (iii); moreover, $a^{*2}$ is isotropic and satifies (i), (ii) and (iii), and we have $(a^{*2})^{*2} = T(a)a$ and $T(a^{*2}) = T(a)^2$.*

Proof. If $a$ and $a^{*2}$ are linearly independent, $\langle \, , \, \rangle$ is degenerate on $la \oplus la^{*2}$ if and only if

$$\langle a, a \rangle \langle a^{*2}, a^{*2} \rangle - \langle a^{*2}, a \rangle^2 = 0,$$

i.e., if

$$T(a)^2 - 4\mathrm{N}_{l/k}(N(a)) = 0,$$

so indeed the two conditions in (iii) are equivalent, provided (ii) holds.

If $F$ is not reduced, every nonzero $a \in F$ satifies (i) and (ii) by Th. 4.2.10. If $N$ is isotropic, we choose $a \neq 0$ with $N(a) = 0$, then (iii) also holds. For anisotropic $N$ we argue as follows: if $\mathrm{char}(k) \neq 2$, then no vector can be orthogonal to itself, so the bilinear form $\langle \, , \, \rangle$ is nondegenerate on any subspace; if $\mathrm{char}(k) = 2$ and $a \neq 0$, then

$$T(a)^2 - 4\,\mathrm{N}_{l/k}(N(a)) = T(a)^2 \neq 0.$$

Now assume $F$ reduced: $F = F(C)_\lambda$ for an octonion algebra $C$ over $k$ and some $\lambda \in l^*$. Let $D$ be a two-dimensional composition subalgebra of $C$. If we have $a \in D$ satisfying (i) and (ii) then (iii) must hold, too. For $a \in C$ we have

$$\begin{aligned}
a^{*2} &= \lambda \bar{a}^2 = \lambda(-a + \langle a, e \rangle e)^2 \\
&= \lambda(a^2 - 2\langle a, e \rangle a + \langle a, e \rangle^2 e) \\
&= \lambda(-\langle a, e \rangle a + (\langle a, e \rangle^2 - N(a))e),
\end{aligned}$$

and

$$T(a) = \langle a^{*2}, a \rangle = \mathrm{N}_{l/k}(\lambda)\langle a, e \rangle \big( \langle a, e \rangle^2 - 3N(a) \big).$$

The conditions (i) and (ii) together are equivalent to the following four conditions:

$$a \notin ke, \quad \langle a, e \rangle \neq 0, \quad \langle a, e \rangle^2 \neq N(a), \quad \langle a, e \rangle^2 \neq 3N(a).$$

If the restriction of $N$ to $D$ is isotropic, we pick $a \in D$ with $\langle a, e \rangle = 1$ and $N(a) = 0$; this satisfies our conditions. If $k$ is finite we may assume this to be the case. So we can now assume that $k$ is infinite and $N$ is anisotropic on $D$. The four conditions require $a \in D$ to lie outside a finite number of lines in $D$. Since $k$ is infinite, such $a$ exist. This proves the first part of the Lemma.

Finally, let $a$ with $N(a) = 0$ satisfy $T(a) \neq 0$. Since $\langle a, a^{*2} \rangle = T(a) \neq 0$, (ii) must hold. Further, (iii) holds. From (4.34) we infer that $(a^{*2})^{*2} = T(a)a$, so $T(a^{*2}) = T(a)^2 \neq 0$; further, $N(a^{*2}) = 0$. Hence $a^{*2}$ is isotropic and satisfies (i), so also (ii) and (iii). □

At the end of § 4.1 we gave some examples of fields over which every normal twisted composition algebra is reduced. These fields have the same property for arbitrary twisted composition algebras. This is clear for the case of finite fields: since every finite extension of a finite field is Galois, every twisted composition algebra over a finite field is normal. Over a complete, discretely valued field with finite residue class field every twisted composition algebra is reduced; in the nonnormal case this follows from the above theorem by the same argument as used at the end of § 4.1 for the normal case.

## 4.3 Twisted Composition Algebras over Split Cubic Extensions

In this section we generalize the notion of twisted composition algebra to the situation where the cubic field extension $l/k$ is replaced by a direct sum of three copies of a field. This generalization will be used in the next section to determine the automorphism groups of eight-dimensional twisted composition algebras. These will turn out to be twisted forms of groups of type $D_4$.

The motivation for the generalization lies in the following situation. Consider as in § 4.1 a normal twisted composition algebra $F = (F, *, N)$ over a cubic cyclic extension field $l$ of $k$. Let $K$ be an extension field of $l$; this is a splitting field of $l$ over $k$, i.e., an extension of $k$ such that $L = K \otimes_k l \cong K \oplus K \oplus K$. See Ch. I, § 16 (in particular ex. 2) in [Ja 64a]. Denote the three primitive idempotents in $L$ by $e_1$, $e_2$ and $e_3$, so $e_1 = (1, 0, 0)$, etc. The action of the Galois automorphism $\sigma$ of $l$ over $k$ is extended $K$-linearly to $L$, i.e., as $\mathrm{id} \otimes \sigma$; it then induces a cyclic permutation of the primitive idempotents, say,

$\sigma(e_i) = e_{i-1}$ (indices to be taken mod 3) (cf. the proof of Th. 8.9 in [Ja 80]). The group $<\sigma>$ plays the role of "Galois group" of $L$ over $K$.

Extend the vector space $F$ over $l$ to the free module $F_K = K \otimes_k F$ over $L$, and the norm $N$, which is a quadratic form over $l$, to a quadratic form over $L$ on $F_K$, also denoted by $N$. Finally, extend the $k$-bilinear product $*$ on $F$ to a $K$-bilinear product $*$ on $F_K$. The conditions (i), (ii) and (iii) of Def. 4.1.1 remain valid on $F_K$. (Since (ii) involves polynomials of degree 4 over $k$, one has difficulties if $k$ does not have at least five elements; these difficulties are avoided if one views $N$ as a polynomial function on $F_K$ which is defined over $k$.) We thus arrive at a notion of twisted composition algebra over the split cubic extension $L$ of $K$; we formalize this in the following definition.

**Definition 4.3.1** Let $K$ be any field. By the *split cubic extension* of $K$ we understand the $K$-algebra $L = K \oplus K \oplus K$. Call its primitive idempotents $e_1 = (1,0,0)$, $e_2 = (0,1,0)$ and $e_3 = (0,0,1)$. Fix the $K$-automorphism $\sigma$ of $L$ by $\sigma(e_i) = e_{i-1}$ $(i = 1,2,3 \bmod 3)$. A *twisted composition algebra* over $L$ and $\sigma$ is a free $L$-module $F$ provided with a $K$-bilinear product $*$ and a non-degenerate quadratic form $N$ over $L$ (this notion being defined in the obvious manner) such that the conditions (i), (ii) and (iii) of Def. 4.1.1 hold.

If $l$ with char$(l) \neq 2,3$ is a cubic field extension of $k$ which is not normal and $F$ is a twisted composition algebra over $l$, we have a normal extension $F' = l' \otimes_l F$, which we identify with $k' \otimes_k F$. If now $K$ is any field extension of $l'$, we have again $K \otimes_k l = K \otimes_{k'} l' = K \oplus K \oplus K$ (with obvious identifications of tensor products), and $F_K = K \otimes_k F = K \otimes_{k'} F'$ with the twisted composition algebra structure induced by that on $F'$ is again a twisted composition algebra over the split extension $L = K \oplus K \oplus K$ of $K$.

The 3-cyclic group generated by $\sigma$ plays the role of "Galois group" of $L$ over $K$. The formulas of Lemmas 4.1.2 and 4.1.3 remain valid in the situation of Def. 4.3.1; we will, in fact, only need (4.4) and (4.6).

We now turn to an explicit determination of the structure of a twisted composition algebra $F$ over $L = K \oplus K \oplus K$. It will turn out that $F$ is the direct sum of three copies of a composition algebra $C$ over $K$, with the product $*$ in $F$ determined in a specific way by the product in $C$.

Put $F_i = e_i F$ for $i = 1,2,3$; these are vector spaces over $K$, and we have a direct sum decomposition

$$F = F_1 \oplus F_2 \oplus F_3$$

of vector spaces over $K$. The formulas in (i) of Def. 4.1.1 with $\lambda = e_i$ show that

$$F_i * F \subseteq F_{i+2} \quad \text{and} \quad F * F_i \subseteq F_{i+1} \quad (i = 1,2,3).$$

It follows that for $1 \leq i, j \leq 3$,

$$F_i * F_j = 0 \quad (j \neq i+1) \quad \text{and} \quad F_i * F_{i+1} \subseteq F_{i+2}.$$

For $x \in F_i$ we have

$$N(x) = N(e_i x) = e_i N(x) \in e_i L = K e_i.$$

So we can define $N_i : F_i \to K$ by

$$N_i(x) e_i = N(x) \qquad (x \in F_i).$$

It is readily verified that $N_i$ is a nondegenerate quadratic form on the vector space $F_i$ over $K$ for $i = 1, 2, 3$. We denote the associated $K$-bilinear form by $N_i(\ ,\ )$. Formulas (ii) and (iii) of Def. 4.1.1 yield

$$N_{i+2}(x * y) = N_i(x) N_{i+1}(y) \qquad (x \in F_i,\ y \in F_{i+1}), \tag{4.37}$$

$$N_i(y * z, x) = N_{i+1}(z * x, y) \qquad (x \in F_i,\ y \in F_{i+1},\ z \in F_{i+2}) \tag{4.38}$$

for $i = 1, 2, 3$.

Put $C = F_1$. Take $a \in F_3$ with $N_3(a) \neq 0$ and $b \in F_2$ with $N_2(b) \neq 0$. In the same way as in Lemma 4.1.4 we derive that the $K$-linear map $f_2 : C \to F_2$, $x \mapsto a * x$, is bijective, and similarly for $f_3 : C \to F_3$, $x \mapsto x * b$.

We define a $K$-bilinear product on $C$ by

$$xy = (a * x) * (y * b) = f_2(x) * f_3(y) \qquad (x, y \in C). \tag{4.39}$$

Using (4.4) and (4.6) one sees that $e = N_3(a)^{-1} N_2(b)^{-1} (b * a)$ is an identity element for this multiplication. Putting $N_0(x) = N_3(a) N_2(b) N_1(x)$ $(x \in C)$, we conclude from (4.37) that

$$N_0(xy) = N_0(x) N_0(y) \qquad (x, y \in C).$$

Thus we have obtained on $C$ a structure of composition algebra over $K$; we denote this by $C_{a,b}(F)$. Set $f_1 = \mathrm{id} : C \to F_1$. We have constructed a $K$-linear bijection

$$f = (f_1, f_2, f_3) : C \oplus C \oplus C \to F,\ (x_1, x_2, x_3) \mapsto (f_1(x_1), f_2(x_2), f_3(x_3)) \tag{4.40}$$

with $f_i(x_i) \in F_i$. Notice that $F$ determines the norm of $C$ up to a $K^*$-multiple, hence it determines $C$ up to isomorphism by Th. 1.7.1.

We encountered a similar situation in the proof of Prop. 4.1.6 (which we gave before stating the proposition itself). In the present case, too, $F$ can have dimension 1, 2, 4 or 8 over $L$. Notice that the composition algebra structure on $C$ determines the quadratic form $N$ and the product $*$ on $F$, provided the $K$-linear bijections $f_2 : C \to F_2$ and $f_3 : C \to F_3$ as well as $N_3(a)$ and $N_2(b)$ are given; to prove this, use (4.37), (4.38), (4.4) and (4.6).

From any composition algebra $C$ over $K$ we can construct a twisted composition algebra $F_s(C)$ over $L = K \oplus K \oplus K$. As an $L$-module we take $F_s(C) = C \oplus C \oplus C$, the product and the norm are defined by

$$(x, y, z) * (u, v, w) = (\bar{y}\bar{w}, \bar{z}\bar{u}, \bar{x}\bar{y}),$$
$$N((x, y, z)) = (N_0(x), N_0(y), N_0(z)),$$

with $N_0$ denoting the norm of $C$. One easily verifies the conditions of Def. 4.1.1, keeping in mind that the "Galois automorphism" $\sigma$ acts on $L$ by $\sigma((\alpha, \beta, \gamma)) = (\beta, \gamma, \alpha)$. Taking $a = (0, 0, 1)$ and $b = (0, 1, 0)$ we reconstruct $C$ from $F_s(C)$ as described above: $C = C_{a,b}(F_s(C))$.

If we start from an arbitrary twisted composition algebra $F$ over the split cubic extension $L$ of $K$, and construct the composition algebra $C = C_{a,b}(F)$ with the aid of $a \in F_3$ and $b \in F_2$ with $N_3(a) = N_2(b) = 1$ (provided these exist), then $F \cong F_s(C)$.

## 4.4 Automorphism Groups of Twisted Octonion Algebras

Let $F$ be a twisted octonion algebra, either normal over the cubic cyclic extension field $l$ of $k$, or nonnormal over the separable but not Galois cubic extension field $l$ of $k$; in the latter case we assume $\operatorname{char} k \neq 2, 3$. Denote the automorphism group of $F$ by $\operatorname{Aut}(F)$. An automorphism of a twisted composition algebra $F_K$ over a split cubic extension $L$ of $K$ is, of course, an $L$-linear bijection that preserves $*$; as in Lemma 4.1.5 one sees that it also leaves $N$ invariant. The group of these automorphisms is denoted by $\operatorname{Aut}(F_K)$. If we take for $K$ an algebraic closure of $l'$, the automorphisms of $F_K = K \otimes_k F$ form a algebraic group $\mathbf{G}$.

Now let $K$ again be any extension field of $l'$. Let $u$ be an automorphism of $F_K$; the $L$-linearity of $u$ implies that it stabilizes every $F_i$. Let $u_i$ be the restriction of $u$ to $F_i$; it is a $K$-linear bijection. As in the preceding section, let $C$ be the octonion algebra $C_{a,b}(F)$ over $K$ defined by $a \in F_3$ and $b \in F_2$ with $N(a)N(b) \neq 0$. We have the linear bijection of (4.39)

$$f = (f_1, f_2, f_3) : C \oplus C \oplus C \to F.$$

From (4.39) we infer that

$$u_1(xy) = u_2(f_2(x)) * u_3(f_3(y)) \qquad (x, y \in C). \tag{4.41}$$

Define $t = (t_1, t_2, t_3) \in \operatorname{GL}(C)^3$ by $t = f^{-1} \circ u \circ f$, i.e., $t_i = f_i^{-1} \circ u_i \circ f_i$. Since $u$ preserves $N$, all $t_i$ lie in $O(N_0)$. According to (4.41) and (4.39), they must satisfy the condition

$$t_1(xy) = t_2(x)t_3(y) \qquad (x, y \in C)$$

(remember that $f_1 = \operatorname{id}$). This means that $(t_1, t_2, t_3)$ is a related triple of rotations of $C$ (cf. the Principle of Triality, Th. 3.2.1), necessarily of spinor norm 1 (cf. § 3.2, in particular Prop. 3.2.2 and Cor. 3.3.3). Conversely, any related triple of rotations $(t_1, t_2, t_3)$ (necessarily of spinor norm 1) of $C$ defines an automorphism $u$ of $F_K$ as above. Thus we get an isomorphism between $\mathrm{Aut}(F_K)$ and the group $\mathrm{RT}(C)$ of related triples of rotations of $C_K$ (see § 3.6):

$$\varphi : G \to \mathrm{RT}(C), \quad u \mapsto f^{-1} \circ u \circ f. \tag{4.42}$$

Taking first $K = l'$, we obtain a composition algebra $C$ over $l'$. Then taking for $K$ the algebraic closure of $l'$, we see that the algebraic groups $\mathbf{G}$ and $\mathbf{RT}(C_K)$ are isomorphic. Now by Prop. 3.7.1 the group $\mathbf{RT}(C_K)$ is defined over $l'$. Also, the isomorphism $f$ of 4.40 is defined over $l'$. We thus obtain on $\mathbf{G}$ a structure of algebraic group over $l'$. By Prop. 3.6.3, $\mathbf{RT}(C_K)$ is isomorphic to the spin group $\mathbf{Spin}(N_0)$. Thus we have shown the following result.

**Proposition 4.4.1** $\mathbf{G}$ *is an algebraic group over $l'$ which is isomorphic to* $\mathbf{Spin}(8)$.

We will see in Th. 4.4.3 that $\mathbf{G}$ is defined over $k$.

In the rest of this section we take for $K$ a separable closure $k_s$ of $k$ containing $l'$. Then $O'(N_0) = \mathrm{SO}(N_0)$. For the case $\mathrm{char}(k) \neq 2$ this follows from the fact that $k_s$ contains all square roots of its elements, so all spinor norms are 1. In all characteristics, one can use a Galois cohomology argument, see [Sp 81, §12.3]. The Galois group $\mathrm{Gal}(k_s/k)$ (which we understand to be the topological Galois group if $k_s$ has infinite degree over $k$) acts on $k_s \otimes_k l$ and on $F_{k_s} = k_s \otimes_k F$ by acting on the first factor. It permutes the idempotents $e_i$ and the components $F_i$; thus we have a homomorphism $\alpha : \mathrm{Gal}(k_s/k) \to S_3$ such that

$$\gamma(e_i) = e_{\alpha(\gamma)(i)} \qquad \big(\gamma \in \mathrm{Gal}(k_s/k)\big).$$

**Lemma 4.4.2** $\alpha(\mathrm{Gal}(k_s/k))$ *has order 3 if $l$ is Galois over $k$, and order 6 if $l$ is not Galois over $k$, so it equals the degree of $l'$ over $k$.*

Proof. The invariants of $\mathrm{Gal}(k_s/k)$ in $k_s \otimes_k l$ form the field $k \otimes_k l = l$, so every idempotent $e_i$ is displaced. This implies that $\alpha(\mathrm{Gal}(k_s/k))$ has order at least 3. The idempotents exist already in $l' \otimes_k l$, which is invariant under $\mathrm{Gal}(k_s/k)$, so their permutation is in fact accomplished by the action of $\mathrm{Gal}(l'/k)$, that is, $\alpha$ can be factored through a homomorphism $\alpha' : \mathrm{Gal}(l'/k) \to S_3$. If $l' = l$, then $\mathrm{Gal}(l'/k)$ has order 3, so then $|\alpha(\mathrm{Gal}(k_s/k))| = 3$. If $l' \neq l$, then $\mathrm{Gal}(l'/k) \cong S_3$ and the kernel of $\alpha'$ in $\mathrm{Gal}(l'/k)$ is a normal subgroup of order at most 2, so it consists of the identity only, whence $|\alpha(\mathrm{Gal}(k_s/k))| = 6$. $\quad\square$

Over $k_s$ the octonion algebra $C_{k_s} = k_s \otimes_k C$ is split, so we may replace $C = C_{a,b}(F)$ by the split octonion algebra. Thus we get from (4.42) an isomorphism, which we also call $\varphi$, of $\mathbf{G}(k_s)$ onto $\mathbf{RT}(C)(k_s)$, where $C$ is the split octonion algebra over $l$. For $\gamma \in \mathrm{Gal}(k_s/k)$ we have the conjugate isomorphism ${}^\gamma\varphi = \gamma \circ \varphi \circ \gamma^{-1}$. Then $z(\gamma) = {}^\gamma\varphi \circ \varphi^{-1}$ is an automorphism of $\mathbf{RT}(C)(k_s)$. This defines a nonabelian 1-cocycle of $\mathrm{Gal}(k_s/k)$ with values in $\mathbf{RT}(C)(k_s)$ (see [Sp 81, § 12.3]).

The automorphism $z(\gamma)$ acts on $\mathbf{RT}(C)(k_s)$ as

$$z(\gamma) : (t_1, t_2, t_3) \mapsto (t'_{\alpha(\gamma)(1)}, t'_{\alpha(\gamma)(2)}, t'_{\alpha(\gamma)(3)}),$$

where $\alpha$ is the homomorphism of $\mathrm{Gal}(k_s/k)$ into $S_3$ as in the above lemma, and $t'_j$ is the image of $t_j$ under some inner automorphism (depending on $j$) of $\mathbf{SO}(N_0)$. In case $\mathrm{char}(k) \neq 2$, $t'_j = t_j$ if $t_j = \pm 1$, so then $z(\gamma)$ permutes the central elements $(1, -1, -1)$ etc. of $\mathbf{RT}(C)(k_s)$ in the same way as $\alpha(\gamma^{-1})$ does; if $\mathrm{char}(k) = 2$, a similar argument with central elements of the Lie algebra of $\mathbf{RT}$ works (see Prop. 3.6.4 and its proof). It follows that the image of $z(\gamma)$ in $\mathrm{Aut}(\mathbf{RT}(C))/\mathrm{Inn}(\mathbf{RT}(C)) \cong S_3$ is $\alpha(\gamma^{-1})$.

$\mathbf{G}$ is a $k_s$-form of $\mathbf{RT}(C)$ which can be obtained by twisting $\mathbf{RT}(C)$ by the cocycle $z$ (see [Sp 81, § 12.3.7]); it is defined over $k$. From Lemma 4.4.2 we infer that it is of type $^3D_4$ or $^6D_4$ (see [Ti]) according to whether the cubic extension $l/k$ is Galois or not. Thus we have proved the following theorem.

**Theorem 4.4.3** $\mathbf{G}$ is defined over $k$. If $l$ is a cubic cyclic extension of $k$ and $F$ a normal twisted octonion algebra over $l$, then $\mathbf{G}$ is a twisted $k$-form of the algebraic group $\mathbf{Spin}(8)$ of type $^3D_4$.

If $l$ is cubic but not Galois over $k$ with $\mathrm{char}(k) \neq 2$ or $3$, and $F$ is a nonnormal twisted octonion algebra over $l$, then $\mathbf{G}$ is a twisted $k$-form of $\mathbf{Spin}(8)$ of type $^6D_4$.

## 4.5 Normal Twisted Octonion Algebras with Isotropic Norm

In this section, $F$ will be a normal twisted octonion algebra over a cubic cyclic field extension $l$ of $k$ (and we will omit the adjective "normal" when speaking about twisted octonion algebras).

From now on we fix $a \in F$ that satisfies the three conditions of Lemma 4.2.12; if $N$ is isotropic, we moreover assume $a$ isotropic. We take $E$ to be the orthogonal complement of $la \oplus la * a$,

$$E = \{ x \in C \mid \langle x, a \rangle = \langle x, a * a \rangle = 0 \}. \tag{4.43}$$

**Lemma 4.5.1** $E * a \subseteq E$ and $a * E \subseteq E$.

Proof. If $x \in E$, then $x * a \in E$, for

$$\langle x * a, a \rangle = \sigma(\langle a * a, x \rangle) = 0,$$

and by (4.1),

$$\langle x * a, a * a \rangle = \sigma(\langle x, a \rangle)\sigma^2(N(a)) = 0.$$

Similarly for $a * x$.                                                                      □

This allows us to define the $\sigma$-linear transformation $t$ in $E$ by

$$t : E \to E, \; x \mapsto x * a. \tag{4.44}$$

**Lemma 4.5.2** *The transformation* $t : E \to E$ *satisfies*

$$t^2(x) = -(a * a) * x \qquad (x \in E),  \tag{4.45}$$
$$t^3(x) = -T(a)x - a * (a * (a * x)) \qquad (x \in E),  \tag{4.46}$$

*and*

$$t^6 + T(a)t^3 + N_{l/k}(N(a)) = 0.$$

Proof. We compute, using Lemma 4.1.3,

$$t^2(x) = (x * a) * a = -(a * a) * x,$$
$$t^3(x) = -((a * a) * x) * a$$
$$= (a * x) * (a * a) - \sigma^2(\langle a * a, a \rangle)x$$
$$= -T(a)x - a * (a * (a * x)) \qquad (\text{since } \langle a * x, a \rangle = \sigma^2(\langle a * a, x \rangle) = 0),$$

so

$$t^3(x) + T(a)x + a * (a * (a * x)) = 0.$$

Apply $t^3$ to both sides of this equation:

$$t^6(x) + T(a)t^3(x) + [\{(a * (a * (a * x))) * a\} * a] * a = 0.$$

Since $(a * y) * a = \sigma^2(N(a))y$, the last term on the left hand side equals $N_{l/k}(N(a))x$, so we get the formula. $\qquad \square$

We call a twisted composition algebra *isotropic* if its norm is isotropic. From now on we assume that $F$ is an isotropic twisted octonion algebra (normal, as is the convention now). We take $a$ as in Lemma 4.2.12 with $N(a) = 0$. We define

$$e_1 = a \qquad \text{and} \qquad e_2 = T(a)^{-1}a * a.  \tag{4.47}$$

Using Lemma 4.1.3 one easily verifies

$$e_1 * e_1 = \lambda_1 e_2,  \tag{4.48}$$
$$e_2 * e_2 = \lambda_2 e_1,  \tag{4.49}$$
$$e_1 * e_2 = e_2 * e_1 = 0,  \tag{4.50}$$

with $\lambda_1 = T(a) \in k^*$ and $\lambda_2 = \lambda_1^{-1}$. Further,

$$N(e_1) = N(e_2) = 0 \quad \text{and} \quad \langle e_1, e_2 \rangle = 1.  \tag{4.51}$$

A straightforward computation now yields

$$T(\xi_1 e_1 + \xi_2 e_2) = \lambda_1 N_{l/k}(\xi_1) + \lambda_2 N_{l/k}(\xi_2).  \tag{4.52}$$

Notice that replacing $a$ by $T(a)^{-1}a * a$, which also satifies the conditions in Lemma 4.2.12 and is isotropic, amounts to interchanging $e_1$ and $e_2$ and also $\lambda_1$ and $\lambda_2$.

Define $D = le_1 \oplus le_2$ and $E = D^\perp$. The restriction of $\langle\ ,\ \rangle$ to $D$ and the restriction to $E$ are both nondegenerate. Define as in (4.44) the $\sigma$-linear transformations

$$t_i : E \to E, \ x \mapsto x * e_i. \tag{4.53}$$

From (4.45) we infer that

$$t_i^2(x) = -\lambda_i e_{i+1} * x \qquad (x \in E, \text{ indices mod } 2). \tag{4.54}$$

Take $E_i = t_i(E)$. Trivially, $t_i(E_i) \subseteq E_i$. Since $N(e_i) = 0$, both $E_i$ are totally isotropic, hence they have dimension $\leq 3$. By (4.7),

$$(e_2 * x) * e_1 + (e_1 * x) * e_2 = \sigma^2(\langle e_1, e_2 \rangle)x = x.$$

Since $e_i * E = t_{i+1}^2(E) \subseteq E$, it follows that $E = E_1 + E_2$. Since $\dim E_i \leq 3$, we must have a direct sum decomposition:

$$E = E_1 \oplus E_2,$$

with both $E_i$ having dimension 3. Using (4.7) again, we see for $x \in E$,

$$(x * e_1) * e_2 + (e_2 * e_1) * x = \sigma^2(\langle x, e_2 \rangle)e_1 = 0.$$

Since $e_2 * e_1 = 0$ by (4.50), we find that $t_2 t_1 = 0$. Similarly, $t_1 t_2 = 0$. Hence

$$t_i(E_j) = 0 \qquad (i \neq j), \tag{4.55}$$

and therefore

$$t_i(E_i) = E_i \qquad (i = 1, 2). \tag{4.56}$$

It follows that $E_i = t_i^2(E)$, so by (4.54)

$$E_i = e_{i+1} * E. \tag{4.57}$$

From (4.46) we know that

$$t_i^3(x) = -\lambda_i x - e_i * (e_i * (e_i * x)).$$

By (4.4), $e_i * E_i = e_i * (E * e_i) = 0$, so

$$t_i^3(x) = -\lambda_i x \qquad (x \in E_i). \tag{4.58}$$

It follows that $\pi_i = -\lambda_{i+1} t_i^3$ is the projection of $E$ on $E_i$.

Since $E_1$ and $E_2$ are totally isotropic and $\langle\ ,\ \rangle$ is nondegenerate on $E$, the $E_i$ are in duality by the isomorphism

$$E_1 \to E_2^*, \ u \mapsto \hat{u},$$

where

$$\hat{u} : E_2 \to l, \ x \mapsto \langle u, x \rangle,$$

and similarly $E_2 \to E_1^*$. For $x_i \in E_i$ we have

$$\langle t_1(x_1), t_2(x_2) \rangle = \sigma(\langle x_1, x_2 \rangle), \tag{4.59}$$

for

$$\langle t_1(x_1), t_2(x_2) \rangle = \langle x_1 * e_1, x_2 * e_2 \rangle = \sigma(\langle e_1 * (x_2 * e_2), x_1 \rangle) = \sigma(\langle x_1, x_2 \rangle),$$

since by (4.54) $e_1 * (x_2 * e_2) = -\lambda_1 t_2^2(t_2(x_2)) = -\lambda_1 t_2^3(x_2) = x_2$. This means that $t_2 = (t_1^*)^{-1}$ and vice versa.

We now compute $x_1 * x_2$ and $x_2 * x_1$ for $x_i \in E_i$. We write $x_i$ as $x_i = t_i(z_i) = z_i * e_i$ with $z_i \in E_i$ and find

$$\begin{aligned} x_1 * x_2 &= (z_1 * e_1) * (z_2 * e_2) \\ &= -((z_2 * e_2) * e_1) * z_1 + \sigma^2(\langle z_1, z_2 * e_2 \rangle)e_1 \quad \text{(by (4.7))}. \end{aligned}$$

Now $(z_2 * e_2) * e_1 = t_1 t_2(z_2) = 0$, so

$$\begin{aligned} x_1 * x_2 &= \sigma^2(\langle t_1^{-1}(x_1), x_2 \rangle)e_1 \\ &= \sigma(\langle x_1, t_2(x_2) \rangle)e_1 \quad \text{(by (4.59))}. \end{aligned}$$

Similarly for $x_2 * x_1$. Thus,

$$x_i * x_j = \sigma(\langle x_i, t_j(x_j) \rangle)e_i \quad (x_i \in E_i, \quad x_j \in E_j, \quad i \neq j). \tag{4.60}$$

Next consider $x * y$ for $x, y \in E_1$. We have

$$\langle x * y, e_1 \rangle = \sigma^2(\langle e_1 * x, y \rangle) = 0,$$

since $e_1 * E_1 = -\lambda_1 t_2^2(E_1) = 0$. Further,

$$\langle x * y, e_2 \rangle = \sigma(\langle y * e_2, x \rangle) = 0,$$

since $E_1 * e_2 = t_2(E_1) = 0$. So $x * y \in E$. Using (4.5) we find

$$t_1(x) * t_1(y) = (x * e_1) * (y * e_1) = -e_1 * (y * (x * e_1)),$$

since $\langle x * e_1, e_1 \rangle = 0$. Also,

$$y * (x * e_1) = -e_1 * (x * y).$$

Now using (4.54) and (4.58), we find

$$t_1(x) * t_1(y) = -\lambda_1 t_2^2(-\lambda_1 t_2^2(x * y)) = -\lambda_1 t_2(x * y),$$

and similarly with $t_2$ and $x, y \in E_2$. Thus,

$$t_i(x) * t_i(y) = -\lambda_i t_{i+1}(x * y) \quad (x, y \in E_i). \tag{4.61}$$

An immediate conclusion is that

$$E_i * E_i \subseteq E_{i+1}. \tag{4.62}$$

To describe the multiplication of elements of $E_i$ it is convenient to introduce a bilinear product

$$\wedge : E_i \times E_i \to E_{i+1} \qquad (i = 1, 2)$$

by defining

$$x_i \wedge y_i = t_i^{-1}(x_i) * t_i(y_i) \qquad (x_i, y_i \in E_i). \tag{4.63}$$

From (4.61) it is immediate that

$$t_i(x_i) \wedge t_i(y_i) = -\lambda_i t_{i+1}(x_i \wedge y_i) \qquad (x_i, y_i \in E_i). \tag{4.64}$$

This wedge product is alternating. For if $x \in E_1$, then

$$
\begin{aligned}
x \wedge x &= (e_2 * x) * (x * e_1) \qquad \big(\text{by (4.54) and (4.58)}\big) \\
&= -((x * e_1) * x) * e_2 \qquad \big(\text{by (4.7), since } \langle e_2, x * e_1 \rangle = 0\big) \\
&= N(x)(e_1 * e_2) \qquad \big(\text{by (4.6)}\big) \\
&= 0 \qquad \big(\text{since } N(x) = 0, \text{ or } e_1 * e_2 = 0\big),
\end{aligned}
$$

and similarly for $x \in E_2$. By linearizing one finds

$$x \wedge y = -y \wedge x \qquad (x, y \in E_i, \quad i = 1, 2). \tag{4.65}$$

For this wedge product we can prove the formulas that are well known for the vector product (cross product) in three-space.

**Lemma 4.5.3** *For $x_i, y_i \in E_i$, $(i = 1, 2)$, one has*

$$(x_1 \wedge y_1) \wedge x_2 = \langle x_1, x_2 \rangle y_1 - \langle y_1, x_2 \rangle x_1, \tag{4.66}$$

$$(x_2 \wedge y_2) \wedge x_1 = \langle x_2, x_1 \rangle y_2 - \langle y_2, x_1 \rangle x_2, \tag{4.67}$$

$$\langle x_1 \wedge y_1, x_2 \wedge y_2 \rangle = \langle x_1, x_2 \rangle \langle y_1, y_2 \rangle - \langle x_1, y_2 \rangle \langle x_2, y_1 \rangle. \tag{4.68}$$

*Further, $x_i \wedge y_i = 0$ for all $y_i$ (or for all $x_i$) implies $x_i = 0$ ($y_i = 0$, respectively), $(i = 1, 2)$.*

Proof. From (4.63) we get

$$
\begin{aligned}
(x_1 \wedge y_1) \wedge x_2 &= t_2^{-1}(t_1^{-1}(x_1) * t_1(y_1)) * t_2(x_2) \\
&= -\lambda_1 (t_1^{-2}(x_1) * y_1) * t_2(x_2) \qquad \big(\text{by (4.61)}\big) \\
&= (t_1(x_1) * y_1) * t_2(x_2) \qquad \big(\text{by (4.58).}\big)
\end{aligned}
$$

Now by (4.7) and (4.59),

$$(t_1(x_1) * y_1) * t_2(x_2) + (t_2(x_2) * y_1) * t_1(x_1) = \sigma^2(\langle t_1(x_1), t_2(x_2) \rangle) y_1$$
$$= \langle x_1, x_2 \rangle y_1,$$

and by (4.59) and (4.60),

$$t_2(x_2) * y_1 = \sigma(\langle t_1(y_1), t_2(x_2) \rangle) e_2 = \sigma^2(\langle y_1, x_2 \rangle) e_2.$$

Using these relations, we find

$$(t_1(x_1) * y_1) * t_2(x_2) = -(t_2(x_2) * y_1) * t_1(x_1) + \langle x_1, x_2 \rangle y_1$$
$$= -\langle y_1, x_2 \rangle (e_2 * t_1(x_1)) + \langle x_1, x_2 \rangle y_1$$
$$= -\langle y_1, x_2 \rangle x_1 + \langle x_1, x_2 \rangle y_1 \qquad \text{(by (4.54) and (4.58))}.$$

This proves the first formula, and the same argument leads to the second one.

For the third formula we proceed as follows:

$$\langle x_1 \wedge y_1, x_2 \wedge y_2 \rangle = \langle t_1^{-1}(x_1) * t_1(y_1), t_2^{-1}(x_2) * t_2(y_2) \rangle$$
$$= (-\lambda_2)(-\lambda_1) \langle t_2(t_1^{-2}(x_1) * y_1), t_1(t_2^{-2}(x_2) * y_2) \rangle$$
$$= \sigma(\langle t_1(x_1) * y_1, t_2(x_2) * y_2 \rangle)$$
$$= \langle y_2, (t_1(x_1) * y_1) * t_2(x_2) \rangle.$$

Now $(t_1(x_1) * y_1) * t_2(x_2)$ was computed above; substituting that expression we find formula (4.68).

The last statement of the Lemma is easily derived from (4.66) and (4.67). □

Define an alternating trilinear function $\langle \, , \, , \, \rangle$ on $E_i$ by

$$\langle x, y, z \rangle = \langle x, y \wedge z \rangle.$$

From 4.59 and 4.64 we obtain

$$\langle t_i(x), t_i(y), t_i(z) \rangle = -\lambda_i \sigma(\langle x, y, z \rangle). \tag{4.69}$$

## 4.6 A Construction of Isotropic Normal Twisted Octonion Algebras

We maintain the convention that all twisted composition algebras are normal. The analysis made in the previous section leads to a construction of isotropic twisted octonion algebras, to be described in the present section. This construction will yield all such algebras.

We first discuss some generalities. Let $V$ be a three-dimensional vector space over the field $l$ and $V'$ its dual space. The bilinear pairing between $V$

and $V'$ is denoted by $\langle\ ,\ \rangle$. There is a vector product $\wedge$ on $V$ with values in $V'$, and one on $V'$ with values in $V$ with the following three properties (where $x, y \in V$, $x', y' \in V'$) :

$$(x \wedge y) \wedge x' = \langle x, x' \rangle y - \langle y, x' \rangle x, \tag{4.70}$$

$$(x' \wedge y') \wedge x = \langle x, x' \rangle y' - \langle x, y' \rangle x', \tag{4.71}$$

$$\langle x' \wedge y', x \wedge y \rangle = \langle x, x' \rangle \langle y, y' \rangle - \langle x, y' \rangle \langle y, x' \rangle. \tag{4.72}$$

Using these properties, one easily verifies that $x \wedge y$ is alternating bilinear on $V$, nonzero if $x$ and $y$ are linearly independent, and that $\langle x, y, z \rangle = \langle x, y \wedge z \rangle$ is an alternating trilinear function on $V$ (i.e., invariant under even permutations of the variables and changing sign under odd permutations); similarly on $V'$. On a three-dimensional space, an alternating trilinear function is unique up to a nonzero factor, viz., it is a multiple of the determinant whose columns are the coordinate vectors of the three variables with respect to some fixed basis. It follows that the vector products on $V$ and on $V'$ are unique up to multiplication by some $\alpha \in l^*$ and $\alpha^{-1}$, respectively.

If $t : V \to V$ is a $\sigma$-linear transformation, where $\sigma$ is an automorphism of $l$, then another alternating trilinear function of $x, y, z$ is $\sigma^{-1}(\langle t(x), t(y), t(z) \rangle)$. Hence it is a multiple of $\langle x, y, z \rangle$. We define $\det(t) \in l$ by

$$\langle t(x), t(y), t(z) \rangle = \det(t)\sigma(\langle x, y, z \rangle) \qquad (x, y, z \in V).$$

If we replace the wedge product $x \wedge y$ on $V$ by $\alpha x \wedge y$, so $\langle\ ,\ \rangle$ by $\alpha\langle\ ,\ \rangle$, then $\det(t)$ changes to $\alpha\sigma(\alpha)^{-1} \det(t)$. We call $\det(t)$ the *determinant* of $t$ with respect to the given choice of the wedge product (or the choice of the alternating trilinear form).

Let $t'$ be the inverse adjoint transformation of $t$ in $V'$, i.e.,

$$\langle t(x), t'(x') \rangle = \sigma(\langle x, x' \rangle) \qquad (x \in V, \quad x' \in V').$$

$t'$ is also $\sigma$-linear.

**Lemma 4.6.1** *For* $x, y \in V$ *we have*

$$t(x) \wedge t(y) = \det(t)t'(x \wedge y). \tag{4.73}$$

*Moreover,* $\det(t') = \det(t)^{-1}$.

Proof. The first formula follows from the equations

$$\langle z, (t')^{-1}(t(x) \wedge t(y)) \rangle = \sigma^2(\langle t(z), t(x) \wedge t(y) \rangle) = \sigma^2(\langle t(z), t(x), t(y) \rangle) =$$

$$\sigma^2(\det(t))\langle z, x \wedge y \rangle.$$

Similarly, we have for $x', y' \in V'$

$$t'(x') \wedge t'(y') = \det(t')t(x' \wedge y').$$

Using (4.70) and the definition of $\det(t)$ we see that

$$(t(x) \wedge t(y)) \wedge (t(x_1) \wedge t(y_1)) = \det(t)t((x \wedge y) \wedge (x_1 \wedge y_1)),$$

which by what we already proved equals

$$\det(t)^2(t'(x \wedge y) \wedge t'(x_1 \wedge y_1)).$$

It follows that

$$t'(x') \wedge t'(y') = \det(t)^{-1}t(x' \wedge y'),$$

where $x' = x \wedge y$, $y' = x_1 \wedge y_1$. Since any element of $V'$ is a wedge product of elements of $V$, the last formula holds for arbitrary $x', y' \in V'$. The second assertion of the lemma follows.                    □

Now assume that $l$ is, as before, a cubic cyclic extension of the field $k$, and that $\sigma$ is a generator of the Galois group. Assume that $t$ is a $\sigma$-linear transformation of $V$ such that $t^3 = -\lambda$ with $\lambda \in k^*$. We also assume that the vector product on $V$ is such that $\det(t) = -\lambda$. This can always be arranged. For if not, then $N_{l/k}(-\lambda^{-1}\det(t)) = 1$, as one sees by computing $\langle t^3(x), t^3(y), t^3(z) \rangle$ in two different ways. By Hilbert's Theorem 90 there exists $\alpha \in l$ such that $\det(t) = -\lambda\alpha^{-1}\sigma(\alpha)$. Replacing the vector product $x \wedge y$ on $V$ by $\alpha x \wedge y$ changes the determinant of $t$ to $\det(t) = -\lambda$. Our assumption determines the vector product on $V$ up to a multiplicative factor $\mu \in k^*$ and the vector product on $V'$ up to $\mu^{-1}$. For the inverse adjoint transformation $t'$ we have $t'^3 = -\lambda^{-1}$ and $\det(t') = -\lambda^{-1}$.

$V$ and $t$ are the ingredients of the construction of a (normal) twisted composition algebra $\mathcal{F}(V, t)$. We are guided by the results of the preceding section. Taking (with the notations of 4.5) $V = E_1$, $V' = E_2$, $t = t_1$ the definition of $\mathcal{F}(V, t)$ is explained by the formulas of that section.

We define $F = \mathcal{F}(V, t) = l \oplus l \oplus V \oplus V'$ and put $e_1 = (1, 0, 0, 0)$, $e_2 = (0, 1, 0, 0)$. We define a product $*$ in $F$ as follows:

$$(\xi_1 e_1 + \xi_2 e_2 + x + x') * (\eta_1 e_1 + \eta_2 e_2 + y + y') =$$
$$\{\lambda^{-1}\sigma(\xi_2)\sigma^2(\eta_2) + \sigma(\langle x, t'(y') \rangle)\}e_1 + \{\lambda\sigma(\xi_1)\sigma^2(\eta_1) + \sigma(\langle t(y), x' \rangle)\}e_2 +$$
$$\sigma(\xi_2)t^{-1}(y) + \sigma^2(\eta_1)t(x) + t'(x') \wedge (t')^{-1}(y') +$$
$$\sigma(\xi_1)(t')^{-1}(y') + \sigma^2(\eta_2)t'(x') + t(x) \wedge t^{-1}(y). \tag{4.74}$$

for $\xi_i, \eta_i \in l$, $x, y \in V$, $x', y' \in V'$. We further define the norm $N$ on $F$ by

$$N(\xi_1 e_1 + \xi_2 e_2 + x + x') = \xi_1\xi_2 + \langle x, x' \rangle \tag{4.75}$$

for $\xi_i \in l$, $x \in V$ and $x' \in V'$.

**Theorem 4.6.2** *With this product and norm $\mathcal{F}(V, t)$ is an isotropic twisted octonion algebra, and all such algebras are of this form. For $z = \xi_1 e_1 + \xi_2 e_2 + x + x'$ the cubic form $T(z) = \langle z * z, z \rangle$ is given by*

$$T(z) = \lambda \, N_{l/k}(\xi_1) + \lambda^{-1} N_{l/k}(\xi_2) + \mathrm{Tr}_{l/k}(\xi_1 \sigma(\langle t(x), x' \rangle)) +$$

$$\mathrm{Tr}_{l/k}(\xi_2 \sigma(\langle x, t'(x') \rangle)) + \langle x, t(x), t^{-1}(x) \rangle + \langle x', t'(x'), t'^{-1}(x') \rangle. \quad (4.76)$$

Proof. It is clear that the product $*$ is $\sigma$-linear in the first variable and $\sigma^2$-linear in the second variable. The verification of the other two requirements for a twisted composition algebra (cf. Def. 4.1.1), viz.,

$$N(x * y) = \sigma(N(x))\sigma^2(N(y)) \qquad (x, y \in F)$$

and

$$\langle x * y, z \rangle = \sigma(\langle y * z, x \rangle) \qquad (x, y, z \in F),$$

and the computation of $T$ are straightforward, so we omit them. That we obtain all twisted composition algebras with isotropic norm in this way, follows from the analysis in the preceding section.                    □

## 4.7 A Related Central Simple Associative Algebra

We continue to consider an isotropic twisted octonion algebra $F = \mathcal{F}(V, t)$. Notations and conventions are as in the preceding section. We introduce the associative algebra $D$ over $k$ consisting of all transformations of $V$ of the form

$$\xi_0 + \xi_1 t + \xi_2 t^2 \qquad (\xi_0, \xi_1, \xi_2 \in l).$$

(We write 1 for the identity here.) We show that this is a cyclic crossed product. (For crossed products, see [Al 61, Ch. V], [ArNT, Ch. VIII, §§ 4 and 5] or [Ja 80, §§ 8.4 and 8.5].) The notation $D$ is used since in the most important case for us it is a division algebra.

**Lemma 4.7.1** *The elements* 1, $t$ *and* $t^2$ *form a basis of* $D$ *over* $l$, *and* $D$ *is isomorphic to the cyclic crossed product* $(l, \sigma, -\lambda)$, *so it is a central simple algebra of degree 3 over* $k$. *If* $\lambda \in N_{l/k}(l^*)$, *this crossed product is isomorphic to the algebra* $M_3(k)$ *of* $3 \times 3$ *matrices over* $k$, *and if* $\lambda \notin N_{l/k}(l^*)$, *it is a division algebra.*

Proof. If

$$\xi_0 + \xi_1 t + \xi_2 t^2 = 0$$

with $\xi_i \in l$, then for all $x \in V$ and $\eta \in l$ we have

$$\eta \xi_0 x + \sigma(\eta)\xi_1 t(x) + \sigma^2(\eta)\xi_2 t^2(x) = 0.$$

Since the automorphisms $1, \sigma, \sigma^2$ are linearly independent over $l$, it follows that all $\xi_i$ are zero. So $1, t, t^2$ is a basis of $D$ over $l$.

The cyclic crossed product $(l, \sigma, -\lambda)$ is the algebra generated by $l$ and an element $u$ such that $u^3 = -\lambda$ and $u\xi = \sigma(\xi)u$ $(\xi \in l)$. It has dimension 3 over $l$. Clearly, there is a homomorphism $(l, \sigma, -\lambda) \to D$ sending $u$ to $t$ and extending the identity map of $l$. Since $D$ has dimension 3 the homomorphism is bijective, hence is an isomorphism. The crossed product is known to be isomorphic to $M_3(k)$ if $-\lambda \in N_{l/k}(l^*)$ and to be a division algebra if $-\lambda \notin N_{l/k}(l^*)$. We may omit the minus sign, since $N_{l/k}(-1) = -1$.    □

**Lemma 4.7.2** *There exists $v_0 \in V$ such that $V = D.v_0$.*

Proof. If $D$ is a division algebra, we can pick any nonzero $v_0 \in V$. For then $D.v_0$ is a nine-dimensional subspace of $V$ over $k$, which must coincide with $V$.

   If $D$ is not a division algebra we have $D \cong M_3(k)$. As a $D$-module, $V$ is isomorphic to the direct sum of three copies of the simple module $k^3$ of $M_3(k)$. But then the $D$-module $V$ is isomorphic to $D$, viewed as a left module over itself. We can then take for $v_0$ the image in $V$ of any invertible element of $D$.    □

On the central simple algebra $D$ over $k$, one has the *reduced norm*, see [Al 61, Ch. VIII, § 11], [Schar, Ch. 8, § 5] or [Weil, Ch. IX, § 2]. It is the unique polynomial function on $D$ which upon extension of $k$ to a splitting field of $D$ becomes the determinant. It is multiplicative, i.e., $N_D(uv) = N_D(u)N_D(v)$ for $u, v \in D$, and $u$ is invertible if and only if $N_D(u) \neq 0$. The following lemma gives a simple characterization of the reduced norm.

**Lemma 4.7.3** *If $A$ is a central simple algebra of degree $n$ over $k$, then the reduced norm $N_A$ is the unique homogeneous polynomial function of degree $n$ on the vector space $A$ over $k$ with $N_A(1) = 1$ which satisfies the conditions: $x \in A$ is invertible if and only if $N_A(x) \neq 0$ and there exists a homogeneous polynomial map $P : A \to A$ of degree $n - 1$ such that*

$$x^{-1} = N_A(x)^{-1}P(x) \qquad (x \in A, N_A(x) \neq 0).$$

Proof. For $A = M_n(k)$ it is known that $x \in A$ is invertible if and only $\det(x) \neq 0$, and then $x^{-1} = \det(x)^{-1} \operatorname{adj}(x)$, where $\operatorname{adj}(x)$ is the adjoint matrix of $x$, i.e., the matrix whose entries are the cofactors of $x$ (see, e.g., [Ja 74, § 2.3]). As to the uniqueness of $N_A$, let $N_A'$ with $N_A'(1) = 1$, in combination with $P'$ also satisfy the conditions. Then $N_A'(x) \neq 0$ if and only if $\det(x) \neq 0$, and

$$\det(x)P'(x) = \operatorname{adj}(x)N_A'(x) \qquad (x \in A, \det(x) \neq 0).$$

Since det is an irreducible polynomial in the entries of the matrix $x$ (see, e.g., [Ja 74, Th. 7.2]), either det divides all entries of adj (viewed as a matrix with polynomial entries), or det divides $N_A'$. The first case being absurd, we must have $N_A' = \det$.

In the case of an arbitrary central simple algebra $A$, work with a Galois splitting field $m$ of $A$, i.e., with a Galois extension $m$ of $k$ such that $m \otimes_k A \cong M_n(k)$. We have a reduced norm $N_A$ over $m$. For any $\sigma \in \mathrm{Gal}(m/k)$, let $\sigma(N_A)$ denote the polynomial obtained by the action of $\sigma$ on the coefficients of $N_A$. Then $\sigma(N_A)$ also satisfies the conditions for a reduced norm on $m \otimes_k A$, so by the uniqueness of this, $\sigma(N_A) = N_A$. This means that $N_A$ has its coefficients in $k$, and hence its restriction to $A$ is a reduced norm on $A$. The uniqueness of the reduced norm on $A$ is immediate from the fact that its extension to $m \otimes_k A$ is a reduced norm on the latter algebra.    $\square$

With the aid of the above lemma it is not hard to compute the reduced norm on $D$ explicitly. (The proof can, in fact, be adapted to any crossed product.)

**Lemma 4.7.4** *The reduced norm of* $u = \xi_0 + \xi_1 t + \xi_2 t^2$ *in* $D$ *is*

$$N_D(u) = N_{l/k}(\xi_0) - \lambda\, N_{l/k}(\xi_1) + \lambda^2\, N_{l/k}(\xi_2) + \lambda\, \mathrm{Tr}_{l/k}(\xi_0 \sigma(\xi_1)\sigma^2(\xi_2)).$$

*This can also be written as* $N_D(u) = \det(A_u)$, *where* $A_u$ *is the matrix*

$$A_u = \begin{pmatrix} \xi_0 & -\lambda\sigma(\xi_2) & -\lambda\sigma^2(\xi_1) \\ \xi_1 & \sigma(\xi_0) & -\lambda\sigma^2(\xi_2) \\ \xi_2 & \sigma(\xi_1) & \sigma^2(\xi_0) \end{pmatrix}.$$

Proof. An element $u = \xi_0 + \xi_1 t + \xi_2 t^2$ has an inverse if and only if the right multiplication by $u$, i.e., $v \mapsto vu$, is bijective. This right multiplication is a linear transformation over $l$ which has matrix $A_u$ as above. Hence $u$ is invertible if and only if $\det(A_u) \neq 0$, and then $u^{-1}$ is the solution $v$ of $A_u v = 1$, so of the form $\det(A_u)^{-1} P(u)$. A straightforward computation yields that

$$\det(A_u) = N_{l/k}(\xi_0) - \lambda\, N_{l/k}(\xi_1) + \lambda^2\, N_{l/k}(\xi_2) + \lambda\, \mathrm{Tr}_{l/k}(\xi_0 \sigma(\xi_1)\sigma^2(\xi_2)).$$

So $\det(A_u) \in k$ and it is a cubic polynomial in coordinates over $k$. $P$ is a map $D \to D$ that is quadratic in coordinates over $k$.    $\square$

In exactly the same way as we did with $D$ we introduce the associative algebra $D'$ of $l$-linear combinations of $1$, $t'$ and $t'^2$, acting on $V'$. This has the same properties as $D$ except that $t'^3 = -\lambda^{-1}$, so in the formulas for the reduced norm one must replace $\lambda$ by $\lambda^{-1}$. We call $D'$ the *opposite algebra* of $D$, a name which is justified by the fact that $D$ and $D'$ are anti-isomorphic as we will see in the following Lemma.

**Lemma 4.7.5** *The mapping*

$$D \to D', \quad u = \xi_0 + \xi_1 t + \xi_2 t^2 \mapsto u' = \xi_0 - \lambda\sigma(\xi_2)t' - \lambda\sigma^2(\xi_1)t'^2$$

*is an anti-isomorphism of* $D$ *onto* $D'$. *It preserves the reduced norm:*

$$N_{D'}(u') = N_D(u) \qquad (u \in D).$$

Proof. Both statements are verified by straightforward explicit computation in the coordinates $\xi_0, \xi_1, \xi_2$. Notice that we can also write $u' = \xi_0 + (t')^{-1}\xi_1 + (t')^{-2}\xi_2$. □

We now return to the twisted composition algebra $\mathcal{F}(V, t)$ of §4.6.

**Lemma 4.7.6** *(i) The nonzero values of the reduced norm $N_D$ on $D$ form a subgroup $N(D)^*$ of $k^*$. We have $N(D')^* = N(D)^*$.*
*(ii) For $v \in V$ we have $T(v) \neq 0$ if and only if $V = D.v$. Then $T(u.v) = N_D(u)T(v)$ for $u \in D$, so the nonzero values of $T$ on $V$ form a coset of $N(D)^*$ in $k^*$.*
*(iii) Similarly, the nonzero values of $T$ on $V'$ form a coset of $N(D)^*$ in $k^*$.*

Proof. $N_D$ is multiplicative, and $N_D(u) \neq 0$ if and only $u$ is invertible, so the nonzero values of $N_D$ form a subgroup of $k^*$. The second point of (i) follows from the preceding lemma. According to Th. 4.6.2, the value of $T(v)$ for $v \in V$ is

$$T(v) = -\lambda^{-1}\langle v, t(v), t^2(v) \rangle.$$

Pick $v_0 \in V$ such that $V = D.v_0$ (cf. Lemma 4.7.2). Write

$$v = \xi_0 v_0 + \xi_1 t(v_0) + \xi_2 t^2(v_0) \qquad (\xi_i \in l).$$

Then

$$T(v) = \det(X)T(v_0),$$

where $X$ is the matrix which expresses $v, t(v), t^2(v)$ in $v_0, t(v_0), t^2(v_0)$:

$$X = \begin{pmatrix} \xi_0 & -\lambda\sigma(\xi_2) & -\lambda\sigma^2(\xi_1) \\ \xi_1 & \sigma(\xi_0) & -\lambda\sigma^2(\xi_2) \\ \xi_2 & \sigma(\xi_1) & \sigma^2(\xi_0) \end{pmatrix}.$$

From Lemma 4.7.4 we infer that $\det(X) = N_D(u)$ with $u = \xi_0 + \xi_1 t + \xi_2 t^2$. Thus we find that $T(v) = T(v_0)N_D(u)$. We see that $T(v) \neq 0$ if and only if $u$ is invertible, which implies the first assertion of (ii). This proves (ii).

(iii) is proved in the same way, also using (i). □

# 4.8 A Criterion for Reduced Twisted Octonion Algebras. Applications

In Th. 4.1.10 we gave criteria for a normal twisted composition algebra to be reduced. The following theorem can be viewed as a sharpening of part (ii) of Th. 4.1.10. Let $F = \mathcal{F}(V, t)$ be as in Th. 4.6.2. Notations and conventions are as before.

**Theorem 4.8.1** *The isotropic twisted octonion algebra $F$ is reduced if and only if there exists $x \in V$, $x \neq 0$, and $u \in D$ such that $T(x) = N_D(u)$.*

Proof. First assume there exists $x \in V$, $x \neq 0$, and $u \in D$ such that $T(x) = N_D(u)$. If $T(x) = 0$, then $F$ is reduced by Th. 4.1.10. If $T(x) = N_D(u) \neq 0$, then by Lemma 4.7.6 $N_D(u) \in T(a)N(D)^*$ for some $a \in V$, so $T(a) \in N(D)^*$. It follows that there exists $y \in V$ with $T(y) = 1$. Now $\langle y * y, y \rangle = T(y) = 1$, whereas $\langle y, y \rangle = 0$ (since $N$ is identically zero on $V$ by (4.75)), so $y$ and $y*y$ are linearly independent. Hence $z = y + y * y \neq 0$ and

$$
\begin{aligned}
z * z &= (y + y * y) * (y + y * y) \\
&= y * y + y * (y * y) + (y * y) * y + (y * y) * (y * y) \\
&= y * y + \sigma(N(y))y + \sigma^2(N(y))y + T(y)y - N(y)y * y \\
&= y * y + y = z.
\end{aligned}
$$

Again we conclude by Th. 4.1.10 that $F$ is reduced.

Conversely, assume $F$ reduced. We are going to show the existence of $x$ and $u$ with the required properties. Since $F$ is reduced there exists by Th. 4.1.10 a nonzero $z = \xi_1 e_1 + \xi_2 e_2 + x + x' \in F$ with $\xi_i \in l$, $x \in V$ and $x' \in V'$ such that $z * z = \alpha z$ for some $\alpha \in l$. By Th. 4.6.2 this amounts to saying that the following system of equations has a nontrivial solution $(\xi_1, \xi_2, x, x', \alpha)$:

$$\lambda^{-1}\sigma(\xi_2)\sigma^2(\xi_2) + \sigma(\langle x, t'(x') \rangle) = \alpha\xi_1 \tag{4.77}$$

$$\lambda\sigma(\xi_1)\sigma^2(\xi_1) + \sigma(\langle t(x), x' \rangle) = \alpha\xi_2 \tag{4.78}$$

$$\sigma(\xi_2)t^{-1}(x) + \sigma^2(\xi_1)t(x) + t'(x') \wedge (t')^{-1}(x') = \alpha x \tag{4.79}$$

$$\sigma(\xi_1)(t')^{-1}(x') + \sigma^2(\xi_2)t'(x') + t(x) \wedge t^{-1}(x) = \alpha x'. \tag{4.80}$$

(a) We first consider the case that there is a solution with $x \neq 0$. We compute

$$T(x) = \langle x, t(x) \wedge t^{-1}(x) \rangle.$$

Using (4.80) we get

$$T(x) = \alpha\langle x, x' \rangle - \sigma(\xi_1)\langle x, (t')^{-1}(x') \rangle - \sigma^2(\xi_2)\langle x, t'(x') \rangle.$$

Applying $\sigma^2$ to both sides of (4.77) and then multiplying the result by $\sigma^2(\xi_2)$ yields

$$\sigma^2(\xi_2)\langle x, t'(x') \rangle = \sigma^2(\alpha\xi_1\xi_2) - \lambda^{-1}N_{l/k}(\xi_2).$$

Further,

$$
\begin{aligned}
\sigma(\xi_1)\langle x, (t')^{-1}(x') \rangle &= \sigma(\xi_1)\sigma^2(\langle t(x), x' \rangle) \\
&= \sigma(\alpha\xi_1\xi_2) - \lambda N_{l/k}(\xi_1) \qquad \text{(by (4.78))}.
\end{aligned}
$$

With these formulas we find

$$T(x) = \lambda \mathrm{N}_{l/k}(\xi_1) + \lambda^{-1} \mathrm{N}_{l/k}(\xi_2) - \sigma(\alpha\xi_1\xi_2) - \sigma^2(\alpha\xi_1\xi_2) + \alpha\langle x, x'\rangle. \quad (4.81)$$

If $\alpha = 0$, then $T(x) = \mathrm{N}_D(u)$ with $u = -\xi_1 t + \lambda^{-1}\xi_2 t^2$ by Lemma 4.7.4.

For the rest of case (a) assume that $\alpha \neq 0$. By (4.11) we have $N(z) = \sigma(\alpha)\sigma^2(\alpha)$. Since $N(z) = \xi_1\xi_2 + \langle x, x'\rangle$ we find that

$$\langle x, x'\rangle = \sigma(\alpha)\sigma^2(\alpha) - \xi_1\xi_2.$$

Inserting this into in (4.81) we find

$$T(x) = \mathrm{N}_{l/k}(\alpha) + \lambda \mathrm{N}_{l/k}(\xi_1) + \lambda^{-1} \mathrm{N}_{l/k}(\xi_2) - \mathrm{Tr}_{l/k}(\alpha\xi_1\xi_2) = \mathrm{N}_D(u),$$

where $u = \alpha - \sigma^2(\xi_1)t + \lambda^{-1}\sigma(\xi_2)t^2$.
(b) Now assume we have a solution $(\xi_1, \xi_2, 0, x', \alpha)$ of the system of equations (4.77)–(4.80) with $\xi_2 \neq 0$. From (4.77) we infer that also $\alpha \neq 0$ and $\xi_1 \neq 0$. Multiplying the opposite sides of (4.77) and (4.78), we get

$$\alpha\lambda^{-1} \mathrm{N}_{l/k}(\xi_2) = \alpha\lambda \mathrm{N}_{l/k}(\xi_1),$$
$$\lambda^{-2} = \mathrm{N}_{l/k}(\xi_1\xi_2^{-1}),$$
$$\lambda = \mathrm{N}_{l/k}(\lambda\xi_1\xi_2^{-1}).$$

This implies that the cyclic crossed product $D$ is isomorphic to the matrix algebra $\mathrm{M}_3(k)$. Hence $\mathrm{N}_D(u)$ runs over all elements of $k$ if $u$ runs over $D$, so for any nonzero $x \in V$ there is $u \in D$ with $T(x) = \mathrm{N}_D(u)$.
(c) Finally, let there be a nonzero solution of the form $(\xi_1, 0, 0, x', \alpha)$. By (4.78) we have $\xi_1 = 0$. By (4.79), $t'(x') \wedge t'^{-1}(x') = 0$. This is only possible if $t'(x') = \xi x'$ for some $\xi \in l$. Since $t'$ is $\sigma$-linear and $t'^3 = -\lambda^{-1}$, we find by computing $t'^3(x')$ that $\lambda^{-1} = \mathrm{N}_{l/k}(-\xi)$. So again $\lambda \in \mathrm{N}_{l/k}(l^*)$ and we can complete the proof as in case (b).    $\square$

We mention, in particular, the following consequence, which we will use in Ch. 8.

**Corollary 4.8.2** *If $F$ is not reduced, then $D$ is a division algebra.*

Proof. If $D$ is not a division algebra, then $D \cong \mathrm{M}_3(k)$. In that case the reduced norm $\mathrm{N}_D$ takes all values in $k$, so for any nonzero $x \in V$ there exists $u \in D$ such that $\mathrm{N}_D(u) = T(x)$. Hence $F$ is reduced by Th. 4.8.1.    $\square$

As an application of the above theorem we can give another class of special fields $k$ over which all twisted octonion algebras are reduced. We assume $\mathrm{char}(k) \neq 2, 3$.

**Theorem 4.8.3** *Assume that $k$ has the following property: If $A$ is a nine-dimensional central simple algebra $A$ whose center $k'$ is either $k$ or a quadratic extension of $k$, then $\mathrm{N}_A(A) = k'$. Let further $l$ be a cubic extension of $k$. Then every twisted octonion algebra over $l$ and $\sigma$ (as in Def. 4.2.1) is reduced.*

Proof. Let $F$ be a twisted octonion algebra over $l$ and $\sigma$. If the norm $N$ of $F$ is isotropic, we take $k' = k$, and if $N$ is anisotropic, we choose a quadratic extension $k'$ of $k$ which makes $N$ isotropic. Extend $l$ to a cubic cyclic extension $l'$ of $k'$, with $\sigma$ (extended to a $k'$-automorphism of $l'$) as generator of $\mathrm{Gal}(l'/k')$. In either case, $F' = k' \otimes_k F$ is a twisted octonion algebra over $l'$ of the form described in Th. 4.6.2. The reduced norm of $l'[t]$ takes all values in $k'$; if $l'[t]$ is a division algebra over $k'$, this follows from the assumption we made about the reduced norm on nine-dimensional division algebras over $k'$, and if $l'[t] \cong \mathrm{M}_3(k')$, this is always the case. By the previous theorem, $F'$ is reduced. By Th. 4.1.10 this is equivalent to the fact that the cubic form $T$ represents zero nontrivially on $F'$. By Lemma 4.2.11, $T$ already represents zero nontrivially on $F$. That implies that $F$ is reduced. □

Here are some examples of fields with the property of the theorem.

(i) *k an algebraic number field*
If $D$ is a central simple algebra over a field $k$, then the nonzero reduced norms of elements of $D$ form a subgroup $\mathrm{N}(D)^*$ of $k^*$. Let $\mathbf{SL}_D$ be the norm one group of $D$, i.e. the group of elements of $D_K$ with reduced norm 1. This is a an algebraic group over $k$. Then the quotient group $k^*/N(D)^*$ can be identified with the Galois cohomology set $H^1(k, \mathbf{SL}_D)$ (see [Se 64, Ch. III, § 3.2]). The reduced norm map will be surjective if and only if the Galois cohomology set is trivial. That this is indeed the case if $k$ is an algebraic number field and $D$ is nine-dimensional follows from the Hasse principle (see [loc.cit., § 4.7, Remarque 1]). This shows that a number field $k$ has the property of the previous theorem. Hence over such a field any twisted composition algebra is reduced.

(ii) As in case (vi) of § 1.10 perfect fields with cohomological dimension $\leq 2$ also have the required property, as follows from [loc.cit., Ch. III, § 3.2]. Examples are finite fields and $p$-adic fields.

## 4.9 More on Isotropic Normal Twisted Octonion Algebras

This section gives some complements to the material of § 4.5 and § 4.6. We prove two somewhat technical lemmas that will be used in Ch. 8. Let $F$ be an isotropic normal twisted octonion algebra and consider an isotropic element $a \in F$ with $T(a) \neq 0$; then $a$ and $a * a$ are linearly independent and the restriction of $\langle \, , \, \rangle$ to $la \oplus la * a$ is nondegenerate (see Lemma 4.2.12). The corresponding subspaces $E = (la \oplus la * a)^\perp$ as in (4.43) and $E_i = t_i(E)$ $\langle i = 1, 2 \rangle$ with $t_i$ as in (4.53) will now be denoted by $E(a)$ and $E_i(a)$, since we are going to vary $a$:

$$E(a) = (la \oplus la * a)^\perp,$$
$$E_1(a) = \{ x * a \mid \langle x, a \rangle = \langle x, a * a \rangle = 0 \},$$
$$E_2(a) = \{ x * (a * a) \mid \langle x, a \rangle = \langle x, a * a \rangle = 0 \}.$$

By (4.57), we also have

$$E_1(a) = \{ (a * a) * x \mid \langle x, a \rangle = \langle x, a * a \rangle = 0 \},$$
$$E_2(a) = \{ a * x \mid \langle x, a \rangle = \langle x, a * a \rangle = 0 \}.$$

Since $(a * a) * (a * a) = T(a)a$ by (4.8), $E_i(a * a) = E_{i+1}(a)$. By (4.4) and (4.6), $a * (a * a) = (a * a) * a = 0$, hence we can also write

$$E_1(a) = \{ x * a \mid \langle x, a * a \rangle = 0 \},$$
$$E_2(a) = \{ a * x \mid \langle x, a * a \rangle = 0 \}.$$

Recall that $E_1(a)$ and $E_2(a)$ are totally isotropic subspaces which are in duality with respect to $\langle \, , \, \rangle$.

**Lemma 4.9.1** *Let $a, b \in F$ be isotropic with $T(a)T(b) \neq 0$. Then $b \in E_1(a)$ if and only if $a \in E_2(b)$.*

Proof. Let $b \in E_1(a)$. By (4.56), we may assume that $b = x * a$ with $x \in E_1(a)$. Pick $z \in E_2(a)$ with $\langle x, z \rangle = 1$. According to (4.62), $b * b \in E_2(a)$, so $\langle z, b * b \rangle = 0$. From (4.7) it follows:

$$b * z = (x * a) * z = -(z * a) * x + \sigma^2(\langle x, z \rangle)a.$$

By (4.55), $z * a = 0$. Hence $a = b * z \in E_2(b)$.

The proof of the converse implication is similar. □

**Lemma 4.9.2** *Assume again $a, b \in F$ to be isotropic with $T(a)T(b) \neq 0$. If $a * b = 0$, then $E_2(a) \cap E_1(b) \neq 0$.*

Proof. We may assume that $F = \mathcal{F}(V, t)$ is as in Th. 4.6.2 with $e_1 = a$, $e_2 = T(a)^{-1}a * a$, $V = E_1(a)$ and $V' = E_2(a)$ (see the part of § 4.5 beginning with equation (4.47)). Using the multiplication rule (4.74) it is straightforward to see that $a * b = 0$ implies that $b = \alpha e_2 + v$, with $\alpha \in k$, $v \in V$. If $v = 0$, then $b$ is a nonzero multiple of $e_2$ and $E_1(b) = E_2(a)$, proving the lemma in that case. So we may assume that $v \neq 0$. Then

$$b * b = \lambda^{-1}\sigma(\alpha)\sigma^2(\alpha)e_1 + \sigma(\alpha)t^{-1}(v) + t(v) \wedge t^{-1}(v).$$

Take $z \in V$, with $\langle z, t(v) \wedge t^{-1}(v) \rangle = 0$, $z \notin kt(v)$, then

$$\langle z, b * b \rangle = \langle z, t(v) \wedge t^{-1}(v) \rangle = 0,$$

and $z * b$ is a nonzero multiple of $t(z) \wedge t^{-1}(v)$ which lies in $V' \cap E_1(b) = E_2(a) \cap E_1(b)$. □

## 4.10 Nonnormal Twisted Octonion Algebras with Isotropic Norm

In this section we briefly discuss analogues of the results of § 4.5 and § 4.6 in the case of a nonnormal twisted octonion algebra. We use the notations of Def. 4.2.1. So $l$ is a non-cyclic cubic extension of $k$. Assume that $\operatorname{char}(k) \neq 2, 3$. Let $F$ be a twisted composition algebra over $l$ and $\sigma$ and assume that the norm $N$ is *isotropic*. The normal extension $F' = l' \otimes_l F$ of $F$, introduced in Prop. 4.2.2, is an isotropic normal twisted composition algebra over $l'$, and $u = \tau \otimes \operatorname{id}$ defines a $\tau$-linear anti-automorphism of $F'$, by Prop. 4.2.4.

Take $a \in F$ with the properties of Lemma 4.2.12. Then $a$ and $a^{*2}$ are fixed by $u$. We carry out the analysis of § 4.5 for $a$, with $F'$ instead of $F$. Notations being as in that section, we have for $x \in F'$

$$u(x * a) = a * u(x), \ u(x * a^{*2}) = a^{*2} * u(x).$$

It follows that for $x \in E_i$

$$u(x * e_i) = e_i * x.$$

Using (4.54) we see that $u$ induces a $\tau$-linear bijection $E_i \to E_{i+1}$. From (4.58) we find that for $x \in E_i$

$$u \circ t_i = t_{i+1}^{-1} \circ u, \tag{4.82}$$

and from (4.63) it then follows that

$$u(x_i \wedge y_i) = u(y_i) \wedge u(x_i) \quad (x_i, y_i \in E_i). \tag{4.83}$$

Next, (4.60) implies that

$$\langle x_1, t_2(x_2) \rangle = \sigma\tau(\langle u(x_2), (t_2 \circ u)(x_1) \rangle),$$

and using (4.82)

$$\langle x_1, x_2 \rangle = \sigma\tau(\langle (u \circ t_2^{-1})(x_2), (t_2 \circ u)(x_1) \rangle) = \sigma\tau(\langle t_1(u(x_2)), t_2(u(x_1)) \rangle),$$

whence

$$\langle u(x_2), u(x_1) \rangle = \tau(\langle x_1, x_2 \rangle). \tag{4.84}$$

From (4.83) and (4.84) we deduce that for $x, y, z \in E_i$

$$\langle u(x), u(y), u(z) \rangle = -\tau(\langle x, y, z \rangle),$$

where the alternating trilinear form $\langle \ , \ , \ \rangle$ is as at the end of § 4.5.

The properties of $u$ which we just established indicate how to modify the construction of § 4.6 in order to deal with nonnormal twisted composition algebras. The notations are as in the beginning of the chapter. Assume we are given a normal twisted composition algebra $F' = \mathcal{F}(V, t)$ over $l'$ and $\sigma$.

**Definition 4.10.1** A *hermitian involution* of $(V, t)$ is a $\tau$-linear bijection $\iota : V \to V'$ with the following properties:

(i) $\langle x, \iota(y) \rangle = \tau(\langle y, \iota(x) \rangle)$   $(x, y \in V)$.

(ii) $\iota \circ t = (t')^{-1} \circ \iota$.

(iii) $\iota^{-1}(x \wedge y) = \iota(y) \wedge \iota(x)$.

Using (4.72) one finds by the same kind of argument as used in the proof of Lemma 4.6.1 that we also have

(iii)' $\iota(x' \wedge y') = \iota^{-1}(y') \wedge \iota^{-1}(x')$   $(x', y' \in V')$.

The next theorem is the analogue of 4.6.2 for nonnormal twisted composition algebras.

**Theorem 4.10.2** *Let $\iota$ be a hermitian involution of $(V, t)$. Then $u(\xi_1 e_1 + \xi_2 e_2 + x + x') = \tau(\xi_1)e_1 + \tau(\xi_2)e_2 + \iota^{-1}(x') + \iota(x)$ defines an involution of $F'$. The fixed point set $\mathrm{Inv}(u)$ is an isotropic nonnormal twisted composition algebra over $l$ and $\sigma$. All such algebras are of this form.*

Proof. The proof that $u$ is an involution (as defined after the proof of Prop. 4.2.4) is straightforward. The second point then follows from Prop. 4.2.4. The last point is a consequence of what was established in the beginning of this section.                                                                    □

The properties of $\iota$ of 4.10.1 can be reformulated. Putting

$$h(x, y) = \langle x, \iota(y) \rangle \; (x, y \in V), \tag{4.85}$$

it follows from property (i) that $h$ is a nondegenerate hermitian form on the $l'$-vector space $V$, relative to $\tau$. Then, using (ii),

$$h(t(x), y) = \langle t(x), \iota(y) \rangle = \sigma(\langle x, (t')^{-1}(\iota(y)) \rangle) = \sigma(\langle x, \iota(t(y)) \rangle) =$$

$$\sigma(h(x, t(y))),$$

which explains the adjective "hermitian" in Def. 4.10.1.

Using the three properties we also find

$$\langle \iota(x), \iota(y), \iota(z) \rangle = \langle \iota^{-1}(y \wedge x), \iota(z) \rangle = \tau(\langle z, y \wedge x \rangle) = -\tau(\langle x, y, z \rangle),$$

from which we see that $\det(\iota) = -1$, the determinant being defined as in § 4.6.

Conversely, given a nondegenerate hermitian form on $V$, there is a unique $\tau$-linear map $\iota$ such that (4.85) holds. The requirements that $t$ be hermitian relative to $h$ and that $\det(\iota) = -1$ give conditions equivalent to those of Def. 4.10.1.

Let $D$ be the cyclic crossed product $(l', \sigma, -\lambda)$, see § 4.7. Its center is $k'$. It is immediate that

$$\xi_0 + \xi_1 t + \xi_2 t^2 \mapsto \tau(\xi_0) + t\tau(\xi_1) + t^2\tau(\xi_2)$$

defines an *involution of the second kind* of $D$, i.e. an anti-automorphism of $D$ which induces a nontrivial automorphism on the center of $D$. (In the present case this is $\tau$.)

Finally, we notice that by property (ii) we have $t^3 = \iota^{-1} \circ (t')^{-3} \circ \iota$. Since $t^3$ and $(t')^{-3}$ both are scalar multiplication by $-\lambda$, we conclude that now $\tau(\lambda) = \lambda$. We will say that in this case the involution of the second kind of $D$ is *hermitian*.

## 4.11 Twisted Composition Algebras with Anisotropic Norm

In this section we will review analogues Th. 4.6.2 for the case of twisted composition algebras with anisotropic norm. We need a complement to Lemma 4.2.12, which we first establish. For the moment, $F$ is an arbitrary twisted composition algebra, as in Def. 4.2.1. The restriction char$(k) \neq 2, 3$ remains in force.. Assume that $b \in F$ has the properties (i), (ii), (iii) of Lemma 4.2.12 and that $N(b) \neq 0$. In particular $D(b) = T(b)^2 - 4\mathrm{N}_{l/k}(N(b)) \neq 0$. The polynomial with coefficients in $k$

$$X^2 + T(b)X + \mathrm{N}_{l/k}(N(b)) \tag{4.86}$$

has two distinct roots $\xi$ and $\eta$, which are nonzero. Assume that they lie in $k$ and put

$$a = (\xi - \eta)^{-1}(N(b)^{-1}\xi b + b^{*2}), \quad a' = N(b)^{-1}b - a = (\eta - \xi)^{-1}(N(b)^{-1}\eta b + b^{*2}).$$

**Lemma 4.11.1** $a$ *is isotropic and* $\langle a, a' \rangle = N(b)^{-1}$. *We have* $a^{*2} = -N(b)\eta^{-1}a'$, $T(a) = -\eta^{-1}$. *Moreover, $a$ has the properties of Lemma 4.2.12. Similar results hold for $a'$.*

Proof. That $a$ is isotropic follows by a direct computation, using that $\xi$ is a root of (4.86). We have $\langle a, b \rangle = (\xi - \eta)^{-1}(2\xi + T(b)) = 1$, since $T(b) = -\xi - \eta$. Hence, $a$ being isotropic, $\langle a, a' \rangle = \langle a, N(b)^{-1}b - a \rangle = N(b)^{-1}$. The formula for $a^{*2}$ follows from (4.34), using that $\mathrm{N}_{l/k}(N(b)) = N(b)N(b)^{*2} = \xi\eta$. The remaining assertions are easy.                    □

Now assume that the norm $N$ is *anisotropic* on $F$. Then $k$ is infinite. We also assume that (with the notations of Prop. 4.2.5) $N(F)^* = M(N)$. By Cor. 4.2.6 this can be achieved by replacing $F$ by an isotope.

**Lemma 4.11.2** *There exists $b \in F$ with $N(b) = 1$ having the three properties of Lemma 4.2.12.*

Proof. As $N$ is anisotropic, $k$ and $l$ are infinite. View $F$ as a vector space over $k$. By Lemma 4.2.12 there exists $c \in F$ with $T(c) \neq 0$, $U(c) = T(c)^2 - 4\mathrm{N}_{l/k}(N(c)) \neq 0$ and $N(c) \neq 0$. View $T$ and $U$ as homogeneous polynomial functions of respective degree 3 and 6, and $N$ as a homogeneous quadratic mapping of $F$ to $l = k^3$. Let $K$ be an algebraic closure of $k$. By homogeneity, there is $d \in K \otimes_k F$ with $T(d) \neq 0$, $U(d) \neq 0$, $N(d) = 1$.

Let $Q$ be the variety in $K \otimes_k F$ defined by the equation $N(x) = 1$. The set $O$ of $x \in Q$ with $T(x) \neq 0$, $U(x) \neq 0$ is an open subset of $Q$ which is nonempty, by what we just saw. By our assumptions there exists $b_0 \in F$ with $N(b_0) = 1$. If $x \in F$, $N(x) \neq 0$, then $\tilde{x} = b_0 - (N(x))^{-1} \langle b_0, x \rangle x \in Q$, as a straightforward calculation shows. Since $k$ is infinite, we can choose $x \in F$ such that $N(x) \neq 0$ and $\tilde{x} \in O$. Then $b = \tilde{x}$ has properties (ii) and (iii) of Lemma 4.2.12, and $N(b) = 1$. Also, if we had $b^{*2} = \xi b$, then since $N(b) = 1$ we had $\xi^2 = 1$ and $T(b) = \langle \xi b, b \rangle = 2\xi$, whence the contradiction $U(b) = 0$. Hence property (i) also holds.    □

Choose $b$ as in the preceding lemma. Since $N$ is anisotropic the polynomial (4.86) has no roots in $k$. We have to distinguish several cases, which we briefly discuss.

Case (A). $F$ is a normal twisted composition algebra.

Let $k_1$ be the quadratic extension of $k$ generated by the roots $\xi$ and $\eta$ of (4.86). Notice that now $\xi\eta = 1$. Denote by $\tau_1$ the nontrivial automorphism of $k_1/k$. Then $l_1 = k_1 \otimes_k l$ is a Galois extension $k$ which is cyclic of degree 6. Viewing $\sigma$ and $\tau_1$ as elements of its Galois group, $\sigma\tau_1$ is a generator of that group. Now $F_1 = l_1 \otimes_l F$ is a normal twisted composition algebra over $l_1$ and $\sigma$, with an isotropic norm.

We perform the analysis of § 4.5 for $F_1$, with $a$ as in Lemma 4.11.1. Then $e_1 = a$, and by the lemma $e_2 = a'$. Denote by $v$ the $\tau_1$-linear map $\tau_1 \otimes \mathrm{id}$ of $F_1 = l_1 \otimes F$. Then $v$ is an automorphism of $F_1$ and $F$ is the space of invariants $\mathrm{Inv}(v)$. Moreover $v(x * e_i) = v(x) * e_{i+1}$ $(x \in F_1, i = 1, 2)$. With the notations of 4.5, $v$ induces a $\tau_1$-linear bijection $E_i \to E_{i+1}$. As in § 4.10 we find

$$v \circ t_i = t_{i+1} \circ v,$$

$$v(x_i \wedge y_i) = v(x_i) \wedge v(y_i),$$

$$\langle v(x_2), v(x_1) \rangle = \tau_1(\langle x_1, x_2 \rangle).$$

These formulas indicate how to modify the construction of 4.6 in the present case. Assume given a twisted composition algebra $F_1 = \mathcal{F}(V_1, t)$ over $l_1$ and $\sigma$.

**Definition 4.11.3** A *unitary involution* of $(V_1, t)$ is a $\tau_1$-linear bijection $\iota_1 : V_1 \to V_1'$ with the following properties:

(i) $\langle x, \iota_1(y) \rangle = \tau(\langle y, \iota_1(x) \rangle)$  $(x, y \in V_1)$.
(ii) $\iota_1 \circ t = t' \circ \iota_1$.
(iii) $\iota_1^{-1}(x \wedge y) = \iota_1(y) \wedge \iota_1(x)$.

We also have

(iii)' $\iota_1(x' \wedge y') = \iota_1^{-1}(y') \wedge \iota_1^{-1}(x')$.

**Theorem 4.11.4** *Let $\iota_1$ be a unitary involution of $(V_1, t)$. Then $v(\xi_1 e_1 + \xi_2 e_2 + x + x') = \tau_1(\xi_2) e_1 + \tau_1(\xi_1) e_2 - \iota_1^{-1}(x') - \iota_1(x)$ defines a $\tau_1$-linear automorphism of $F_1$. The fixed point set $\mathrm{Inv}(v)$ is a normal twisted composition algebra over $l$ and $\sigma$, with $N(F)^* = M(N)$. All anisotropic normal twisted composition algebras with the last property are of this form.*

Proof. The proofs of the assertions about $v$ and $F$ are straightforward. The last point follows from what was established in the beginning of this section. □

Again, there is a reformulation of the properties of Def. 4.11.3. Define

$$h(x, y) = \langle x, \iota_1(y) \rangle \quad (x, y \in V_1).$$

Then $h$ is a nondegenerate hermitian form on $V_1$, relative to $\tau_1$. We now have

$$h(t(x), t(y)) = \sigma(h(x, y)),$$

which explains the adjective "unitary". Also, $\det(\iota_1) = -1$.

The cyclic crossed product occurring in the present case is $D = (l_1, \sigma, \xi)$, where $\xi$ is as before. Notice that $\eta = \tau_1(\xi) = \xi^{-1}$. The central simple algebra $D$ over $k_1$ has the involution of the second kind

$$\xi_0 + \xi_1 t + \xi_2 t^2 \mapsto \tau_1(\xi_0) + t^{-1}\tau_1(\xi_1) + t^{-2}\tau_1(\xi_2).$$

We now call the involution *unitary*.

Case (B). $F$ is a nonnormal twisted composition algebra and the normal composition algebra $F'$ is anisotropic.

In this case (4.86) has no root in $k'$. Let $k_1$ and $\tau_1$ be as in Case (A). Now $l_1' = k_1 \otimes_k l'$ is a Galois extension of $k$ whose group is $S_2 \times S_3$. We view $\sigma, \tau$ and $\tau_1$ as automorphisms of $l_1'$. Put $F_1' = k_1 \otimes_k F'$. By 4.6.2 we may assume that is of the form $F_1' = \mathcal{F}(V_1, t)$, where $V_1$ is a vector space over $l_1'$ and $t$ is $\sigma$-semilinear. By 4.10.2 we have a $\tau$-linear hermitian involution $\iota$ on $(V_1, t)$ and an involution on $V_1$, whereas by case (A) we have a $\tau_1$-linear unitary involution on $(V_1, t)$ and an automorphism $v$ of $F_1'$. Then

$$F = \mathrm{Inv}(u, v) = \mathrm{Inv}(u) \cap \mathrm{Inv}(v).$$

In the present case the cyclic crossed product $D$ is a central simple algebra over the field $k_1' = k_1 \otimes_k k'$. It has two commuting involutions of the second kind: a $\tau$-linear one which is hermitian, and a $\tau_1$-linear one which is unitary.

Case (C). $F$ is a nonnormal twisted composition algebra and (4.86) has roots in $k'$.

We can now take $\xi$ and $\eta$ in $k'$, hence $a \in F'$. Proceeding as in 4.10 we have

$$u(x * e_i) = e_{i+1} * x \quad (x \in E_i),$$

$$u \circ t_i = t_i^{-1} \circ u.$$

It follows that $u$ defines $\tau$-linear bijections $\iota$ and $\iota'$ of $V$ and $V'$, respectively. Moreover, (4.83) holds, and we have the counterpart of (4.84)

$$\langle \iota(x_1), \iota'(x_2) \rangle = \tau(\langle x_1, x_2 \rangle).$$

Assume again that $F' = \mathcal{F}(V, t)$, with $V$ a vector space over $l'$. Then we have a $\tau$-linear automorphism $\iota$ of order 2 such that

$$\iota \circ t = t^{-1} \circ \iota.$$

Moreover, $\iota'$ being as in the last equation,

$$\iota'(x \wedge y) = \iota(x) \wedge \iota(y) \quad (x, y \in V).$$

With these notations,

$$u(\xi_1 e_1 + \xi_2 e_2 + x + x') = \tau(\xi_2)e_1 + \tau(\xi_1)e_2 + \iota(x) + \iota'(x)$$

defines a $\tau$-linear involution of $F'$, such that $F$ is the fixed point set $\mathrm{Inv}(u)$. Any twisted composition algebra of Case (C) can be obtained in this way.

The cyclic crossed product $D = (l', \sigma, -\lambda)$ has the unitary involution of the second kind

$$\xi_0 + \xi_1 t + \xi_2 t^2 \mapsto \tau(\xi_0) + t\tau(\xi_1) + t^2\tau(\xi_2)$$

## 4.12 Historical Notes

*Twisted composition algebras* were introduced by T.A. Springer in the normal case, with a view to a good description of nonreduced Albert algebras (see Ch. 6). The theory was first exposed in a course at the University of Göttingen in the summer of 1963 (see [Sp 63]). The generalization to the nonnormal case is due to F.D. Veldkamp (in an unpublished manuscript). Independently, it was also given in [KMRT, §36].

# 5. J-algebras and Albert Algebras

In this chapter we discuss a class of Jordan algebras which includes those that are usually named exceptional central simple Jordan algebras or Albert algebras. Our interest in Albert algebras is motivated by their connections with exceptional simple algebraic groups of type $E_6$ and $F_4$, a topic we will deal with in Ch. 7. They also play a role in a description of algebraic groups of type $E_7$ and $E_8$, but we leave that aspect aside. We will not enter into the general theory of Jordan algebras, but use an ad hoc characterization of the algebras under consideration by simple axioms, which are somewhat reminiscent of those for composition algebras. We call these algebras J-algebras, since they are in fact a limited class of Jordan algebras; see Remark 5.1.7.

In this and the following chapters, fields will always be assumed to have characteristic $\neq 2, 3$. The assumption characteristic $\neq 3$ is for technical reasons and could possibly be removed. However, characteristic $\neq 2$ is essential for our approach to Jordan algebras as (nonassociative) algebras with a binary product. If one wants to include all characteristics, it is necessary to use quadratic Jordan algebras as introduced by K. McCrimmon [McC]; see also [Ja 69], [Ja 81] or [Sp 73].

## 5.1 J-algebras. Definition and Basic Properties

Let $k$ be a field with $\operatorname{char}(k) \neq 2, 3$ and let $C$ be a composition algebra over $k$. For fixed $\gamma_i \in k^*$, let $A = \mathrm{H}(C; \gamma_1, \gamma_2, \gamma_3)$ be the set of $(\gamma_1, \gamma_2, \gamma_3)$-hermitian $3 \times 3$ matrices

$$
x = h(\xi_1, \xi_2, \xi_3; c_1, c_2, c_3) = \begin{pmatrix} \xi_1 & c_3 & \gamma_1^{-1}\gamma_3\bar{c}_2 \\ \gamma_2^{-1}\gamma_1\bar{c}_3 & \xi_2 & c_1 \\ c_2 & \gamma_3^{-1}\gamma_2\bar{c}_1 & \xi_3 \end{pmatrix} \tag{5.1}
$$

with $\xi_i \in k$ and $c_i \in C$ $(i = 1, 2, 3)$; here $^-$ denotes conjugation as in § 1.3. We define a product in $A$ which is different from the standard matrix product:

$$
xy = \frac{1}{2}(x \cdot y + y \cdot x) = \frac{1}{2}((x+y)^2 - x^2 - y^2), \tag{5.2}
$$

where the dot indicates the standard matrix product and the square is the usual one with respect to the standard product (which coincides with the square with respect to the newly defined product). This multiplication is not associative. Together with the usual addition of matrices and multiplication by elements of $k$, it makes $A$ into a commutative, nonassociative $k$-algebra with the $3 \times 3$ identity matrix $e$ as identity element. We introduce a quadratic norm $Q$ on $A$ by

$$
\begin{aligned}
Q(x) &= \tfrac{1}{2}\operatorname{tr}(x^2) \\
&= \tfrac{1}{2}(\xi_1^2 + \xi_2^2 + \xi_3^2) + \gamma_3^{-1}\gamma_2 N(c_1) + \gamma_1^{-1}\gamma_3 N(c_2) + \gamma_2^{-1}\gamma_1 N(c_3)
\end{aligned} \tag{5.3}
$$

for $x = h(\xi_1, \xi_2, \xi_3; c_1, c_2, c_3) \in A$, and the associated bilinear form

$$
\langle x, y \rangle = Q(x+y) - Q(x) - Q(y) = \operatorname{tr}(xy) \qquad (x, y \in A).
$$

This bilinear form is nondegenerate. Of special interest is the case that $C$ is an octonion algebra; we then call $A = \mathrm{H}(C; \gamma_1, \gamma_2, \gamma_3)$ an *Albert algebra*. More generally, $A$ is called an *Albert algebra* if $k' \otimes_k A$ is isomorphic to such a matrix algebra $\mathrm{H}(C'; \gamma_1, \gamma_2, \gamma_3)$ for some field extension $k'$ of $k$ and some octonion algebra $C'$ over $k'$. In Prop. 5.1.6 we will prove the relation $x^2(xy) = x(x^2y)$, which is typical for Jordan algebras. The reader will have no difficulty in verifying the following three rules:

$$
Q(x^2) = Q(x)^2 \qquad \text{if } \langle x, e \rangle = 0, \tag{5.4}
$$

$$
\langle xy, z \rangle = \langle x, yz \rangle, \tag{5.5}
$$

$$
Q(e) = \frac{3}{2}. \tag{5.6}
$$

We will, conversely, consider a class of algebras with a quadratic norm $Q$ that satisfies (5.4), (5.5) and (5.6). This class will turn out to contain, besides the algebras related to the algebras $A = \mathrm{H}(C; \gamma_1, \gamma_2, \gamma_3)$ introduced above, one other type of algebras; see Prop. 5.3.5, the remark that follows it, and the classification in Th. 5.4.5.

**Definition 5.1.1** Let $k$ be a field of characteristic $\neq 2, 3$. A *J-algebra* over $k$ is a finite-dimensional commutative, not necessarily associative, $k$-algebra $A$ with identity element $e$ together with a nondegenerate quadratic form $Q$ on $A$ such that the conditions (5.4), (5.5) and (5.6) are satisfied. $Q$ is called the *norm* of $A$, and the associated bilinear form $\langle \, , \, \rangle$ will often be called the *inner product*.

A *J-subalgebra* is a nonsingular (with respect to $Q$) linear subspace which contains $e$ and is closed under multiplication.

An *isomorphism* $t : A \to A'$ of J-algebras over $k$ is a bijective linear transformation which preserves multiplication: $t(xy) = t(x)t(y)$ $(x, y \in A)$.

**Remark 5.1.2** It will be shown in Prop. 5.3.10 that in a J-algebra of dimension $> 2$ the norm $Q$ is already determined by the linear structure and the product, and the same holds for the cubic form det that will be introduced in Prop. 5.1.5. As a consequence, an isomorphism necessarily leaves $Q$ and det invariant in dimension $> 2$.

We begin the study of J-algebras with a lemma that gives a linearized version of (5.4).

**Lemma 5.1.3** *If* $\langle x, e \rangle = \langle y, e \rangle = \langle z, e \rangle = \langle u, e \rangle = 0$, *then*

$$2\langle xy, zu \rangle + 2\langle xz, yu \rangle + 2\langle xu, yz \rangle = \langle x, y \rangle\langle z, u \rangle + \langle x, z \rangle\langle y, u \rangle + \langle x, u \rangle\langle y, z \rangle.$$

Proof. By substituting $\lambda x + \mu y + \nu z + \varrho u$ for $x$ in (5.4), writing both sides out as polynomials in $\lambda, \mu, \nu$ and $\varrho$, and equating the coefficients of $\lambda\mu\nu\varrho$ on either side, we immediately get the formula. Here we use that the degree of the polynomials is 4 and that $|k| > 4$. ☐

**Proposition 5.1.4** *If $A$ is a J-algebra over a field $k$ and $l$ is any extension field of $k$, then $l \otimes_k A$, with the extension of the product and the quadratic form, is a J-algebra over $l$.*

Proof. The linearized version of (5.4) that we proved in the Lemma is in fact equivalent to (5.4) itself. This multilinear version clearly also holds in $l \otimes_k A$. Similarly for (5.5). ☐

By the Hamilton-Cayley Theorem, every element in a $3 \times 3$ matrix algebra over a field satisfies a cubic equation. It is conceivable that a similar result holds in a "matrix algebra" $H(C; \gamma_1, \gamma_2, \gamma_3)$. In fact, it does in all J-algebras.

**Proposition 5.1.5** *Every element $x$ in a J-algebra $A$ satisfies a cubic equation*

$$x^3 - \langle x, e \rangle x^2 - (Q(x) - \frac{1}{2}\langle x, e \rangle^2)x - \det(x)e = 0, \qquad (5.7)$$

*called its Hamilton-Cayley equation. Here det is a cubic form on $A$.*

Proof. With the aid of equation (5.5) we derive from Lemma 5.1.3 (with $z = u = x$)

$$\langle x^3 - Q(x)x, y \rangle = 0 \qquad (x, y \in A, \ \langle x, e \rangle = \langle y, e \rangle = 0).$$

Since $\langle \ , \ \rangle$ is nondegenerate, this implies that

$$x^3 - Q(x)x = \kappa(x)e \qquad (x \in A, \ \langle x, e \rangle = 0), \qquad (5.8)$$

where $\kappa$ is a cubic form with values in $k$. We can write any $x \in A$ as $x = x' + \frac{1}{3}\langle x, e \rangle e$ with $\langle x', e \rangle = 0$. Substitution of $x - \frac{1}{3}\langle x, e \rangle e$ in equation (5.8) yields after some computation

$$x^3 - \langle x, e \rangle x^2 - (Q(x) - \frac{1}{2}\langle x, e \rangle^2)x - \det(x)e = 0 \qquad (x \in A),$$

where det is a cubic form on $A$ with values in $k$. ☐

If $\det(x) \neq 0$, then

$$x^{-1} = \det(x)^{-1}(x^2 - \langle x, e \rangle x - (Q(x) - \frac{1}{2}\langle x, e \rangle^2)e)$$

satisfies $xx^{-1} = e$; we will come back to this in Lemma 5.2.3. The polynomial

$$\chi_x(T) = T^3 - \langle x, e \rangle T^2 - (Q(x) - \frac{1}{2}\langle x, e \rangle^2)T - \det(x) \qquad (5.9)$$

is called the *characteristic polynomial* of $x$, and $\det(x)$ the *determinant* of $x$; the cubic form det is called the *determinant function* on $A$, or just the *determinant* of $A$.

By taking the inner product of the left hand side of (5.7) with $e$, one finds

$$0 = \langle x^3, e \rangle - \langle x, e \rangle\langle x^2, e \rangle - (Q(x) - \frac{1}{2}\langle x, e \rangle^2)\langle x, e \rangle - \det(x)\langle e, e \rangle$$

$$= \langle x^2, x \rangle - 3Q(x)\langle x, e \rangle + \frac{1}{2}\langle x, e \rangle^3 - 3\det(x).$$

Hence

$$\det(x) = \frac{1}{3}\langle x^2, x \rangle - Q(x)\langle x, e \rangle + \frac{1}{6}\langle x, e \rangle^3. \qquad (5.10)$$

Notice that $\det(e) = 1$.

With the aid of (5.10) one easily computes $\det(x)$ for an element $x = h(\xi_1, \xi_2, \xi_3; c_1, c_2, c_3)$ of $H(C; \gamma_1, \gamma_2, \gamma_3)$ as in (5.1):

$$\det(x) = \xi_1\xi_2\xi_3 - \gamma_3^{-1}\gamma_2\xi_1 N(c_1) - \gamma_1^{-1}\gamma_3\xi_2 N(c_2) - \gamma_2^{-1}\gamma_1\xi_3 N(c_3) + N(c_1 c_2, \bar{c}_3).$$
$$(5.11)$$

Here $N(\,,\,)$ denotes the bilinear form associated with the norm $N$ on $C$.

The cubic form det uniquely determines a symmetric trilinear form $\langle\,,\,,\,\rangle$ with

$$\langle x, x, x \rangle = \det(x).$$

We have

$$6\langle x, y, z \rangle = \det(x + y + z) - \det(x + y) - \det(y + z) - \det(x + z) +$$

$$\det(x) + \det(y) + \det(z).$$

We derive some consequences of the Hamilton-Cayley equation. Replacing $x$ by $x + y + z$ in (5.7) yields

$$(x+y+z)^3 - \langle x+y+z, e \rangle(x+y+z)^2 - (Q(x+y+z) - \frac{1}{2}\langle x+y+z, e \rangle^2)(x+y+z) =$$

$$\det(x + y + z)e.$$

Collecting in both sides terms which are linear in each of the variables $x$, $y$ and $z$, we obtain the following formula.

$$x(yz) + y(xz) + z(xy) = \langle x, e \rangle yz + \langle y, e \rangle xz + \langle z, e \rangle xy +$$

$$\frac{1}{2}(\langle y,z\rangle - \langle y,e\rangle\langle z,e\rangle)x + \frac{1}{2}(\langle x,z\rangle - \langle x,e\rangle\langle z,e\rangle)y +$$

$$\frac{1}{2}(\langle x,y\rangle - \langle x,e\rangle\langle y,e\rangle)z + 3\langle x,y,z\rangle e. \tag{5.12}$$

Replacing $z$ by $x$ in this equation we find

$$2x(xy) + x^2y = 2\langle x,e\rangle xy + \langle y,e\rangle x^2 +$$

$$(\langle x,y\rangle - \langle x,e\rangle\langle y,e\rangle)x + (Q(x) - \frac{1}{2}\langle x,e\rangle^2)y + 3\langle x,x,y\rangle e. \tag{5.13}$$

From equation (5.12) one easily derives a formula that expresses the symmetric trilinear form associated with det in the inner product and the product. Namely, take the inner product of either side of (5.12) with $e$, apply condition (5.5) several times and use $\langle e,e\rangle = 3$. After rearrangement and dividing by 3 one finds:

$$3\langle x,y,z\rangle = \langle xy,z\rangle - \frac{1}{2}\langle x,e\rangle\langle y,z\rangle - \frac{1}{2}\langle y,e\rangle\langle x,z\rangle - \frac{1}{2}\langle z,e\rangle\langle x,y\rangle +$$

$$\frac{1}{2}\langle x,e\rangle\langle y,e\rangle\langle z,e\rangle. \tag{5.14}$$

We can now prove the Jordan identity.

**Proposition 5.1.6** *In any J-algebra A the Jordan identity holds:*

$$x^2(xy) = x(x^2y).$$

Proof. It suffices to prove that $x^2(xy)$ and $x(x^2y)$ have equal inner products with any $z \in A$. In view of (5.5) this amounts to showing that

$$\langle xy, x^2z\rangle = \langle xz, x^2y\rangle \qquad (x,y,z \in A). \tag{5.15}$$

This relation is immediate from (5.5) if $y = e$ or $z = e$, so it suffices to prove it for the case $\langle y,e\rangle = \langle z,e\rangle = 0$. Under these assumptions, we take the inner product of either side of equation (5.13) with $xz$ and, using (5.5), find the relation

$$2\langle x(xy), xz\rangle + \langle x^2y, xz\rangle = 2\langle x,e\rangle\langle xy,xz\rangle + \langle x,y\rangle\langle x,xz\rangle +$$

$$Q(x)\langle y,xz\rangle - \frac{1}{2}\langle x,e\rangle^2\langle y,xz\rangle + 3\langle x,x,y\rangle\langle x,z\rangle.$$

Replacing $3\langle x,x,y\rangle$ in the right hand side by the expression that follows from equation (5.14), we arrive at the formula

$$\langle x^2y, xz\rangle = -2\langle x(xy), xz\rangle + 2\langle x,e\rangle\langle xy,xz\rangle + \langle x,y\rangle\langle x,xz\rangle +$$

$$Q(x)\langle y,xz\rangle - \frac{1}{2}\langle x,e\rangle^2\langle y,xz\rangle + \langle x,xy\rangle\langle x,z\rangle - \langle x,e\rangle\langle x,y\rangle\langle x,z\rangle$$

for $\langle y,e\rangle = \langle z,e\rangle = 0$. It is straightforward to verify that the right hand side of this equation is symmetric in $y$ and $z$. So the left hand side is symmetric in $y$ and $z$, too, which just amounts to (5.15). $\qquad\square$

**Remark 5.1.7** A commutative algebra over a field $k$ of characteristic $\neq 2$ in which the Jordan identity holds is called a (commutative) *Jordan algebra*. So the above proposition says that every J-algebra is a Jordan algebra. A consequence of the Jordan identity is power associativity: $x^m x^n = x^{m+n}$ $(m, n \geq 1)$; see, e.g., [Ja 68, Ch. I, Th. 8] or [Schaf, Ch. IV, §1, p. 92]. For J-algebras, power associativity follows more easily, as we show in the following corollary.

**Corollary 5.1.8** *For any $x$ in a J-algebra $A$, the subalgebra $k[x]$ generated by $x$ is a homomorphic image of $k[T]/\chi_x(T)$, where $\chi_x$ is the characteristic polynomial of $x$ (see equation (5.9)). Consequently, $A$ is power associative.*

Proof. By substituting $x$ for $y$ in the Jordan identity we find that $x^2 x^2 = x^4$. By the Hamilton-Cayley equation (5.7), every element of $k[x]$ can be written in the form $\xi_0 e + \xi_1 x + \xi_2 x^2$ (but notice that $e$, $x$ and $x^2$ need not be linearly independent). The product of two such elements is associative since $(x^l x^m) x^n = x^{l+m+n}$ for $l, m, n \leq 2$, as follows from the Jordan identity. In other words, $k[x]$ is the homomorphic image of the associative algebra $k[T]/\chi_x(T)$.                                                                    □

**Remark 5.1.9** The word "subalgebra" above is meant in the sense of the theory of nonassociative algebras, so as a linear subspace containing $e$ and closed under multiplication. For a J-subalgebra we also required in Def. 5.1.1 that the restriction of the norm $Q$ to it is nondegenerate; we will see in Prop. 5.3.8, that this need not be the case with $k[x]$.

## 5.2 Cross Product. Idempotents

With the aid of the symmetric trilinear form $\langle \, , \, , \, \rangle$ associated with det and the bilinear inner product we introduce a *cross product* $\times$ that will be used frequently in future computations: in a J-algebra $A$, we define $x \times y$ $(x, y \in A)$ to be the element such that

$$\langle x \times y, z \rangle = 3 \langle x, y, z \rangle \qquad (z \in A). \tag{5.16}$$

The cross product is evidently symmetric. In the following lemma we express it in terms of the ordinary product, and collect some formulas that will be useful in later computations.

**Lemma 5.2.1** *The following formulas hold for the cross product.*

(i) $x \times y = xy - \frac{1}{2}\langle x, e \rangle y - \frac{1}{2}\langle y, e \rangle x - \frac{1}{2}\langle x, y \rangle e + \frac{1}{2}\langle x, e \rangle \langle y, e \rangle e$;

(ii) $x(x \times x) = \det(x)e$;

(iii) $(x_1 \times x_2) \times y = \frac{1}{2}(x_1 x_2)y - \frac{1}{2}x_1(x_2 y) - \frac{1}{2}x_2(x_1 y) + \frac{1}{4}\langle x_1, y \rangle x_2 + \frac{1}{4}\langle x_2, y \rangle x_1$;

(iv) $(x \times x) \times (x \times x) = \det(x)x$;

*(v)* $4(x \times y) \times (z \times u) + 4(x \times z) \times (y \times u) + 4(x \times u) \times (y \times z) =$
$$3\langle x, y, z \rangle u + 3\langle x, y, u \rangle z + 3\langle x, z, u \rangle y + 3\langle y, z, u \rangle x;$$

*(vi)* $4x \times (y \times (x \times x)) = \langle x, y \rangle x \times x + \det(x)y;$

*(vii)* $\det(x \times x) = \det(x)^2.$

Proof. (i) By equation (5.14), $\langle x \times y, z \rangle$ equals the inner product with $z$ of the right hand side of the formula in (i), for all $z$. Hence (i) holds.

(ii) Using the above formula to compute $x \times x$, one finds

$$x(x \times x) = x^3 - \langle x, e \rangle x^2 - (Q(x) - \frac{1}{2}\langle x, e \rangle^2)x$$
$$= \det(x)e \quad \text{(by the Hamilton-Cayley equation (5.7))}.$$

(iii) A straightforward computation using (i) and equation (5.14) yields

$$(x \times x) \times y = x^2 y - \frac{1}{2}\langle y, e \rangle x^2 - \langle x, e \rangle xy +$$

$$\frac{1}{2}\langle x, e \rangle\langle y, e \rangle x - \frac{1}{2}(Q(x) - \frac{1}{2}\langle x, e \rangle^2)y - \frac{3}{2}\langle x, x, y \rangle e.$$

Using equation (5.13), one reduces this to

$$(x \times x) \times y = \frac{1}{2}x^2 y - x(xy) + \frac{1}{2}\langle x, y \rangle x.$$

Linearizing this one obtains the formula in (iii).

(iv) Replace $x_1$ and $x_2$ by $x$, and $y$ by $x \times x$ in the formula of (iii). Then using (ii), the Hamilton-Cayley equation and power associativity, one easily gets the result.

(v) This follows by linearizing the previous formula.

(vi) and (vii) In (v), replace $x$, $z$ and $u$ by $x \times x$; this yields

$$4[(x \times x) \times (x \times x)] \times [y \times (x \times x)] = 3\langle x \times x, x \times x, y \rangle x \times x + \det(x \times x)y. \quad (5.17)$$

The left hand side equals $4\det(x)x \times (y \times (x \times x))$ by (iv). Further,

$$3\langle x \times x, x \times x, y \rangle = \langle (x \times x) \times (x \times x), y \rangle = \det(x)\langle x, y \rangle,$$

by (5.16) and (iv). Replacing $y$ by $x \times x$ in this formula we get

$$\det(x \times x) = \det(x)^2,$$

i.e., (vii) holds. Using these last two equations to replace terms in the right hand side of (5.17), we find

$$4\det(x)x \times (y \times (x \times x)) = \det(x)\langle x, y \rangle x \times x + \det(x)^2 y.$$

This yields formula (vi) for $\det(x) \neq 0$; by continuity for the Zariski topology over an algebraic closure of $k$ it holds everywhere (cf. the end of the proof of Prop. 3.3.4). □

An important role in developing the theory of J-algebras will be played by idempotent elements, i.e., elements $u$ such that $u^2 = u$.

**Lemma 5.2.2** *If $u \neq 0, e$ is an idempotent in $A$, then $\det(u) = 0$ and $Q(u) = \frac{1}{2}$ or $Q(u) = 1$; in addition, $\langle u, e \rangle = 2Q(u)$. Further, $e - u$ is an idempotent with $u(e - u) = 0$, $\langle u, e - u \rangle = 0$ and $Q(e - u) = \frac{3}{2} - Q(u)$. So if $A$ contains an idempotent $\neq 0, e$, then it contains an idempotent $u$ with $Q(u) = \frac{1}{2}$.*

Proof. The Hamilton-Cayley equation reads for an idempotent $u$:

$$(1 - \langle u, e \rangle - Q(u) + \frac{1}{2}\langle u, e \rangle^2)u = \det(u)e.$$

Since an idempotent $\neq 0, e$ can not be a multiple of $e$, we find $\det(u) = 0$ and

$$1 - \langle u, e \rangle - Q(u) + \frac{1}{2}\langle u, e \rangle^2 = 0. \tag{5.18}$$

Using (5.5) we see that

$$Q(u) = \frac{1}{2}\langle u, u \rangle = \frac{1}{2}\langle u^2, e \rangle = \frac{1}{2}\langle u, e \rangle,$$

so $\langle u, e \rangle = 2Q(u)$. Substituting this in equation (5.18), we get

$$Q(u)^2 - \frac{3}{2}Q(u) + \frac{1}{2} = 0.$$

This yields $Q(u) = \frac{1}{2}$ or $Q(u) = 1$. The rest of the proof is straightforward. $\square$

An idempotent $u$ with $Q(u) = \frac{1}{2}$ is called a *primitive idempotent*. This name is in agreement with the usual terminology. For let $u$ be an idempotent with $Q(u) = \frac{1}{2}$, and suppose we could decompose

$$u = u_1 + u_2 \qquad \left(u_i^2 = u_i, \ u_1 u_2 = 0\right).$$

Then we would have

$$1 = \langle u, e \rangle = \langle u_1, e \rangle + \langle u_2, e \rangle$$

and $\langle u_i, e \rangle = 2Q(u_i) = 1$ or $2$, which leads to a contradiction. On the other hand we will see in Prop. 5.3.7 that in most cases $e - u$ is a sum of two orthogonal primitive idempotents.

We next prove an addition to Prop. 5.1.5.

**Lemma 5.2.3** *Let $A$ be a J-algebra. For $x \in A$, there exists $x^{-1} \in A$ having the properties $xx^{-1} = e$ and $x(x^{-1}y) = x^{-1}(xy)$ $(y \in A)$ if and only if $\det(x) \neq 0$. Such an element $x^{-1}$ is unique, viz.,*

$$x^{-1} = \det(x)^{-1}(x^2 - \langle x, e \rangle x - (Q(x) - \frac{1}{2}\langle x, e \rangle^2)e).$$

Proof. If $\det(x) \neq 0$, then

$$x^{-1} = \det(x)^{-1}(x^2 - \langle x, e \rangle x - (Q(x) - \frac{1}{2}\langle x, e \rangle^2)e)$$

satisfies $xx^{-1} = e$ by the Hamilton-Cayley equation (5.7). From the Jordan identity it follows that $x(x^{-1}y) = x^{-1}(xy)$ $(y \in A)$. If $z$ satisfies $xz = e$ and $x(zy) = z(xy)$ $(y \in A)$, then $x^2z = x$ and $x^3z = x^2$. Hence by the Hamilton-Cayley equation,

$$\det(x)z = x^2 - \langle x, e \rangle x - (Q(x) - \frac{1}{2}\langle x, e \rangle^2 e,$$

which shows uniqueness.

Now suppose $x^{-1}$ would exist for $x \neq 0$ with $\det(x) = 0$. Then

$$x^2 - \langle x, e \rangle x - (Q(x) - \frac{1}{2}\langle x, e \rangle^2)e = 0.$$

Taking the inner product with $e$ we find that $Q(x) = \frac{1}{2}\langle x, e \rangle^2$. Hence $x^2 = \langle x, e \rangle x$. Then

$$x = x(x^{-1}x) = x^{-1}x^2 = \langle x, e \rangle e.$$

It follows that $x$ is a nonzero multiple of $e$. Since $\det(e) = 1$ we obtain a contradiction. $\qquad\qquad\Box$

The element $x^{-1}$ as in the Lemma is called the *J-inverse* or just *inverse* of $x$. It is not an inverse in the general sense of nonassociative algebras as we defined at the end of § 1.3, since $x^{-1}(xy) = y$ need not hold for $y \notin k[x]$.

## 5.3 Reduced J-algebras and Their Decomposition

A J-algebra is said to be *reduced* provided it contains an idempotent $\neq 0, e$. By Lemma 5.2.2 it also contains a primitive idempotent.

We consider a reduced J-algebra $A$ over $k$ and fix a primitive idempotent $u$ in $A$. Define

$$E = (ke \oplus ku)^{\perp} = \{ x \in A \mid \langle x, e \rangle = \langle x, u \rangle = 0 \}.$$

The restriction of $Q$ to $ke \oplus ku$ is nondegenerate, hence the same holds for $E$. From $x \in E$ we infer that $ux \in E$, for $\langle ux, e \rangle = \langle x, u \rangle = 0$ and $\langle ux, u \rangle = \langle x, u \rangle = 0$. So we can define the linear transformation

$$t : E \to E, \ x \mapsto ux. \tag{5.19}$$

**Lemma 5.3.1** *$t$ is symmetric with respect to $\langle\ ,\ \rangle$, and $t^2 = \frac{1}{2}t$. We have*

$$E = E_0 \oplus E_1 \qquad with \quad E_0 \perp E_1,$$

*where*

$$E_i = \{\, x \in E \mid t(x) = \frac{1}{2}ix \,\}.$$

Proof. Let $x, y \in E$. The symmetry of $t$ follows from property (5.5):

$$\langle t(x), y \rangle = \langle ux, y \rangle = \langle x, uy \rangle = \langle x, t(y) \rangle.$$

Using Lemma 5.2.2, we get from equations (5.13) and (5.14)

$$2u(ux) + ux = 2ux,$$

from which $t^2 = \frac{1}{2}t$ follows. This implies that the possible eigenvalues of $t$ are $0$ and $\frac{1}{2}$, and the symmetry of $t$ implies that the eigenspaces are orthogonal and span $E$. □

$E_0$ and $E_1$ are called, respectively, the *zero space* and *half space* of $u$. The restrictions of $Q$ to $E_0$ and to $E_1$ are nondegenerate.

**Lemma 5.3.2** *The following rules hold for the product in $E$.*
*(i)    For $x, y \in E_0$,*

$$xy = \frac{1}{2}\langle x, y \rangle(e - u).$$

*(ii)   For $x, y \in E_1$,*

$$xy = \frac{1}{4}\langle x, y \rangle(e + u) + x \circ y \quad with\ x \circ y \in E_0.$$

*(iii) If $x \in E_0$ and $y \in E_1$, then $xy \in E_1$.*

Proof. First we derive a formula for arbitrary $x, y \in E$. Replacing $z$ by $u$ in equation (5.12) we get

$$u(xy) + x(uy) + y(ux) = xy + \frac{1}{2}\langle x, y \rangle u + 3\langle u, x, y \rangle e \qquad (x, y \in E).$$

Now $3\langle u, x, y \rangle = \langle u, x \times y \rangle$ and in the latter expression we replace $x \times y$ by the expression given in Lemma 5.2.1. This leads to the equation

$$u(xy) + x(uy) + y(ux) = xy + \frac{1}{2}\langle x, y \rangle u + (\langle ux, y \rangle - \frac{1}{2}\langle x, y \rangle)e \qquad (x, y \in E).$$
$$(5.20)$$

For $x, y \in E_0$ this equation reduces to

$$u(xy - \frac{1}{2}\langle x, y \rangle e) = xy - \frac{1}{2}\langle x, y \rangle e.$$

The only elements $z$ of $A$ satisfying the relation $uz = z$ are the multiples of $u$, so

$$xy - \frac{1}{2}\langle x, y \rangle e = \kappa u$$

for some $\kappa \in k$. Taking inner products with $u$, we get

$$\langle xy, u \rangle - \frac{1}{2}\langle x, y \rangle \langle e, u \rangle = \kappa \langle u, u \rangle.$$

Since $\langle xy, u \rangle = \langle x, uy \rangle = 0$, we find $\kappa = -\frac{1}{2}\langle x, y \rangle$. Hence

$$xy = \frac{1}{2}\langle x, y \rangle (e - u),$$

which proves (i). Next, we consider $x, y \in E_1$. We define $x \circ y$ by

$$x \circ y = xy - \frac{1}{4}\langle x, y \rangle (e + u) \qquad (x, y \in E_1). \tag{5.21}$$

Using that $\langle x \circ y, u \rangle = \frac{1}{2}\langle x, y \rangle$ we find that $\langle x \circ y, u \rangle = 0$. Likewise, $\langle x \circ y, e \rangle = 0$. Hence $x \circ y \in E$. From equation (5.20) we infer that $u(xy) = \frac{1}{2}\langle x, y \rangle u$, so

$$u(x \circ y) = u(xy - \frac{1}{4}\langle x, y \rangle (e + u)) = u(xy) - \frac{1}{2}\langle x, y \rangle u = 0.$$

Hence $x \circ y \in E_0$. This proves (ii). Finally, let $x \in E_0$ and $y \in E_1$. Then

$$\langle xy, u \rangle = \langle y, ux \rangle = 0 \quad \text{and} \quad \langle xy, e \rangle = \langle x, y \rangle = 0,$$

so $xy \in E$. Equation (5.20) yields $u(xy) + \frac{1}{2}xy = xy$, so $u(xy) = \frac{1}{2}xy$. This proves that $xy \in E_1$. $\qquad\square$

In the following lemma we collect a number of formulas involving products between elements of $E_0$ and $E_1$.

**Lemma 5.3.3** *The following formulas hold for $x, x_1, x_2 \in E_0$ and $y \in E_1$.*

(i) $x(xy) = \frac{1}{4}Q(x)y$;

(ii) $x_1(x_2 y) + x_2(x_1 y) = \frac{1}{4}\langle x_1, x_2 \rangle y$;

(iii) $y \circ xy = \frac{1}{4}Q(y)x$;

(iv) $(y \circ y)y = \frac{1}{4}Q(y)y$;

(v) $(y_1 \circ y_2)y_3 + (y_2 \circ y_3)y_1 + (y_3 \circ y_1)y_2 = \frac{1}{8}\langle y_1, y_2 \rangle y_3 + \frac{1}{8}\langle y_2, y_3 \rangle y_1 + \frac{1}{8}\langle y_3, y_1 \rangle y_2$;

(vi) $Q(xy) = \frac{1}{4}Q(x)Q(y)$;

(vii) $Q(y \circ y) = \frac{1}{4}Q(y)^2$.

Proof. (i) For $x \in E_0$ and $y \in E_1$ we have by (5.13),

$$2x(xy) + x^2 y = Q(x)y + 3\langle x, x, y \rangle e.$$

According to equation (5.16) and Lemma 5.2.1,

$$3\langle x, x, y \rangle = \langle x, x \times y \rangle = \langle x, xy \rangle = 0,$$

since $x \in E_0$ and $xy \in E_1$. By Lemma 5.3.2 (i), $x^2 = Q(x)(e - u)$. So we get

$$2x(xy) + Q(x)y - \frac{1}{2}Q(x)y = Q(x)y,$$

from which (i) follows. (ii) follows from (i) by linearizing.
(iii) Interchanging $x$ and $y$ in formula (5.13) and using (5.16) and Lemma 5.2.1 again, we find

$$xy^2 + 2y(xy) = Q(y)x + \langle x, y^2 \rangle e.$$

Using $y^2 = y \circ y + \frac{1}{2}Q(y)(e + u)$, we get

$$x(y \circ y + \frac{1}{2}Q(y)(e + u)) + 2y(xy) = Q(y)x + \langle x, y^2 \rangle e.$$

By Lemma 5.3.2 (i),

$$x(y \circ y) = \frac{1}{2}\langle x, y \circ y \rangle (e - u) = \frac{1}{2}\langle x, y^2 - \frac{1}{2}Q(y)(e+u) \rangle (e - u) = \frac{1}{2}\langle x, y^2 \rangle (e - u).$$

Substituting this into the above formula and rearranging terms, we get

$$2y(xy) - \frac{1}{2}\langle x, y^2 \rangle u = \frac{1}{2}Q(y)x + \frac{1}{2}\langle x, y^2 \rangle e.$$

By (5.5) this yields

$$y(xy) - \frac{1}{4}\langle y, xy \rangle (e + u) = \frac{1}{4}Q(y)x,$$

from which (iii) follows.
(iv) The Hamilton-Cayley equation (see Prop. 5.1.5) for $y \in E$ reads

$$y^3 - Q(y)y - \det(y)e = 0.$$

Now for $y \in E_1$,

$$3\det(y) = \langle y, y \times y \rangle = \langle y, y^2 - Q(y)e \rangle = \langle y, y^2 \rangle =$$

$$\langle y, y \circ y - \frac{1}{2}Q(y)(e + u) \rangle = 0,$$

so the Hamilton-Cayley equation for $y \in E_1$ becomes

$$y^3 = Q(y)y.$$

From this we derive

$$(y \circ y)y = (y^2 - \frac{1}{2}Q(y)(e + u))y = Q(y)y - \frac{1}{2}Q(y)y - \frac{1}{4}Q(y)y = \frac{1}{4}Q(y)y,$$

thus obtaining (iv). (v) is obtained by linearizing.
(vi) From Lemma 5.1.3 we derive

$$4Q(xy) + \langle x^2, y^2 \rangle = 2Q(x)Q(y).$$

Computing $x^2$ and $y^2$ with the aid of Lemma 5.3.2, we get

$$4Q(xy) + \frac{1}{2}Q(x)Q(y)\langle e - u, e + u \rangle = 2Q(x)Q(y).$$

Since $\langle e - u, e + u \rangle = 2$, we arrive at the formula of (vi).
(vii) By (vi) and (iv),

$$\frac{1}{4}Q(y \circ y)Q(y) = Q((y \circ y)y) = Q(\frac{1}{4}Q(y)y) = \frac{1}{16}Q(y)^3,$$

so $Q(y \circ y) = \frac{1}{4}Q(y)^2$ follows for $Q(y) \neq 0$. By Zariski continuity (over an algebraic closure of $k$), this holds for all $y \in E_1$. □

Statement (i) in the above lemma can be interpreted in terms of Clifford algebras (cf. § 3.1). This will be used later on.

**Corollary 5.3.4** *Let* $Cl(Q; E_0)$ *be the Clifford algebra of the restriction of* $Q$ *to* $E_0$. *The map* $\varphi : E_0 \to \mathrm{End}(E_1)$ *defined by*

$$\varphi(x)(y) = 2xy \qquad (x \in E_0,\ y \in E_1)$$

*can be extended to a representation of* $Cl(Q; E_0)$ *in* $E_1$, *i.e., a homomorphism of* $Cl(Q; E_0)$ *into* $\mathrm{End}(E_1)$.

Proof. By (i) in the Lemma,

$$\varphi(x)^2(y) = Q(x)y \qquad (x \in E_0,\ y \in E_1),$$

so the extension of $\varphi$ to a homomorphism of the tensor algebra $T(E_0)$ into $\mathrm{End}(E_1)$ respects the defining relations for $Cl(Q; E_0)$. □

**Proposition 5.3.5** *Let* $A$ *be a reduced J-algebra. Consider a primitive idempotent* $u$ *in* $A$, *and let* $E$, $E_0$ *and* $E_1$ *be as before, with respect to* $u$.
*(i)* $E_0 = 0$ *if and only if* $A$ *is 2-dimensional; then* $A = ku \oplus k(e - u)$, *an orthogonal direct sum.*
*(ii) If* $E_1 = 0$, *then*

$$A = ku \oplus k(e - u) \oplus E_0.$$

For $\lambda, \lambda', \mu, \mu' \in k$, $x, x' \in E_0$ product and norm given by

$$(\lambda u + \mu(e - u) + x)(\lambda' u + \mu'(e - u) + x') =$$

$$= \lambda \lambda' u + (\mu \mu' + \frac{1}{2} q(x, x'))(e - u) + \mu x' + \mu' x,$$

and

$$Q(\lambda + \mu(e - u) + x) = \frac{1}{2}\lambda^2 + \mu^2 + q(x),$$

where $q$ is a nondegenerate quadratic form on $E_0$ with associated bilinear form $q(\ ,\ )$. Conversely, for any vector space $E_0$ (possibly 0) with a nondegenerate quadratic form $q$, the above formulas define a J-algebra $A$.

Proof. If $E_0 = 0$, then $y \circ y = 0$ for $y \in E_1$, so $Q(y) = 0$ by Lemma 5.3.3 (iv). Since the restriction of $Q$ to $E_1$ is nondegenerate, this implies $E_1 = 0$. Hence $A$ is the 2-dimensional algebra $ku \oplus k(e - u)$.

If $E_1 = 0$, then $A$ is an orthogonal direct sum of vector spaces:

$$A = ku \oplus k(e - u) \oplus E_0.$$

The product is determined by Lemma 5.3.2 (i), if we take for $q$ the restriction of $Q$ to $E_0$. Conversely, any vector space $E_0$ with a nondegenerate quadratic form $q$ yields a J-algebra $A$ of dimension equal to $\dim E_0 + 2$ as above; it is straightforward to verify the axioms (5.4)–(5.6).    □

We call a J-algebra as in (ii) of the above proposition a *J-algebra of quadratic type*. Such a J-algebra $A$ is closely related to the Jordan algebra of the quadratic form $q$ as in [Ja 68, p. 14]; in fact, $A$ is the algebra direct sum of a one-dimensional algebra $ku$ and the subalgebra $k(e - u) \oplus E_0$, the latter being the Jordan algebra of $q$ with $e - u$ as identity element (and also a J-algebra if we multiply $Q$ by $\frac{3}{2}$).

**Lemma 5.3.6** *If $E_1 \neq 0$, hence also $E_0 \neq 0$, then there exists $x_1 \in E_0$ with $Q(x_1) = \frac{1}{4}$. In fact, one can take $x_1 = Q(y)^{-1} y \circ y$ for any $y \in E_1$ with $Q(y) \neq 0$; then $x_1 y = \frac{1}{4}y$.*

Proof. Since the restriction of $Q$ to $E_1$ is nondegenerate, there exists $y \in E_1$ such that $Q(y) \neq 0$. Then $x_1 = Q(y)^{-1} y \circ y \in E_0$. By Lemma 5.3.3 (vii), $Q(x_1) = \frac{1}{4}$. From (iv) of that same lemma we infer $x_1 y = \frac{1}{4}y$.    □

In the discussion after Lemma 5.2.2 we claimed that in most cases $e - u$ is a sum of two orthogonal primitive idempotents if $u$ is a primitive idempotent. We can now prove the precise result.

**Proposition 5.3.7** *If $A$ is a reduced J-algebra and $u$ is a primitive idempotent in $A$, then $e - u$ is a sum of two orthogonal primitive idempotents unless*

$$A = ku \oplus k(e - u) \oplus E_0$$

*and $Q$ does not represent 1 on $E_0$. The latter condition is independent of the choice of $u$ such that the corresponding half space $E_1$ is zero.*

Proof. If $E_1 \neq 0$, then there exists $x_1 \in E_0$ with $Q(x_1) = \frac{1}{4}$ by the above lemma. Then $\frac{1}{2}(e - u) + x_1$ and $\frac{1}{2}(e - u) - x_1$ are primitive idempotents with sum $e - u$. If $E_1 = 0$, then

$$A = ku \oplus k(e - u) \oplus E_0.$$

Consider $a = \lambda u + \mu(e - u) + a_1$ with $a_1 \in E_0$, and $b = e - u - a$, so $b = -\lambda u + (1 - \mu)(e - u) - a_1$ and $a + b = e - u$. One easily verifies that $a$ and $b$ are both idempotent if and only if $\lambda = 0$, $\mu = \frac{1}{2}$ and $Q(a_1) = \frac{1}{4}$. Then indeed $Q(a) = Q(b) = \frac{1}{2}$ and $ab = 0$.

The restriction of $Q$ to $ku \oplus k(e - u)$ is independent of $u$, so by Witt's Theorem the same is true for the restriction to the orthogonal complement $E_0$.                                                                          □

We saw in Prop. 5.1.5 that every element satisfies a cubic equation, the Hamilton-Cayley equation. We compare this with the minimum equation. Let $a \in A$ and denote by $m_a$ its minimum polynomial. So $k[a] \cong k[T]/m_a(T)$ and $m_a$ divides $\chi_a$.

**Proposition 5.3.8** *The polynomials $m_a$ and $\chi_a$ have the same roots in a common splitting field. Hence $m_a = \chi_a$ if $\chi_a$ has three distinct roots in $a$. For $a \notin ke$, the restriction of the norm $Q$ of $A$ to $k[a]$ is nondegenerate if and only if not all roots of $\chi_a$ are equal. If $\chi_a$ has a root in $k$, then $k[a]$ contains a primitive idempotent if and only if not all roots of $\chi_a$ are equal.*

Proof. Upon replacing $k$ by a splitting field of $\chi_a$, we may assume that $\chi_a$ splits in $k$. If $\dim k[a] = 1$, then $a = \lambda e$ for some $\lambda \in k$. Then $m_a(T) = T - \lambda$, and one easily computes that $\chi_a(T) = (T - \lambda)^3$.

Next assume $\dim k[a] = 2$. If $m_a(T) = (T - \lambda)^2$, consider $x = a - \lambda e$. This satisfies $x^2 = 0$ and $x \neq 0$, so $m_x(T) = T^2$. It follows that $\chi_x(T) = T^3 - \langle x, e \rangle T^2$, so $Q(x) = \frac{1}{2}\langle x, e \rangle^2$. But $Q(x) = \frac{1}{2}\langle x, x \rangle = \frac{1}{2}\langle x^2, e \rangle = 0$, hence also $\langle x, e \rangle = 0$. This implies that $\langle a, e \rangle = 3\lambda$ and $Q(a) = \frac{3}{2}\lambda^2$. Since $\det(a) = \lambda^3$, we find that $\chi_a(T) = (T - \lambda)^3$. It is easily verified that in $k[a] = k[x]$ there is no idempotent $\neq e$, and that the restriction of $Q$ to this subspace is degenerate.

If $m_a(T) = (T - \lambda)(T - \mu)$ with $\lambda \neq \mu$, then

$$k[a] \cong k[T]/(T - \lambda)(T - \mu) \cong k \oplus k,$$

so $k[a]$ contains orthogonal idempotents $u$ and $e - u$; we may assume $u$ to be primitive. $Q$ is nondegenerate on $k[a]$ (see Lemma 5.2.2). If $a = \alpha u + \beta(e - u)$, then $m_a(T) = (T - \alpha)(T - \beta)$, so $\alpha = \lambda$ and $\beta = \mu$, or $\alpha = \mu$ and $\beta = \lambda$. One easily computes that $\chi_a(T) = (T - \alpha)(T - \beta)^2$, which is $(T - \lambda)(T - \mu)^2$ or $(T - \lambda)^2(T - \mu)$.

Now let $\dim k[a] = 3$. Then $m_a = \chi_a$. If $\chi_a$ has three distinct roots, then $k[a] \cong k \oplus k \oplus k$, i.e., $k[a]$ is spanned by three idempotents $u_1$, $u_2$, $u_3$ with $u_i u_j = 0$ for $i \neq j$. These must be orthogonal with respect to $Q$ since $\langle u_i, u_j \rangle = \langle e, u_i u_j \rangle = 0$ if $i \neq j$. Hence the restriction of $Q$ to $k[a]$ is nondegenerate.

If $\chi_a(T) = (T - \lambda)(T - \mu)^2$ with $\lambda \neq \mu$, then $k[a] \cong k \oplus k[x]$ for some $x \neq 0$ with $x^2 = 0$. So $k[a]$ contains an idempotent. The restriction of $Q$ to $k[x]$ is degenerate, and since the two components $k$ and $k[x]$ in the direct sum decomposition $k[a] = k \oplus k[x]$ are ideals generated by orthogonal idempotents, the restriction of $Q$ to $k[a]$ is degenerate.

If $\chi_a(T) = (T - \lambda)^3$, then $k[a] = k[x]$ for some $x$ with $x^2 \neq 0$ and $x^3 = 0$. With arguments as in the case $m_a(T) = (T - \lambda)^2$ treated above, one sees that the restriction of $Q$ to the subspace $kx \oplus kx^2$ is identically zero and that this subspace is orthogonal to $ke$. Thus we find that the restriction of $Q$ to $k[a]$ is degenerate. It is also straightforward to verify that $k[x]$ contains no idempotents $\neq e$.

Finally, suppose no longer that $k$ is necessarily a splitting field of $\chi_a$, but that $\chi_a$ has a root in $k$. Then either all three roots lie in $k$, and then the statement about the existence of a primitive idempotent in $k[a]$ follows from the above analysis. Or there are two distinct roots in a quadratic extension $l$ of $k$ which are not in $k$ itself. Then $k[a] \cong k \oplus l$, which contains an idempotent. $\square$

**Corollary 5.3.9** *If $k$ is algebraically closed and $\dim_k A > 2$, then $m_x = \chi_x$ for $x$ in a nonempty Zariski open subset of $A$.*

Proof. We first construct a primitive idempotent $u_1 \in A$. To this end, we pick an element $x$ with $\langle x, e \rangle = 0$ and $Q(x) \neq 0$. Then

$$\chi_x(T) = T^3 - Q(x)T - \det(x) \quad \text{with} \quad Q(x) \neq 0,$$

which does not have three equal roots. By Prop. 5.3.8, $k[x]$ contains a primitive idempotent $u_1$. Prop. 5.3.7 implies that $e - u_1$ is the sum of two orthogonal primitive idempotents $u_2$ and $u_3$, since $Q$ represents all values on the subspace $(ku_1 \oplus k(e - u_1))^\perp$ if $k$ is algebraically closed. An element $y = \eta_1 u_1 + \eta_2 u_2 + \eta_3 u_3$ with three distinct $\eta_i$ has characteristic polynomial $\chi_y$ with three distinct roots, viz., the $\eta_i$, so then $\chi_y = m_y$. The $x \in A$ such that $\chi_x$ has three distinct roots are characterized by the fact that the discriminant of $\chi_x$ is not zero, so these form a Zariski open set. $\square$

In Rem. 5.1.2 we indicated that the norm $Q$ and the cubic form det on a J-algebra are determined by the algebra structure, provided it is of dimension $> 2$. We will now prove this.

**Proposition 5.3.10** *On a J-algebra $A$ over $k$ with $\dim_k(A) > 2$, the quadratic form $Q$ satisfying conditions (5.4)–(5.6) and the cubic form* det *are determined by the algebra structure of $A$, i.e. the structure of vector space over $k$ and the product. Hence for an isomorphism $t : A \to A'$ of J-algebras over $k$ of dimension $> 2$ we have: $Q'(t(x)) = Q(x)$ and $\det'(t(x)) = \det(x)$ $(x \in A)$.*

Proof. It suffices to prove this for algebraically closed $k$. There is a nonempty Zariski open subset $S$ of $A$ such that $\chi_x$ is the minimum polynomial of $x$ for $x \in S$. The coefficients of $\chi_x$ determine $\det(x)$ and $Q(x)$. So the polynomials det and $Q$ are determined on $S$ and therefore on all of $A$.    □

**Remark 5.3.11** If $\dim_k(A) = 2$ and $A$ is reduced, then there are two orthogonal idempotents $u_1$ and $u_2$ such that $A = ku_1 \oplus ku_2$ (cf. Prop. 5.3.5). One of these idempotents is primitive and the other is not: $Q(u) = \frac{1}{2}$ and $Q(e-u) = 1$ for $u = u_1$ or $u_2$. For $x = \xi u + \eta(e-u)$ we have $Q(x) = \frac{1}{2}\xi^2 + \eta^2$; with the Hamilton-Cayley equation one finds $\det(x) = \xi\eta^2$. Hence in this case there are two possibilities for $Q$ and det, i.e., these are not determined by the algebra structure.

**Corollary 5.3.12** *If $\det(x) \neq 0$ then $x$ has a J-inverse $x^{-1}$ and $\det(x^{-1}) = \det(x)^{-1}$.*

Proof. For the first point see Lemma 5.2.3. It suffices to prove the equality for algebraically closed $k$. We first assume that $\dim_k(A) > 2$. Let $V$ be the Zariski open set $\{x \in A \mid \det(x) \neq 0\}$, and $W$ the Zariski open set on which $m_x = \chi_x$ (see Cor. 5.3.9). On $V \cap W$, $\chi_x(T) = T^3 \ldots - \det(x)$ is the unique cubic polynomial which has $x$ as a root, and similarly for $\chi_{x^{-1}}(T) = T^3 \ldots - \det(x^{-1})$. But $k[x]$ is associative, so $x^{-1}$ is also a root of

$$- \det(x)^{-1} T^3 \chi_x(T^{-1}) = T^3 \ldots - \det(x)^{-1},$$

so $\det(x^{-1}) = \det(x)^{-1}$ for $x \in V \cap W$. By Zariski continuity, the relation holds on all of $V$.

Now let $\dim_k(A) = 2$. For $x = \xi u + \eta(e - u)$ as in the above remark, $\det(x) = \xi\eta^2$. If $\xi\eta \neq 0$, then $\det(x^{-1}) = \xi^{-1}\eta^{-2} = \det(x)^{-1}$.    □

# 5.4 Classification of Reduced J-algebras

We continue with the determination of the structure of reduced J-algebras. This will lead to the result that besides the J-algebras of quadratic type, which we found in Prop. 5.3.5, there is only one other type of reduced J-algebra, viz., the matrix algebras $H(C; \gamma_1, \gamma_2, \gamma_3)$ we introduced at the beginning of § 5.1; see Th. 5.4.5. We again fix a primitive idempotent $u$ and assume that $E_1 \neq 0$. Further we fix $x_1 \in E_0$ with $Q(x_1) = \frac{1}{4}$.

**Lemma 5.4.1** *Consider the linear mapping*

$$s : E_1 \to E_1, \; y \mapsto x_1 y.$$

*(i)  $s$ is symmetric with respect to $\langle \, , \, \rangle$ and $s^2 = \frac{1}{16}$.*
*(ii)  $E_1 = E_+ \oplus E_-$ with $E_+ \perp E_-$, where $E_+$ and $E_-$ are the eigenspaces of $s$ for the eigenvalues $\frac{1}{4}$ and $-\frac{1}{4}$, respectively.*
*(iii)  If $\dim E_0 > 1$, then both $E_+ \neq 0$ and $E_- \neq 0$.*
*(iv)  If $\dim E_0 = 1$, then $E_1 = E_+$ or $E_1 = E_-$.*

Proof. (i) The symmetry of $s$ follows from (5.5). The second statement of (i) is a consequence of Lemma 5.3.3 (i).
(ii) From (i) it follows that $s$ has eigenvalues $\frac{1}{4}$ and $-\frac{1}{4}$, and that $E_1$ is the orthogonal direct sum of the corresponding eigenspaces $E_+$ and $E_-$.
(iii) Pick $x \in E_0$ with $\langle x, x_1 \rangle = 0$ and $Q(x) \neq 0$. If $E_+ \neq 0$, pick $y \in E_+$ with $Q(y) \neq 0$. By Lemma 5.3.3 (ii),

$$x_1(xy) = -x(x_1 y) = -\frac{1}{4}xy,$$

so $xy \in E_-$. Further, $Q(xy) \neq 0$ by Lemma 5.3.3 (vi), so $xy \neq 0$. Hence $E_+ \neq 0$ implies $E_- \neq 0$. Similarly, $E_- \neq 0$ implies $E_+ \neq 0$.
(iv) Suppose $E_+ \neq 0$ and $E_- \neq 0$. Pick $y \in E_+$ with $Q(y) \neq 0$ and $z \in E_-$ with $Q(z) \neq 0$. By Lemma 5.3.3 (iii),

$$y \circ y = 4y \circ x_1 y = Q(y)x_1.$$

Hence $(y \circ y)z = -\frac{1}{4}Q(y)z$. By Lemma 5.3.3 (v),

$$2(y \circ z)y + (y \circ y)z = \frac{1}{4}Q(y)z + \frac{1}{4}\langle y, z \rangle y.$$

Since $\langle y, z \rangle = 0$, we find

$$2(y \circ z)y = \frac{1}{2}Q(y)z,$$

which shows that $y \circ z \neq 0$. On the other hand,

$$\langle x_1, y \circ z \rangle = \langle x_1, yz \rangle = \langle x_1 y, z \rangle = \frac{1}{4}\langle y, z \rangle = 0.$$

This implies that $\dim E_0 > 1$. Now (iv) follows.    □

In case (iv) of the above lemma we may assume that $E_1 = E_+$; otherwise, we take $-x_1$ instead of $x_1$. We can now replace $u$ by another primitive idempotent, viz., $u' = \frac{1}{2}(e - u) - x_1$: one easily verifies that indeed $u'^2 = u'$ and $Q(u') = \frac{1}{2}$. Further, it is straightforward that $A$ is an orthogonal direct sum

$$A = ku' \oplus k(e - u') \oplus E_0',$$

where $E_0' = k(\frac{1}{2}u' + x_1) \oplus E_+$ satisfies $u'E_0' = 0$. This shows that $A$ is a J-algebra of quadratic type if $\dim E_0 = 1$; see Prop. 5.3.5.

This being dealt with, we sharpen our assumptions: $A$ is a reduced J-algebra with a primitive idempotent $u$ such that $\dim E_0 > 1$ and $E_1 \neq 0$. Then, by (iii) of the above lemma, $E_1 = E_+ \oplus E_-$ with both $E_+ \neq 0$ and $E_- \neq 0$. We will see that under these assumptions $A$ is isomorphic to a J-algebra $H(C; \gamma_1, \gamma_2, \gamma_3)$ of Hermitian $3 \times 3$ matrices over a composition algebra $C$, as introduced in the beginning of § 5.1. To start with, we define the vector space

$$C = x_1^{\perp} \cap E_0 = \{\, x \in E_0 \mid \langle x, x_1 \rangle = 0 \,\}.$$

In Prop. 5.4.4, $C$ will be given the structure of a composition algebra; before that, we prove two technical lemmas.

**Lemma 5.4.2** *For $y_+, z_+ \in E_+$ and $y_-, z_- \in E_-$ we have:*

(i)   $y_+ \circ z_+ = \frac{1}{2}\langle y_+, z_+ \rangle x_1$;
(ii)  $y_- \circ z_- = -\frac{1}{2}\langle y_-, z_- \rangle x_1$;
(iii) $y_+ \circ y_- \in C$;
(iv)  $(y_+ \circ y_-)y_+ = \frac{1}{4}Q(y_+)y_-$;
(v)   $(y_+ \circ y_-)y_- = \frac{1}{4}Q(y_-)y_+$;
(vi)  $Q(y_+) = Q(y_-) = 0$ *if* $y_+ \circ y_- = 0$ *and* $y_+ \neq 0 \neq y_-$;
(vii) $Q(y_+ \circ y_-) = \frac{1}{4}Q(y_+)Q(y_-)$.

Proof. (i) Using Lemma 5.3.3 (iii), we find $y_+ \circ y_+ = 4y_+ \circ x_1 y_+ = Q(y_+)x_1$. By linearizing we find the result for $y_+ \circ z_+$. Similarly for (ii).
(iii) $\langle x_1, y_+ \circ y_- \rangle = \langle x_1, y_+ y_- \rangle = \langle x_1 y_+, y_- \rangle = \frac{1}{4}\langle y_+, y_- \rangle = 0$.
(iv) By Lemma 5.3.3 (v),

$$2(y_+ \circ y_-)y_+ + (y_+ \circ y_+)y_- = \frac{1}{4}Q(y_+)y_- + \frac{1}{4}\langle y_+, y_- \rangle y_+.$$

Using (i) and the fact that $E_+ \perp E_-$, we easily get the result. The proof of (v) is similar.
(vi) This is immediate from (iv) and (v).
(vii) By Lemma 5.3.3 (vi),

$$Q((y_+ \circ y_-)y_+) = \frac{1}{4}Q(y_+ \circ y_-)Q(y_+).$$

Using (iv) we find

$$\frac{1}{16}Q(y_+)^2Q(y_-) = \frac{1}{4}Q(y_+ \circ y_-)Q(y_+),$$

so

$$Q(y_+ \circ y_-) = \frac{1}{4}Q(y_+)Q(y_-),$$

if $Q(y_+) \neq 0$. By continuity for the Zariski topology (over an algebraic closure of $k$), the relation holds everywhere. It is not hard, by the way, to prove (vii) for $Q(y_+) = 0$ directly with the aid of Lemma 5.3.3 (i), and (iv) of the present lemma:

$$\frac{1}{4}Q(y_+ \circ y_-)y_+ = (y_+ \circ y_-)((y_+ \circ y_-)y_+) = 0.$$

$\square$

Now we fix $a_+ \in E_+$ and $a_- \in E_-$ with $Q(a_+)Q(a_-) \neq 0$.

**Lemma 5.4.3** *The maps*

$$e_+ : C \to E_+, \quad x \mapsto 4Q(a_-)^{-1}xa_-,$$
$$e_- : C \to E_-, \quad x \mapsto 4Q(a_+)^{-1}xa_+,$$

*are linear isomorphisms. Their inverses are, respectively,*

$$E_+ \to C, \quad y_+ \mapsto a_- \circ y_+ = a_-y_+,$$
$$E_- \to C, \quad y_- \mapsto a_+ \circ y_- = a_+y_-.$$

Proof. From (v) of the above lemma one sees that

$$E_+ \to C, \quad y_+ \mapsto a_- \circ y_+,$$

is injective, and from Lemma 5.3.3 (iii) that it is surjective. That $e_+$ is its inverse, follows from either of these. Similarly for $e_-$ with (iv) of the above lemma instead of (v). $\square$

We now make $C$ into a composition algebra.

**Proposition 5.4.4** *The vector space $C$ with the product $\diamond$ defined by*

$$x \diamond x' = e_+(x)e_-(x')$$
$$= 16Q(a_+)^{-1}Q(a_-)^{-1}(xa_-)(x'a_+) \quad (x, x' \in C)$$

*and with the norm $N$ defined by*

$$N(x) = 4Q(a_+)^{-1}Q(a_-)^{-1}Q(x) \quad (x \in C)$$

*is a composition algebra.*

Proof. We have to verify the conditions of Def. 1.2.1. First, $\varepsilon = a_+a_-$ is an identity element for the multiplication:

$$e_-(\varepsilon) = 4Q(a_+)^{-1}(a_+a_-)a_+ = a_-$$

by Lemma 5.4.2 (iv), hence

$$x \diamond \varepsilon = e_+(x)e_-(\varepsilon) = 4Q(a_-)^{-1}(xa_-)a_- = x \quad (x \in C)$$

by Lemma 5.3.3 (iii), and similarly $\varepsilon \diamond x = x$. Using Lemma 5.4.2 (vii) and Lemma 5.3.3 (vi), we derive

$$Q(x \diamond x') = Q(e_+(x)e_-(x'))$$
$$= \frac{1}{4}Q(e_+(x))Q(e_-(x'))$$
$$= \frac{1}{4}Q(4Q(a_-)^{-1}xa_-)Q(4Q(a_+)^{-1}x'a_+)$$
$$= \frac{1}{4}\cdot16Q(a_-)^{-2}\cdot16Q(a_+)^{-2}\cdot\frac{1}{4}Q(x)Q(a_-)\cdot\frac{1}{4}Q(x')Q(a_+)$$
$$= 4Q(a_+)^{-1}Q(a_-)^{-1}Q(x)Q(x').$$

From this it is immediate that $N$ permits composition. $N$ is nondegenerate since the restriction of $Q$ to $C$ is so. $\qquad\square$

We denote the bilinear form associated with $N$ by $N(\,,\,)$ to distinguish it from the bilinear form $\langle\,,\,\rangle$ associated with $Q$.

**Theorem 5.4.5** *A reduced J-algebra $A$ over a field $k$ of characteristic $\neq 2, 3$ with identity element $e$ and quadratic form $Q$ is of one of the following types, and conversely, all such algebras are reduced J-algebras.*
*(i) $A = ku \oplus k(e - u) \oplus E_0$ (orthogonal direct sum), where $u$ is a primitive idempotent, $ux = 0$ for $x \in E_0$, and $xx' = \frac{1}{2}\langle x, x'\rangle(e - u)$ for $x, x' \in E_0$. Here $E_0$ can be any vector space (possibly 0), and the restriction of $Q$ to $E_0$ can be any nondegenerate quadratic form on it.*
*(ii) $A \cong \mathrm{H}(C; \gamma_1, \gamma_2, \gamma_3)$, the algebra of $3 \times 3$ $(\gamma_1, \gamma_2, \gamma_3)$-hermitian matrices over the composition algebra $C$ over $k$. In this case $\dim A = 6, 9, 15$ or $27$. A J-algebra of type (i) cannot be isomorphic to one of type (ii).*

Proof. We maintain the assumption that we have a primitive idempotent $u$ such that $\dim E_0 > 1$ and $E_1 \neq 0$, and fix $x_1 \in E_0$ with $Q(x_1) = \frac{1}{4}$; so $E_1 = E_+ \oplus E_-$. We know already that in all other cases $A$ is of type (i).

The elements $u_1 = u$, $u_2 = \frac{1}{2}(e - u) + x_1$ and $u_3 = \frac{1}{2}(e - u) - x_1$ are three orthogonal primitive idempotents whose sum is $e$. Using Lemma 5.4.3, we see that every $x \in A$ can be written in a unique way as

$$x = x(\xi_1, \xi_2, \xi_3; c_1, c_2, c_3) = \qquad\qquad (5.22)$$

$$= \xi_1 u_1 + \xi_2 u_2 + \xi_3 u_3 + 2Q(a_+)^{-1}Q(a_-)^{-1}c_1 + 4Q(a_+)^{-1}\bar{c}_2 a_+ + 4Q(a_-)^{-1}\bar{c}_3 a_-$$

with $\xi_1, \xi_2, \xi_3 \in k$ and $c_1, c_2, c_3 \in C$; as usual, $\bar{\phantom{c}}$ denotes conjugation in the composition algebra $C$. As we know, $\bar{c}_2 a_+ \in E_-$ and $\bar{c}_3 a_- \in E_+$. We claim that

$$x^2 = \{\xi_1^2 + Q(a_-)N(c_2) + Q(a_+)N(c_3)\}u_1 +$$
$$\{\xi_2^2 + Q(a_+)^{-1}Q(a_-)^{-1}N(c_1) + Q(a_+)N(c_3)\}u_2 +$$
$$\{\xi_3^2 + Q(a_+)^{-1}Q(a_-)^{-1}N(c_1) + Q(a_-)N(c_2)\}u_3 +$$
$$2Q(a_+)^{-1}Q(a_-)^{-1}[(\xi_2 + \xi_3)c_1 + Q(a_+)Q(a_-)\bar{c}_3 \diamond \bar{c}_2] +$$
$$4Q(a_+)^{-1}[(\xi_1 + \xi_3)c_2 + Q(a_-)^{-1}\bar{c}_1 \diamond \bar{c}_3]a_+ +$$
$$4Q(a_-)^{-1}[(\xi_1 + \xi_2)c_3 + Q(a_+)^{-1}\bar{c}_2 \diamond \bar{c}_1]a_-. \qquad (5.23)$$

We postpone the proof of this formula and first complete the argument that leads us to type (ii). To this end, we compare the expression for $x^2$ with the square of an element in $\mathrm{H}(C; \gamma_1, \gamma_2, \gamma_3)$. Using the notation as in (5.1), we get by a straightforward computation:

$$h(\xi_1, \xi_2, \xi_3; c_1, c_2, c_3)^2 = h(\eta_1, \eta_2, \eta_3; d_1, d_2, d_3),$$

where

$$\eta_1 = \xi_1^2 + \gamma_1^{-1}\gamma_3 N(c_2) + \gamma_2^{-1}\gamma_1 N(c_3)$$
$$\eta_2 = \xi_2^2 + \gamma_3^{-1}\gamma_2 N(c_1) + \gamma_2^{-1}\gamma_1 N(c_3)$$
$$\eta_3 = \xi_3^2 + \gamma_3^{-1}\gamma_2 N(c_1) + \gamma_1^{-1}\gamma_3 N(c_2)$$
$$d_1 = (\xi_2 + \xi_3)c_1 + \gamma_2^{-1}\gamma_3 \bar{c}_3 \diamond \bar{c}_2$$
$$d_2 = (\xi_1 + \xi_3)c_2 + \gamma_3^{-1}\gamma_1 \bar{c}_1 \diamond \bar{c}_3$$
$$d_3 = (\xi_1 + \xi_2)c_3 + \gamma_1^{-1}\gamma_2 \bar{c}_2 \diamond \bar{c}_1$$

with $\diamond$ denoting the product in $C$. Now, if we choose

$$\gamma_1 = 1, \quad \gamma_2 = Q(a_+)^{-1} \quad \text{and} \quad \gamma_3 = Q(a_-),$$

then the bijective map $\varphi : A \to \mathrm{H}(C; \gamma_1, \gamma_2, \gamma_3)$ given by

$$\varphi(x(\xi_1, \xi_2, \xi_3; c_1, c_2, c_3)) = h(\xi_1, \xi_2, \xi_3; c_1, c_2, c_3)$$

satisfies $\varphi(x^2) = \varphi(x)^2$. Since both algebras are commutative, it follows that $\varphi$ is an isomorphism. One sees that $\varphi$ maps $u_1 = u$, $x_1$, $a_+$ and $a_-$ to, respectively, $h(1, 0, 0; 0, 0, 0)$, $h(0, \frac{1}{2}, -\frac{1}{2}; 0, 0, 0)$, $h(0, 0, 0; 0, \frac{1}{4}Q(a_+), 0)$ and $h(0, 0, 0; 0, 0, \frac{1}{4}Q(a_-))$, hence $u_2$ to $h(0, 1, 0; 0, 0, 0)$ and $u_3$ to $h(0, 0, 1; 0, 0, 0)$.

Conversely, we saw already in § 5.1 that the algebras $\mathrm{H}(C; \gamma_1, \gamma_2, \gamma_3)$ satisfy the axioms for J-algebras.

Now we give the postponed proof of formula (5.23). We multiply the right hand side of (5.22) by itself and consider separately the squares and products that arise; we only deal with the less trivial ones. Notice that the idempotents $u_1, u_2, u_3$ act on $c \in C$, $y_+ \in E_+$ and $y_- \in E_-$ as follows:

$$u_1 c = 0, \quad u_1 y_+ = \tfrac{1}{2}y_+, \quad u_1 y_- = \tfrac{1}{2}y_-;$$
$$u_2 c = \tfrac{1}{2}c, \quad u_2 y_+ = \tfrac{1}{2}y_+, \quad u_2 y_- = 0;$$
$$u_3 c = \tfrac{1}{2}c, \quad u_3 y_+ = 0, \quad u_3 y_- = \tfrac{1}{2}y_-.$$

(a) Consider $16Q(a_+)^{-2}Q(a_-)^{-1}c_1(\bar{c}_2 a_+)$; by (iii) of Lemma 5.3.3, this can be written as

$$64Q(a_+)^{-2}Q(a_-)^{-2}\{a_- (c_1 a_-)\}(\bar{c}_2 a_+).$$

In formula (5.12) we substitute $x = a_-$, $y = c_1 a_-$, $z = \bar{c}_2 a_+$; this yields

$$\{a_- (c_1 a_-)\}(\bar{c}_2 a_+) + \{(\bar{c}_2 a_+)(c_1 a_-)\}a_- + \{a_- (\bar{c}_2 a_+)\}(c_1 a_-) =$$

$$\frac{1}{2}\langle a_-, \bar{c}_2 a_+ \rangle c_1 a_- + 3\langle a_-, c_1 a_-, \bar{c}_2 a_+ \rangle. \tag{5.24}$$

By (5.5),

$$\frac{1}{2}\langle a_-, \bar{c}_2 a_+ \rangle c_1 a_- = \frac{1}{2}\langle a_+ a_-, \bar{c}_2 \rangle c_1 a_- = \frac{1}{2}\langle \varepsilon, c_2 \rangle c_1 a_-.$$

(For the second step we use the fact that the restriction of $\langle\ ,\ \rangle$ to $C$ is a multiple of the bilinear form $N(\ ,\ )$ associated with the norm $N$ on $C$, so $\langle \varepsilon, \bar{c}_2 \rangle = \langle \varepsilon, c_2 \rangle$.) Next,

$$
\begin{aligned}
3\langle a_-, c_1 a_-, \bar{c}_2 a_+ \rangle &= \langle a_- \times c_1 a_-, \bar{c}_2 a_+ \rangle \\
&= \langle a_-(c_1 a_-), \bar{c}_2 a_+ \rangle \qquad \text{(by (i) of Lemma 5.2.1)} \\
&= \langle \frac{1}{4} Q(a_-) c_1, \bar{c}_2 a_+ \rangle \qquad \text{(by (iii) of Lemma 5.3.3)} \\
&= 0 \qquad \text{(since } E_0 \perp E_1\text{).}
\end{aligned}
$$

By the definition of $\diamond$ (see Prop. 5.4.4),

$$\{(\bar{c}_2 a_+)(c_1 a_-)\} a_- = \frac{1}{16} Q(a_+) Q(a_-)(c_1 \diamond \bar{c}_2) a_-.$$

Further, by (5.21), (ii) of Lemma 5.4.2 and (5.5)

$$a_-(\bar{c}_2 a_+)\}(c_1 a_-) = \{\frac{1}{4}\langle a_-, \bar{c}_2 a_+ \rangle(e + u) + a_- \circ (\bar{c}_2 a_+)\}(c_1 a_-) =$$

$$\{\frac{1}{4}\langle a_-, \bar{c}_2 a_+ \rangle(e + u) - \frac{1}{2}\langle a_-, \bar{c}_2 a_+ \rangle x_1\}(c_1 a_-) =$$

$$\frac{1}{2}\langle \bar{c}_2, a_+ a_- \rangle\{\frac{1}{2}(e + u) - x_1\}(c_1 a_-) =$$

$$\frac{1}{2}\langle \bar{c}_2, \varepsilon \rangle(u_1 + u_3)(c_1 a_-) =$$

$$\frac{1}{4}\langle c_2, \varepsilon \rangle c_1 a_-.$$

Substituting all this into equation (5.24), we find

$$\{a_-(c_1 a_-)\}(\bar{c}_2 a_+) + \frac{1}{16} Q(a_+) Q(a_-)(c_1 \diamond \bar{c}_2) a_- + \frac{1}{4}\langle c_2, \varepsilon \rangle c_1 a_- =$$

$$\frac{1}{2}\langle c_2, \varepsilon \rangle c_1 a_-,$$

so

$$\{a_-(c_1 a_-)\}(\bar{c}_2 a_+) = \frac{1}{4}\langle c_2, \varepsilon \rangle c_1 a_- - \frac{1}{16} Q(a_+) Q(a_-)(c_1 \diamond \bar{c}_2) a_-.$$

Thus we find

$$16Q(a_+)^{-2}Q(a_-)^{-1}c_1(\bar{c}_2 a_+) =$$
$$16Q(a_+)^{-2}Q(a_-)^{-2}\langle c_2, \varepsilon \rangle c_1 a_- - 4Q(a_+)^{-1}Q(a_-)^{-1}(c_1 \diamond \bar{c}_2)a_- =$$
$$4Q(a_+)^{-1}Q(a_-)^{-1}N(c_2,\varepsilon)c_1 a_- - 4Q(a_+)^{-1}Q(a_-)^{-1}(c_1 \diamond \bar{c}_2)a_- =$$
$$4Q(a_+)^{-1}Q(a_-)^{-1}\{c_1 \diamond (N(c_2,\varepsilon)\varepsilon - \bar{c}_2)\}a_- =$$
$$4Q(a_+)^{-1}Q(a_-)^{-1}(c_1 \diamond c_2)a_-.$$

In the same way one computes that

$$16Q(a_+)^{-1}Q(a_-)^{-2}c_1(\bar{c}_3 a_-) = 4Q(a_+)^{-1}Q(a_-)^{-1}(c_3 \diamond c_1)a_+.$$

(b) $32Q(a_+)^{-1}Q(a_-)^{-1}(\bar{c}_2 a_+)(\bar{c}_3 a_-) = 2\bar{c}_3 \diamond \bar{c}_2$ by the definition of $\diamond$.

(c) Finally,

$$16Q(a_+)^{-2}(\bar{c}_2 a_+)^2 = 16Q(a_+)^{-2}(\frac{1}{2}Q(\bar{c}_2 a_+)(e + u) - Q(\bar{c}_2 a_+)x_1)$$
$$= 16Q(a_+)^{-2}\frac{1}{4}Q(c_2)Q(a_+)(u_1 + u_3)$$
$$= Q(a_-)N(c_2)(u_1 + u_3),$$

and similarly

$$16Q(a_-)^{-2}(\bar{c}_3 a_-)^2 = Q(a_+)N(c_3)(u_1 + u_2).$$

The remaining computations needed to prove formula (5.23) are left to the reader.

Finally, we show that a reduced J-algebra as in (i) and one as in (ii) cannot be isomorphic. This will be done by showing that the cubic form det is reducible in case (i), but irreducible in case (ii). In case (i), consider an element $z = \lambda u + \mu(e - u) + x$ with $x \in E_0$. Using the Hamilton-Cayley equation (5.7), one finds by a straightforward computation that

$$\det(z) = \lambda(\mu^2 - Q(x)),$$

which is reducible.

For case (ii), we may assume that $k$ is algebraically closed. Consider $A = H(C; \gamma_1, \gamma_2, \gamma_3)$ and its subspace $V = H(k; \gamma_1, \gamma_2, \gamma_3)$ consisting of the elements $h(\xi_1, \xi_2, \xi_3; c_1, c_2, c_3)$ with all $c_i \in k$. The polynomial det is homogeneous of degree three, so if it were reducible on $A$, the factors would be homogeneous of degree at most two. Hence the restriction of det to $V$, which is the common $3 \times 3$ determinant, would have to be reducible or identically zero. It is known that this is not the case; see, e.g., [Ja 74, Th. 7.2].    □

We already named reduced J-algebras of type (i) in the Theorem *J-algebras of quadratic type*; we call those of type (ii) *proper J-algebras*, since our main interest is in this type of J-algebras, or rather in those which are isomorphic to $H(C; \gamma_1, \gamma_2, \gamma_3)$ with an octonion algebra $C$ (these are the reduced Albert algebras).

**Corollary 5.4.6** *A reduced J-algebra is proper if and only if the determinant polynomial of $A$ is absolutely irreducible. The image of a J-algebra $A$ under an isomorphism is of the same type as $A$ itself, that is, of quadratic type or proper according to whether $A$ is of quadratic type or proper.*

## 5.5 Further Properties of Reduced J-algebras

The structure theory for J-algebras in the previous section is based on the existence of an idempotent, i.e., it holds for reduced algebras only. We now look for conditions that ensure a J-algebra is reduced, and in that case we classify the primitive idempotents. This being done, we will prove that in a proper reduced J-algebra, i.e., in $H(C; \gamma_1, \gamma_2, \gamma_3)$, the composition algebra $C$ is independent (up to isomorphism) of the choice of the primitive idempotent $u$ and of the choices of $x_1$, $a_+$ and $a_-$.

**Theorem 5.5.1** *In a J-algebra $A$, an element $x$ satisfies $x \times x = 0$ if and only if either $x$ is a multiple of a primitive idempotent (and then $\langle x, e \rangle \neq 0$) or $x^2 = 0$ (and then $\langle x, e \rangle = 0$). If $A$ contains $a \neq 0$ with $a^2 = 0$, then it contains a primitive idempotent $u$ with $ua = 0$.*

*So a J-algebra is reduced if and only if it contains $x \neq 0$ with $x \times x = 0$.*

Proof. Let $x \times x = 0$. If $\langle x, e \rangle = 0$, we infer from Lemma 5.2.1 (i),

$$x^2 - Q(x)e = 0.$$

Hence

$$0 = \langle x^2 - Q(x)e, e \rangle = 2Q(x) - 3Q(x) = -Q(x),$$

so $x^2 = 0$. Conversely, if $x^2 = 0$, the Hamilton-Cayley equation ( 5.7) implies that $Q(x) = \frac{1}{2}\langle x, e \rangle^2$. Since $Q(x) = \frac{1}{2}\langle x, x \rangle = \langle x^2, e \rangle = 0$, we conclude that $Q(x) = \langle x, e \rangle = 0$, whence $x \times x = 0$ by part (i) of Lemma 5.2.1.

If $x \times x = 0$ and $\langle x, e \rangle \neq 0$, we may assume that $\langle x, e \rangle = 1$ and then Lemma 5.2.1 (i) yields

$$x^2 - x - (Q(x) - \frac{1}{2})e = 0.$$

Taking the inner product of both sides of this equation with $e$, we find

$$\langle x, x \rangle - 1 - 3Q(x) + \frac{3}{2} = 0,$$

hence $Q(x) = \frac{1}{2}$ and therefore $x^2 = x$, so $x$ is a primitive idempotent.

Conversely, if $x$ is a primitive idempotent then $\langle x, e \rangle = 2Q(x) = 1$, so $x \times x = 0$ by Lemma 5.2.1 (i).

Let $a^2 = 0$, $a \neq 0$. As we saw, this implies $Q(a) = 0$ and $\langle a, e \rangle = 0$. Since the restriction of $Q$ to $e^\perp$ is nondegenerate, there exist $b \in e^\perp$ with $Q(b) = 0$ and $\langle a, b \rangle = 1$. From equation (5.13) we infer

$$2a(ab) = a \quad \text{(since } a \times a = 0),\tag{5.25}$$
$$2b(ab) = -ab^2 + b + \langle a, b^2 \rangle e \quad (b \times b = b^2 \text{ by Lemma 5.2.1 (i))},\tag{5.26}$$
$$2a(ab^2) = \langle a, b^2 \rangle a.\tag{5.27}$$

Equation (5.12) with $x = a$, $y = b$ and $z = ab$ yields

$$a(b(ab)) + b(a(ab)) + (ab)^2 = \frac{3}{2}ab + \frac{1}{2}\langle a, b^2 \rangle a,$$

since by Lemma 5.2.1 (i),

$$3\langle a, b, ab \rangle = \langle a \times ab, b \rangle = \langle a(ab) - \frac{1}{2}a, b \rangle = 0 \quad \text{(by equation (5.25))}.$$

Computing $a(b(ab))$ and $b(a(ab))$ with the aid of equations (5.25)–(5.27), we derive from this

$$(ab)^2 = \frac{1}{2}ab + \frac{1}{4}\langle a, b^2 \rangle a.$$

Now one easily verifies that $u = e + \langle a, b^2 \rangle a - 2ab$ is idempotent. Further, $\langle u, e \rangle = 1$, so $u$ is primitive. By (5.25), $ua = 0$. $\qquad\square$

**Theorem 5.5.2** *A J-algebra A is reduced if and only if the cubic form* det *represents zero nontrivially on A.*

Proof. If $A$ contains a primitive idempotent $u$, then $u \times u = 0$ by the previous theorem, so $\det(u) = 0$. Conversely, let $x \in A$ be nonzero with $\det(x) = 0$. By Lemma 5.2.1 (iv), either $x \times x = 0$ or $y = x \times x \neq 0$ and $y \times y = 0$. By the previous theorem again, $A$ is reduced. $\qquad\square$

**Proposition 5.5.3** *Let $A$ be a reduced J-algebra and let $u \in A$ be a fixed primitive idempotent. The primitive idempotents in $A$ are the elements*

(i) $t = (Q(y) + 1)^{-1}(u + \frac{1}{2}Q(y)(e - u) + y \circ y + y)$ $\qquad (y \in E_1, Q(y) \neq -1)$;

(ii) $t = \frac{1}{2}(e - u) + x + y$ $\qquad (x \in E_0, Q(x) = \frac{1}{4}, y \in E_1, xy = \frac{1}{4}y, Q(y) = 0)$.

*In type (ii), the condition $Q(y) = 0$ can be replaced by $y \circ y = 0$. The primitive idempotents of type (i) are characterized by the fact that $\langle t, u \rangle \neq 0$, those of type (ii) by $\langle t, u \rangle = 0$.*

*The elements $t \in A$ with $t^2 = 0$ are, up to a scalar factor, of the form*

(i) $t = u - \frac{1}{2}(e - u) + y \circ y + y$ $\qquad (y \in E_1, Q(y) = -1)$;

(ii) $t = x + y$ $\qquad (x \in E_0, y \in E_1, Q(x) = Q(y) = 0, xy = 0, y \circ y = 0)$.

*In case (ii), $Q(y) = 0$ already follows from $y \circ y = 0$. An element $t$ with $t^2 = 0$ is of type (i) if $\langle t, e \rangle \neq 0$ and of type (ii) if $\langle t, e \rangle = 0$.*

Proof. The notations are as in §5.3. To find the primitive idempotents and the nilpotents of order 2, we describe the elements $t \in A$ with $t \times t = 0$. Let

$$t = \xi e + \eta u + x + y \qquad (\xi, \eta \in k, \ x \in E_0, \ y \in E_1).$$

A straightforward computation yields

$$t \times t = (\xi^2 + \xi\eta - \frac{1}{2}Q(y))e + (-\xi\eta - Q(x) + \frac{1}{2}Q(y))u + (y \circ y - (\xi + \eta)x) + (2xy - \xi y).$$

So $t \times t = 0$ is equivalent to the following set of equations:

$$\xi^2 + \xi\eta - \frac{1}{2}Q(y) = 0,$$

$$-\xi\eta - Q(x) + \frac{1}{2}Q(y) = 0,$$

$$y \circ y - (\xi + \eta)x = 0,$$

$$2xy - \xi y = 0 \qquad (\xi, \eta \in k, \ x \in E_0, \ y \in E_1).$$

We distinguish three cases.
(i) $\xi + \eta \neq 0$. Replacing $t$ by a nonzero multiple, we may assume that $\xi + \eta = 1$. Then we must have

$$t = u + \frac{1}{2}Q(y)(e - u) + y \circ y + y \qquad (y \in E_1).$$

Using part (iv) of Lemma 5.3.3 one sees that the set of equations is satisfied, whence $t \times t = 0$. From $\langle t, e \rangle = 1 + Q(y)$ it follows by Th. 5.5.1 that $t^2 = 0$ if $Q(y) = -1$ and that otherwise $(Q(y) + 1)^{-1}t$ is a primitive idempotent.
(ii) $\xi + \eta = 0$, $\xi \neq 0$. In this case we may assume that $\xi = \frac{1}{2}$ and we find

$$t = \frac{1}{2}(e - u) + x + y,$$

where $Q(x) = \frac{1}{4}$, $xy = \frac{1}{4}y$, $y \circ y = 0$ and $Q(y) = 0$. We may drop either of the last two conditions. For if $y \circ y = 0$, then

$$0 = Q(y \circ y) = \frac{1}{4}Q(y)^2,$$

so $Q(y) = 0$. Conversely, if $Q(x) = \frac{1}{4}$, $xy = \frac{1}{4}y$ and $Q(y) = 0$, choose $x_1 = x$. In the decomposition of $E$ with respect to this $x_1$ we have $y \in E_+$. By Lemma 5.4.2 (i), $y \circ y = Q(y)x_1 = 0$. Since $\langle t, e \rangle = 1$, these elements are primitive idempotents.
(iii) $\xi = \eta = 0$. Then

$$t = x + y \qquad (x \in E_0, \ y \in E_1, \ Q(x) = Q(y) = 0, \ xy = 0, \ y \circ y = 0).$$

As in (ii), $y \circ y = 0$ implies $Q(y) = 0$. These elements are nilpotent.

The statement about a primitive idempotent $t$ being of type (i) if $\langle t, u \rangle \neq 0$ and of type (ii) if $\langle t, u \rangle = 0$ follows from the fact that

$$\langle t, u \rangle = (Q(y) + 1)^{-1} \neq 0$$

for $t$ of type (i) and $\langle t, u \rangle = 0$ for $t$ of type (ii). Similarly for $t^2 = 0$.    □

From the classification of primitive idempotents we easily derive the following lemma.

**Lemma 5.5.4** *If $u$ and $t$ are primitive idempotents in a J-algebra, then there exist primitive idempotents $v_0 = u, v_1, \ldots, v_{n-1}, v_n = t$ with $n \leq 3$ such that $\langle v_{i-1}, v_i \rangle = 0$ for $1 \leq i \leq n$.*

Proof. Let $\langle t, u \rangle \neq 0$; by the above proposition,

$$t = (Q(b) + 1)^{-1}(u + \frac{1}{2}Q(b)(e - u) + b \circ b + b) \qquad (b \in E_1, \ Q(b) \neq -1).$$

We have $\langle t, u \rangle = (Q(b) + 1)^{-1}$. We distinguish two cases.
(i) $\langle t, u \rangle \neq 1$, i.e., $Q(b) \neq 0$. A primitive idempotent $v$ with $\langle u, v \rangle = 0$ is of the form

$$v = \frac{1}{2}(e - u) + x + y \qquad (x \in E_0, \ Q(x) = \frac{1}{4}, \ y \in E_1, \ xy = \frac{1}{4}y, \ Q(y) = 0).$$

Then

$$\langle t, v \rangle = (Q(b) + 1)^{-1}(\frac{1}{2}Q(b) + \langle b \circ b, x \rangle + \langle b, y \rangle).$$

If we choose $x = -Q(b)^{-1}b \circ b$, $y = 0$, then $\langle t, v \rangle = 0$. So in this case the result holds with $n \leq 2$.
(ii) $\langle t, u \rangle = 1$, so $Q(b) = 0$. Pick

$$v = \frac{1}{2}(e - u) + x \qquad (x \in E_0, \ Q(x) = \frac{1}{4}).$$

Then $\langle t, v \rangle = \langle b \circ b, x \rangle$. If this equals 1, replace $x$ by $-x$, so we may assume that $\langle t, v \rangle \neq 1$. Hence we can go from $v$ to $t$ in at most two steps, by (i). □

**Remark 5.5.5** The elements $x$ with $x \times x = 0$ have a geometric characterization. Assume that $k$ is algebraically closed and that $A$ is a proper J-algebra over $k$. The cubic polynomial function det is irreducible (see Cor. 5.4.6). It defines an irreducible cubic hypersurface $S$ in the projective space $\mathbf{P}(A)$ (whose points are the one-dimensional subspaces of $A$). It follows from (5.16) that the singular points of $S$ are the lines $kx$, where $x \neq 0$ and $x \times x = 0$.

## 5.6 Uniqueness of the Composition Algebra

In Th. 5.4.5 we saw that a proper reduced J-algebra is isomorphic to an algebra $H(C; \gamma_1, \gamma_2, \gamma_3)$. We show in this section that the composition algebra $C$ depends only on $A$ and not on the choice of an idempotent $u$ nor on the choices of $x_1$, $a_+$ and $a_-$. So we may call $C$ the *composition algebra associated with A*. This will be followed by the result that $C$ depends only on the cubic form det.

**Theorem 5.6.1** *If $A$ is a proper reduced J-algebra, then the composition algebra $C$ such that $A \cong H(C; \gamma_1, \gamma_2, \gamma_3)$ is uniquely determined up to isomorphism.*

Proof. We fix a primitive idempotent $u$, and for any other primitive idempotent $t$ we consider

$$E_0' = \{\, x' \in A \,|\, \langle x', e \rangle = \langle x', t \rangle = 0,\ tx' = 0 \,\}$$
$$= \{\, x' \in A \,|\, \langle x', e \rangle = 0,\ tx' = 0 \,\},$$

since $\langle x', t \rangle = \langle tx', e \rangle = 0$ if $tx' = 0$. In $E_0'$ we choose $x_1'$ with $Q(x_1') = \frac{1}{4}$, and then

$$C' = \{\, x' \in E_0' \,|\, \langle x', x_1' \rangle = 0 \,\}.$$

By Witt's Theorem, the restriction of the quadratic form $Q$ to the orthogonal complement $C'$ of $x_1'$ in $E_0'$ is unique up to isometry. The norm $N$ on $C'$ is a multiple of $Q$, so it is unique up to similarity. By Th. 1.7.1, this implies uniqueness of the composition algebra up to isomorphism. So we have to show for one choice of $x_1'$ in $E_0'$ only that the restriction of $Q$ to $C'$ is similar to the restriction of $Q$ to $C$.

By Lemma (5.5.4) it suffices to prove this for a primitive idempotent $t$ with $\langle t, u \rangle = 0$. So we may assume by Prop. 5.5.3 that $t$ is of the form

$$t = \frac{1}{2}(e - u) + a + b \qquad \left( a \in E_0,\ Q(a) = \frac{1}{4},\ b \in E_1,\ ab = \frac{1}{4}b,\ Q(b) = 0 \right).$$

We recall that the last condition can be replaced by $b \circ b = 0$. The zero space $E_0'$ of $t$ consists of the elements $x' = \xi e + \eta u + x + y$ $\left( \xi, \eta \in k,\ x \in E_0 \right.$ and $\left. y \in E_1 \right)$ which satisfy $\langle x', e \rangle = 0$ and $tx' = 0$. Writing out these two conditions, we get the following five equations in $\xi$, $\eta$, $x$ and $y$.

$$3\xi + \eta = 0, \tag{5.28}$$

$$\xi + \langle a, x \rangle + \frac{1}{2}\langle b, y \rangle = 0, \tag{5.29}$$

$$-\xi - \langle a, x \rangle + \frac{1}{2}\langle b, y \rangle = 0, \tag{5.30}$$

$$\frac{1}{2}x + \xi a + b \circ y = 0, \tag{5.31}$$

$$\frac{1}{4}y + ay + (\xi + \frac{1}{2}\eta)b + bx = 0. \tag{5.32}$$

Equation (5.28) yields $\eta = -3\xi$. The equations (5.29) and (5.30) give

$$\langle b, y \rangle = 0 \qquad \text{and} \qquad \xi = -\langle a, x \rangle.$$

From (5.31) we get

$$x = -2\xi a - 2b \circ y.$$

If $x$ satisfies this equation, then $\xi = -\langle a, x \rangle$, for

$$
\begin{aligned}
\langle a, x \rangle &= -2\xi \langle a, a \rangle - 2\langle b \circ y, a \rangle \\
&= -\xi - 2\langle by, a \rangle \\
&= -\xi - 2\langle ab, y \rangle \\
&= -\xi - \frac{1}{2}\langle b, y \rangle \\
&= -\xi \quad \big(\text{since } \langle b, y \rangle = 0\big).
\end{aligned}
$$

From (5.32) we infer, using (5.31),

$$ay = -\frac{1}{4}y + \xi b + 2(b \circ y)b.$$

By Lemma 5.3.3 (v),

$$(b \circ b)y + 2(b \circ y)b = \frac{1}{4}Q(b)y + \frac{1}{4}\langle b, y \rangle b.$$

Now $b \circ b = 0$, $Q(b) = 0$ and $\langle b, y \rangle = 0$, so $(b \circ y)b = 0$. Hence

$$ay = -\frac{1}{4}y + \xi b.$$

All this together shows that $x' \in E_0'$ if and only if

$$x' = \xi(e - 3u - 2a) - 2b \circ y + y \qquad \big(\langle b, y \rangle = 0, \; ay = -\frac{1}{4}y + \xi b\big). \qquad (5.33)$$

If we pick $\xi = \frac{1}{4}$ and $y = \frac{1}{2}b$, then indeed $\langle b, y \rangle = Q(b) = 0$ and

$$ay = \frac{1}{2}ab = \frac{1}{8}b = -\frac{1}{8}b + \frac{1}{4}b = -\frac{1}{4}y + \xi b.$$

Further, $b \circ y = \frac{1}{2}y \circ y = 0$. So if we choose $x_1' = \frac{1}{4}(e - 3u - 2a) + \frac{1}{2}b$, then $x_1' \in E_0'$, and one easily verifies that $Q(x_1') = \frac{1}{4}$. For $x' \in E_0'$ as in (5.33),

$$\langle x', x_1' \rangle = 2\xi + \langle a, b \circ y \rangle + \frac{1}{2}\langle b, y \rangle.$$

Since

$$\langle a, b \circ y \rangle = \langle a, by \rangle = \langle ab, y \rangle = \frac{1}{4}\langle b, y \rangle = 0,$$

we find $\langle x', x_1' \rangle = 2\xi$. Hence

$$C' = \{\, x' = -2b \circ y + y \mid y \in E_1,\ \langle b, y \rangle = 0,\ ay = -\frac{1}{4}y \,\}.$$

If we choose $x_1 = a$, then $b \in E_+$ and $y \in E_-$, so automatically $\langle b, y \rangle = 0$. It follows that

$$C' = \{\, x' = -2b \circ y + y \mid y \in E_- \,\}. \tag{5.34}$$

Using Lemma 5.4.2 (vii), we find for $x' \in C'$,

$$Q(x') = 4Q(b \circ y) + Q(y) = Q(b)Q(y) + Q(y) = Q(y), \tag{5.35}$$

so we see that $C'$ and $E_-$ are isometric. Since the latter is similar to $C$ (with multiplier $\frac{1}{4}Q(a_+)$), we conclude that $C$ and $C'$ are similar. $\qquad\square$

We now prove that the associated composition algebra depends only on the determinant.

**Theorem 5.6.2** *If $A$ and $B$ are proper reduced $J$-algebras over $k$ with determinants $\det_A$ and $\det_B$, respectively, then the associated composition algebras are isomorphic if and only if there exists a linear transformation $t : A \to B$ such that $\det_B(t(x)) = \alpha \det_A(x)$ $(x \in A)$ for some $\alpha \in k^*$.*

Proof. Let $A = \mathrm{H}(C; \gamma_1, \gamma_2, \gamma_3)$ and $B = \mathrm{H}(C'; \gamma_1', \gamma_2', \gamma_3')$, and let $s : C \to C'$ be an isomorphism of composition algebras. For $x = h(\xi_1, \xi_2, \xi_3; c_1, c_2, c_3) \in A$ we have by equation (5.11):

$$\det_A(x) = \xi_1\xi_2\xi_3 - \gamma_3^{-1}\gamma_2\xi_1 N(c_1) - \gamma_1^{-1}\gamma_3\xi_2 N(c_2) - \gamma_2^{-1}\gamma_1\xi_3 N(c_3) + \langle c_1 c_2, \bar{c}_3 \rangle,$$

and similarly for $\det_B$. The linear transformation

$$t : A \to B,\ h(\xi_1, \xi_2, \xi_3; c_1, c_2, c_3) \mapsto h(\lambda_1\xi_1, \lambda_2\xi_2, \lambda_3\xi_3; s(c_1), s(c_2), s(c_3))$$

with $\lambda_i = (\gamma_{i+1}'\gamma_{i+2})^{-1}\gamma_{i+2}'\gamma_{i+1}$ (indices mod 3) satisfies $\det_B(t(x)) = \det_A(x)$.

Conversely, we will prove that if det is given up to a nonzero scalar factor, then $C$ is determined up to isomorphism. By Th. 1.7.1 it suffices to show that the norm $N$ of $C$ is determined up to a nonzero scalar factor. Take $v \neq 0$ in $A$ with $v \times v = 0$. Consider the quadratic form $F_v$ on $A$ defined by

$$F_v(x) = \langle v, x, x \rangle \qquad (x \in A).$$

We will show that $F_v$ determines $N$ up to a scalar factor. Two cases must be distinguished.
(a) $v$ is a primitive idempotent.
We then decompose $A$ as in § 5.3 with respect to $v$ instead of $u$. By equation (5.14),

$$3F_v(x) = \langle vx, x \rangle - Q(x) - \langle x, e \rangle \langle v, x \rangle + \frac{1}{2}\langle x, e \rangle^2 \qquad (x \in A).$$

For $x = \xi e + \eta v + a + b$ $(\xi, \eta \in k, a \in E_0, b \in E_1)$ this leads after some simple computations to

$$3F_v(x) = \xi^2 - Q(a).$$

This shows that the radical $R_v$ of $F_v$ is $kv \oplus E_1$. The quadratic form induced by $F_v$ on $A/R_v$ is equivalent to the restriction of $F_v$ to $S_v = ke \oplus E_0$, a complement of $R_v$ in $A$. This restriction is given by

$$3F_v(\xi e + a) = \xi^2 - Q(a) \qquad (\xi \in k, a \in E_0).$$

We see that $S_v$ is the orthogonal direct sum of a hyperbolic plane, viz. $ke \oplus kx_1$ for $x_1 \in E_0$ with $Q(x_1) = \frac{1}{4}$, and $C$, so $F_v$ determines the restriction of $Q$ to $C$ by Witt's Theorem. The latter in turn is a scalar multiple of the norm $N$ on $C$.

(b) $v^2 = 0$.

By Th. 5.5.1 we can choose a primitive idempotent $u$ with $uv = 0$. Decompose $A$ with respect to $u$. From the classification of elements with square zero in Prop. 5.5.3 we infer that $v \in E_0$ with $Q(v) = 0$. Using (5.14) we find for $x = \xi e + \eta u + a + b$ $(\xi, \eta \in k, a \in E_0, b \in E_1)$:

$$3F_v(x) = \langle v, x^2 \rangle - \langle x, e \rangle \langle v, x \rangle = -(\xi + \eta)\langle v, a \rangle + \langle v, b \circ b \rangle.$$

The radical of $F_v$ is easily seen to be

$$R_v = \{\, \xi(e - u) + a + b \mid \xi \in k, a \in E_0, \langle v, a \rangle = 0, b \in E_1, vb = 0 \,\}.$$

The isotropic element $v$ is contained in a hyperbolic plane in $E_0$. So there exist $x_1 \in E_0$ with $Q(x_1) = \frac{1}{4}$ and $c \in C = x_1^{\perp} \cap E_0$ with $Q(c) = -\frac{1}{4}$ such that $v$ is a nonzero multiple of $x_1 + c$, say $v = x_1 + c$.

Decompose $E_1 = E_+ \oplus E_-$ with respect to $x_1$ and write $b = b_+ + b_-$ with $b_\pm \in E_\pm$. Then $vb = \frac{1}{4}(b_+ - b_-) + cb_+ + cb_-$, so $vb = 0$ if and only if $cb_- = -\frac{1}{4}b_+$ and $cb_+ = \frac{1}{4}b_-$. From $b_- = 4cb_+$ it follows by Lemma 5.3.3 (i) that $cb_- = 4c(cb_+) = Q(c)b_+ = -\frac{1}{4}b_+$. So $vb = 0$ is equivalent to $b_- = 4cb_+$, hence

$$R_v = \{\, \xi(e - u) + a + b_+ + 4cb_+ \mid \xi \in k, a \in E_0, \langle v, a \rangle = 0, b_+ \in E_+ \,\}.$$

A complementary subspace of $R_v$ is

$$S_v = \{\, \xi e + \eta v_1 + b_- \mid \xi, \eta \in k, b_- \in E_- \,\}$$

for some fixed $v_1 \in E_0$, $\langle v, v_1 \rangle = 1$, $Q(v_1) = 0$. The quadratic form induced by $F_v$ on $A/R_v$ is equivalent to the restriction of $F_v$ to $S_v$, which is given by

$$3F_v(\xi e + \eta v_1 + b_-) = -\xi\eta - \frac{1}{2}Q(b_-).$$

$S_v$ is the orthogonal direct sum of the hyperbolic plane $ke \oplus kv_1$ and $E_-$. Hence $F_v$ determines the restriction of $Q$ to $E_-$ up to equivalence. Since the latter is equivalent to a scalar multiple of the norm $N$ on $C$, $F_v$ determines $N$ up to a scalar factor.    □

## 5.7 Norm Class of a Primitive Idempotent

In a proper reduced J-algebra the isometry class of the restriction of $Q$ to $E_0$ is *not* independent of the choice of the primitive idempotent $u$. For $x = \xi x_1 + c$ $(\xi \in k,\, c \in C)$,

$$Q(x) = \frac{1}{4}\xi^2 + \alpha N(c),$$

where $\alpha = \frac{1}{4}Q(a_+)Q(a_-)$ (cf. Prop. 5.4.4). The isometry class of this form depends on the class of $\alpha$ in $k^*$ modulo the subgroup

$$N(C)^* = \{\, N(c) \,|\, c \in C,\, N(c) \neq 0 \,\}.$$

This class is denoted by $\kappa(\alpha)$ and is called the *norm class* of $\alpha$:

$$\kappa(\alpha) = \alpha N(C)^* \in k^*/N(C)^*.$$

$\kappa(\alpha)$ depends on the primitive idempotent $u$, but we claim that it is independent of the choice of $x_1 \in E_0$ with $Q(x_1) = \frac{1}{4}$. For a different choice of $x_1$ in $E_0$, say $x_1'$, leads to another composition algebra $C'$ with norm $N'$ and identity element $\varepsilon'$. For $x = \xi'x_1' + c'$ $(\xi' \in k,\, c' \in C')$, let $Q(x) = \frac{1}{4}\xi'^2 + \alpha'N'(c')$. By Witt's Theorem, there exists a linear transformation $t : C \to C'$ such that $\alpha N(c) = \alpha'N'(t(c))$ $(c \in C)$. Take $c \in C$ with $t(c) = \varepsilon'$, then $\alpha N(c) = \alpha'$, which shows that $\alpha'$ has the same norm class as $\alpha$. We may therefore call $\kappa(\alpha)$ the *norm class* of $u$, denoted by $\kappa(u)$.

**Proposition 5.7.1** *Let $A$ be a proper reduced J-algebra, let $u$ be a primitive idempotent in $A$ and $x_1 \in E_0$ with $Q(x_1) = \frac{1}{4}$. Set*

$$T = (ke \oplus ku \oplus kx_1)^{\perp} = C \oplus E_1.$$

*The set of norm classes of the primitive idempotents in $A$ coincides with*

$$\{\, \kappa(Q(t)) \,|\, t \in T,\, Q(t) \neq 0 \,\}.$$

Proof. By Witt's Theorem, the restriction of the norm $Q$ to $T$ is independent of the special choice of $u$ and $x_1$ in $A$. We fix $u$ and $x_1$. We are going to compute the norm classes $\kappa(v)$ for the different primitive idempotents $v$. First assume $\langle v, u \rangle \neq 0$, so by Prop. 5.5.3,

$$v = (Q(b) + 1)^{-1}(u + \frac{1}{2}Q(b)(e - u) + b \circ b + b) \qquad (b \in E_1,\, Q(b) \neq -1).$$

We determine the zero space $E_0'$ of $v$. Writing out the equation $vx' = 0$ for $x' = \xi e + \eta u + x + y$ with $\xi, \eta \in k$, $x \in E_0$ and $y \in E_1$, i.e.,

$$(u + \frac{1}{2}Q(b)(e - u) + b \circ b + b)(\xi e + \eta u + x + y) = 0,$$

we arrive at the following four equations in $\xi$, $\eta$, $x$ and $y$.

$$\frac{1}{2}Q(b)\xi + \frac{1}{2}\langle b \circ b, x \rangle + \frac{1}{4}\langle b, y \rangle = 0, \tag{5.36}$$

$$-\frac{1}{2}Q(b)\xi + \xi + \eta - \frac{1}{2}\langle b \circ b, x \rangle + \frac{1}{4}\langle b, y \rangle = 0, \tag{5.37}$$

$$\frac{1}{2}Q(b)x + \xi b \circ b + b \circ y = 0, \tag{5.38}$$

$$(\frac{1}{4}Q(b) + \frac{1}{2})y + (\xi + \frac{1}{2}\eta)b + (b \circ b)y + bx = 0. \tag{5.39}$$

Multiplying equation (5.38) by $2b$, we find with the aid of Lemma 5.3.3 (iv) and (v)

$$Q(b)bx + \frac{1}{2}\xi Q(b)b + \frac{1}{4}Q(b)y + \frac{1}{4}\langle b, y \rangle b - (b \circ b)y = 0. \tag{5.40}$$

Adding equations (5.36) and (5.37), we find

$$\frac{1}{4}\langle b, y \rangle = -\frac{1}{2}(\xi + \eta). \tag{5.41}$$

From (5.39) we infer

$$(b \circ b)y = -(\frac{1}{4}Q(b) + \frac{1}{2})y - \xi b - \frac{1}{2}\eta b - bx = 0.$$

If we substitute this and (5.41) into equation (5.40) and then divide this by $\frac{1}{2}(Q(b) + 1)$ (recall that $Q(b) \neq -1$), we find

$$y = -2bx - \xi b. \tag{5.42}$$

Conversely, using Lemmas 5.3.2 and 5.3.3, we see that if $y$ has this form the equations (5.36)–(5.39) hold if and only if

$$(-Q(b) + 1)\xi + \eta - \langle b \circ b, x \rangle = 0. \tag{5.43}$$

The condition $\langle x', e \rangle = 0$ is equivalent to

$$3\xi + \eta = 0. \tag{5.44}$$

Thus we find that the elements of $E_0'$ are of the form

$$x' = \xi e - 3\xi u + x - (2bx + \xi b) \qquad (\xi \in k, \ x \in E_0), \tag{5.45}$$

where $\xi$ and $x$ have to satisfy

$$-(Q(b) + 2)\xi - \langle b \circ b, x \rangle = 0. \tag{5.46}$$

The norm $Q$ on $E'_0$ is easily computed:

$$Q(x') = (Q(b) + 3)\xi^2 + (Q(b) + 1)Q(x) + 2\xi\langle b \circ b, x \rangle. \tag{5.47}$$

If $Q(b) = -2$, then by equation (5.46) $\langle b \circ b, x \rangle = 0$. In this case we choose $x_1 = \frac{1}{4}b \circ b$, then $Q(x_1) = \frac{1}{4}$ and $x \in x_1^{\perp} \cap E_0 = C$ by (5.46). Take $x'_1 = \frac{1}{2}(e - 3u - b)$; according to equation (5.45) with $\xi = \frac{1}{2}$ and $x = 0$, $x'_1 \in E'_0$ and $Q(x'_1) = \frac{1}{4}$. Since $\langle x', x'_1 \rangle = \xi$, the subspace $C' = (x'_1)^{\perp} \cap E'_0$ consists of the elements

$$x' = x - 2bx \qquad (x \in C). \tag{5.48}$$

For such an element,

$$Q(x') = (Q(b) + 1)Q(x) \tag{5.49}$$

by (5.47), so $\kappa(v) = \kappa(Q(b) + 1)\kappa(u)$.

If $Q(b) \neq -2, -1$, we get from (5.46): $\xi = -(Q(b) + 2)^{-1}\langle b \circ b, x \rangle$. Thus we find as elements of $E'_0$:

$$x'(x) = -(Q(b) + 2)^{-1}\langle b \circ b, x \rangle(e - 3u - b) + x - 2bx \qquad (x \in E_0).$$

An easy computation shows

$$Q(x'(x)) = (Q(b) + 1)\{Q(x) - (Q(b) + 2)^{-2}\langle b \circ b, x \rangle^2\},$$

and hence

$$\langle x'(x), x'(y) \rangle = (Q(b) + 1)\{\langle x, y \rangle - 2(Q(b) + 2)^{-2}\langle b \circ b, x \rangle\langle b \circ b, y \rangle\}.$$

Now, in addition to $Q(b) \neq -2, -1$, we assume $Q(b) \neq 0$. In this case we choose

$$x_1 = Q(b)^{-1}b \circ b \qquad \text{and} \qquad x'_1 = \frac{1}{2}Q(b)^{-1}(Q(b) + 1)^{-1}(Q(b) + 2)x'(b \circ b).$$

$Q(x_1) = Q(x'_1) = \frac{1}{4}$, as one easily verifies. $C = \{x \in E_0 \mid \langle b \circ b, x \rangle = 0\}$ and

$$C' = \{x'(x) \in E'_0 \mid x \in C\}. \tag{5.50}$$

For $x \in C$,

$$Q(x'(x)) = (Q(b) + 1)Q(x), \tag{5.51}$$

so again we find $\kappa(v) = \kappa(Q(b) + 1)\kappa(u)$.

From equations (5.49) and (5.51) we conclude that the possible norm classes of primitive idempotents include all $\kappa(Q(b) + 1)\kappa(u)$ with $b \in E_1$, $Q(b) \neq -1, 0$. We may drop the condition $Q(b) \neq 0$ here, since $\kappa(u)$ is also a norm class. With the notations of § 5.4 it follows from Prop. 5.4.4 that $\kappa(u) = \kappa(Q(a_+)Q(a_-))$. We conclude that the possible norm classes of primitive idempotents we have found so far are the norm classes of the nonzero elements of the form

$$Q(a_+)Q(a_-) + Q(a_-)N(c_2) + Q(a_+)N(c_3) \qquad (c_i \in C).$$

These are all $\kappa(Q(t))$ with $t \in T$, $Q(t) \neq 0$.

Now consider any primitive idempotent $v$. Pick a primitive idempotent $u$ such that $\langle v, u \rangle = 0$. We remarked already that a change of $u$ does not affect the possible values of $Q$ on $T$. For such $v$ it follows from the proof of Th. 5.6.1 that $\kappa(v) = \kappa(Q(a_+))\kappa(u) = \kappa(Q(a_-))$, so $\kappa(v) = \kappa(Q(t))$ for some $t \in T$ with $Q(t) \neq 0$.    □

## 5.8 Isomorphism Criterion. Classification over Some Fields

If two proper reduced J-algebras are isomorphic, the quadratic forms $Q$ and $Q'$ must be equivalent by Prop. 5.3.10, and by Th. 5.6.1 the associated composition algebras are isomorphic. We will now show the converse.

**Theorem 5.8.1** *Two proper reduced J-algebras $A$ and $A'$ with isomorphic associated composition algebras are isomorphic if and only if the quadratic forms $Q$ on $A$ and $Q'$ on $A'$ are equivalent. If this is the case and if $u \in A$ and $u' \in A'$ are primitive idempotents, there exists an isomorphism of $A$ onto $A'$ which carries $u$ to $u'$ if and only if $u$ and $u'$ have the same norm classes: $\kappa(u) = \kappa(u')$.*

Proof. Assume $Q$ and $Q'$ are equivalent. Prop. 5.7.1 implies that we can choose primitive idempotents $u \in A$ and $u' \in A'$ such that their norm classes are the same. This implies that the restrictions of $Q$ to $E_0$ and of $Q'$ to $E_0'$ are equivalent, hence the same holds for the restrictions of these quadratic forms to $E_1$ and $E_1'$ by Witt's Theorem. So we can pick $a_+ \in E_1$ and $a_+' \in E_1'$ such that $Q(a_+) = Q'(a_+') \neq 0$. Take $x_1 = \frac{1}{4}Q(a_+)^{-1}a_+ \circ a_+ \in E_0$; by Lemma 5.3.6, $Q(x_1) = \frac{1}{4}$ and $a_+ \in E_+$. Similarly with $a_+'$ in $A'$. Choose any $a_- \in E_-$ with $Q(a_-) \neq 0$ and $a_-' \in E_-'$ with $Q'(a_-') \neq 0$. Since

$$\kappa(Q(a_+)Q(a_-)) = \kappa(u) = \kappa(u') = \kappa(Q'(a_+')Q'(a_-')),$$

we may replace $a_-'$ by some $a_-'c'$ so as to make $Q(a_-) = Q'(a_-')$; here $c' \in C'$, the orthogonal complement of $x_1'$ in $E_0'$, which as a composition algebra is isomorphic to the composition algebra $C$ of $A$.

Now it follows from the proof of Th. 5.4.5 that $A$ and $A'$ are isomorphic to the same algebra $H(C; \gamma_1, \gamma_2, \gamma_3)$, viz. with

$$\gamma_1 = 1, \quad \gamma_2 = Q(a_+)^{-1} \quad \text{and} \quad \gamma_3 = Q(a_-).$$

Under these isomorphisms the primitive idempotents $u$ and $u'$ are both mapped upon the matrix $h(1,0,0;0,0,0)$ with 1 in the left upper corner and zeros elsewhere. Thus we have found an isomorphism of $A$ onto $A'$ which carries $u$ to $u'$. $\qquad\square$

**Corollary 5.8.2** *If $C$ is a split composition algebra, there is only one isomorphism class of proper reduced J-algebras with $C$ as associated composition algebra. The automorphism group of such a J-algebra is transitive on primitive idempotents.*

Proof. If $C$ is split, its norm form $N$ takes on all values in $k$. So the quadratic forms $Q$ and $Q'$ on any two proper reduced J-algebras which have $C$ as associated composition algebra are necessarily equivalent (see equation (5.3) for the form of $Q$ and $Q'$). Since $k^*/N(C)^*$ has only one element in this case, there is only one norm class of primitive idempotents. $\qquad\square$

The above theorem reduces the classification of proper reduced J-algebras over a given field $k$ to a problem about quadratic forms over $k$. We will discuss the situation for some special fields. We only consider Albert algebras, so the associated composition algebras are octonion algebras, since that is the case we are most interested in. We make use of the classification of octonion algebras over special fields that is given in § 1.10.

*(i) $k$ algebraically closed.*
By Th. 5.5.2, $A$ is reduced. There is only the split octonion algebra in this case, so Cor. 5.8.2 implies that all Albert algebras over $k$ are isomorphic and that $\mathrm{Aut}(A)$ is transitive on primitive idempotents.

*(ii) $k = \mathbb{R}$, the field of the reals.*
The cubic form det represents zero nontrivially over the quadratic extension $\mathbb{C}$ of $\mathbb{R}$, hence so it does over $\mathbb{R}$ itself by Lemma 4.2.11. So Th. 5.5.2 implies that $A$ is reduced. There are two isomorphism classes of octonion algebras $C$, the split algebra and the Cayley numbers. All reduced Albert algebras with split $C$ are isomorphic and in that case $\mathrm{Aut}(A)$ is transitive on primitive idempotents. For the Cayley numbers, $N$ is positive definite and takes on all positive values, so $k^*/N(C)^* = \{ \pm 1 \}$. In this case there are two isomorphism classes of reduced Albert algebras, for as we see from equation (5.3) there are two inequivalent possibilities for the quadratic form $Q$, viz., the positive definite form with all $\gamma_i = 1$ and the indefinite form with, e.g., $\gamma_1 = \gamma_2 = 1$, $\gamma_3 = -1$. In the positive definite case, all primitive idempotents have norm class 1, so $\mathrm{Aut}(A)$ is transitive on them. In the indefinite case, the norm class of a primitive idempotent can be 1 or $-1$, so then there are two transitivity classes of primitive idempotents under the action of $\mathrm{Aut}(C)$.

*(iii) k a finite field.*
According to a theorem of Chevalley [Che 35, p. 75] (see also [Gre, Th. (2.3)],
[Lang, third ed., 1993, p. 214, ex. 7], [Se 70, §2.2, Th. 3] or [Se 73, p. 5]), the
cubic form det represents zero nontrivially over a finite field, so $A$ is reduced
by Th. 5.5.2. $C$ must be the split octonion algebra, so there is one possibility
for $A$, with $\mathrm{Aut}(A)$ acting transitively on the primitive idempotents.

*(iv) k a complete, discretely valuated field with finite residue class field.*
The cubic form det represents zero nontrivially (see [Sp 55, remark after
Prop. 2], or [De] or [Le]), so $A$ is reduced. $C$ is split, so there is one isomorphism class of Albert algebras $A$ and $\mathrm{Aut}(A)$ acts transitively on the
primitive idempotents.

*(v) k an algebraic number field.*
We know that every twisted composition algebra over such a field $k$ is reduced
(see the end of § 4.8). In the next chapter we will show that this implies that
every Albert algebra over $k$ is reduced (see Cor. 6.3.4). As in (v) of § 1.10 we
use Hasse's Theorem on the classification of quadratic forms (see [O'M, §66]).
There is only one possibility for $k_v \otimes_k A$ at each finite or complex infinite
place $v$ by (iii) and (i) above. At each real place there are three possibilities as
we saw in (ii), so we get $3^r$ isomorphism classes of Albert algebras, $r$ denoting
the number of real places of $k$. For $k = \mathbb{Q}$ this leads to three isomorphism
classes of Albert algebras, just as in the real case.

## 5.9 Isotopes. Orbits of the Invariance Group of the Determinant

In this last section of Ch. 5 we intend to prove a transitivity result for the linear transformations in a proper reduced J-algebra that leave the determinant
invariant, a result we need in Ch. 7. For this purpose, we develop a procedure
to construct from a J-algebra a new J-algebra with different identity element
and different norm. We further characterize automorphisms of J-algebras by
the fact that they leave $e$ and the determinant invariant, also for use in Ch. 7.

    $A$ is as in the previous sections. We first give two special cases of formula
(v) in Lemma 5.2.1, which will be frequently used in this section.

$$4(x \times x) \times (x \times y) = \det(x)y + 3\langle x, x, y \rangle x \qquad (5.52)$$

$$2(x \times x) \times (y \times y) + 4(x \times y) \times (x \times y) =$$
$$3\langle x, x, y \rangle y + 3\langle x, y, y \rangle x, \qquad (5.53)$$

where $x, y \in A$. From the first formula we derive a simple but important
lemma.

**Lemma 5.9.1** *If $a, x \in A$ and $\det(a) \neq 0$, then $a \times x = 0$ implies $x = 0$.*

Proof. Since $a \times x = 0$ we have $3\langle a, a, x \rangle = \langle a, a \times x \rangle = 0$. From (5.52) we obtain $\det(a)x = 0$. Since $\det(a) \neq 0$, we must have $x = 0$.    □

Let $a$ be an element of $A$ with $\det(a) = \lambda \neq 0$. On the vector space $A$ we define a symmetric bilinear form $\langle \ , \ \rangle_a$ by

$$\langle x, y \rangle_a = -6\lambda^{-1}\langle x, y, a \rangle + 9\lambda^{-2}\langle x, a, a \rangle\langle y, a, a \rangle \qquad (x, y \in A). \quad (5.54)$$

Notice that

$$\langle x, a \rangle_a = 3\lambda^{-1}\langle x, a, a \rangle, \quad (5.55)$$

whence $\langle a, a \rangle = 3$.

The form is nondegenerate. We have

$$\langle x, y \rangle_a = \langle x, -2\lambda^{-1}y \times a + 3\lambda^{-2}\langle y, a, a \rangle a \times a \rangle.$$

If this is zero for all $x$, then $y \times a = \alpha a \times a$ for some $\alpha \in k$, so $(y - \alpha a) \times a = 0$. Then $y = \alpha a$ by the above Lemma, but $\langle x, a \rangle_a$ is not identically zero, so $y = 0$.

Let $Q_a$ be the nondegenerate quadratic form on $A$ whose associated bilinear form is $\langle \ , \ \rangle_a$:

$$Q_a(x) = -3\lambda^{-1}\langle x, x, a \rangle + \frac{9}{2}\lambda^{-2}\langle x, a, a \rangle^2 \qquad (x \in A). \quad (5.56)$$

By (5.55) we can also write this as

$$Q_a(x) = -3\lambda^{-1}\langle x, x, a \rangle + \frac{1}{2}\langle x, a \rangle_a^2 \qquad (x \in A). \quad (5.57)$$

Further, we define a new product on $A$, for which the notation $._a$ will be used:

$$x._a y = 4\lambda^{-1}(x \times a) \times (y \times a) + \frac{1}{2}(\langle x, y \rangle_a - \langle x, a \rangle_a\langle y, a \rangle_a)a \qquad (x, y \in A). \quad (5.58)$$

**Proposition 5.9.2** *Let $a \in A$ with $\det(a) = \lambda \neq 0$. The algebra $A_a$ which has the vector space structure of $A$, the norm $Q_a$ as in (5.56) and the product $._a$ defined by (5.58), is a J-algebra with $a$ as identity element. The determinant of $A_a$ is*

$$\det_a(x) = \lambda^{-1}\det(x) \qquad (x \in A).$$

*For $a = e$ we get the original J-algebra structure of $A$ (with $Q_e = Q$ and $\det_e = \det$).*

*$A_a$ is reduced or proper if and only if $A$ is reduced or proper, respectively. If $A$ is proper and reduced (hence so is $A_a$), then the composition algebras associated with $A_a$ are isomorphic to those associated with $A$.*

Proof. We first observe that $Q_a(a) = \frac{3}{2}$. Using (5.52) we derive from (5.58):

$$a._a x = 4\lambda^{-1}(a \times a) \times (x \times a) + \frac{1}{2}(\langle a, x \rangle_a - \langle a, a \rangle_a \langle x, a \rangle_a)a = x \qquad (x \in A),$$

so $a$ is an identity element of $A_a$. To verify (5.5) for $A_a$, we have to show that $\langle x._a y, z \rangle_a$ is symmetric in $x$, $y$ and $z$. From (5.54) we find

$$\langle x._a y, z \rangle_a = -6\lambda^{-1}\langle x._a y, z, a \rangle + 9\lambda^{-2}\langle x._a y, a, a \rangle\langle z, a, a \rangle.$$

Substituting (5.58) into this, we get:

$$\langle x._a y, z \rangle_a = -24\lambda^{-2}\langle (x \times a) \times (y \times a), z, a \rangle + \frac{3}{2}\lambda^{-1}\langle x, y \rangle_a \langle z, a, a \rangle$$

$$-\frac{3}{2}\lambda^{-1}\langle x, a \rangle_a \langle y, a \rangle_a \langle z, a, a \rangle + 36\lambda^{-3}\langle (x \times a) \times (y \times a), a, a \rangle\langle z, a, a \rangle.$$

The first term on the right hand side equals $-24\lambda^{-2}\langle x \times a, y \times a, z \times a \rangle$, which is symmetric. The third term on the right is symmetric by (5.55). For the second term we find, using (5.54),

$$\frac{3}{2}\lambda^{-1}\langle x, y \rangle_a \langle z, a, a \rangle = -9\lambda^{-2}\langle x, y, a \rangle\langle z, a, a \rangle +$$

$$\frac{27}{2}\lambda^{-2}\langle x, a, a \rangle\langle y, a, a \rangle\langle z, a, a \rangle.$$

As to the fourth term, we have

$$4\langle (x \times a) \times (y \times a), a, a \rangle = 4\langle (x \times a) \times (a \times a), y, a \rangle.$$

Applying (5.52) to the right hand side we get:

$$4\langle (x \times a) \times (y \times a), a, a \rangle = \lambda\langle x, y, a \rangle + 3\langle x, a, a \rangle\langle y, a, a \rangle, \qquad (5.59)$$

so

$$36\lambda^{-3}\langle (x \times a) \times (y \times a), a, a \rangle\langle z, a, a \rangle = 9\lambda^{-2}\langle x, y, a \rangle\langle z, a, a \rangle +$$

$$27\lambda^{-3}\langle x, a, a \rangle\langle y, a, a \rangle\langle z, a, a \rangle.$$

We see that the contribution of the second plus the fourth term is symmetric. This proves the symmetry of $\langle x._a y, z \rangle_a$ in $x$, $y$ and $z$, and hence (5.5).

Now to the proof of (5.4). We use the notation $x^{.2}$ for $x._a x$. By (5.57),

$$Q_a(x^{.2}) = -3\lambda^{-1}\langle x^{.2}, x^{.2}, a \rangle + \frac{1}{2}\langle x^{.2}, a \rangle_a^2.$$

If $\langle x, a \rangle_a = 0$, we get using (5.58) for $-3\lambda^{-1}\langle x^{.2}, x^{.2}, a \rangle$:

$$-3\lambda^{-1}\langle 4\lambda^{-1}(x \times a) \times (x \times a) + Q_a(x)a, 4\lambda^{-1}(x \times a) \times (x \times a) + Q_a(x)a, a \rangle =$$

$$-16\lambda^{-3}\langle((x\times a)\times(x\times a))\times((x\times a)\times(x\times a)),a\rangle$$

$$-24\lambda^{-2}Q_a(x)\langle(x\times a)\times(x\times a),a,a\rangle-3Q_a(x)^2.$$

By Lemma 5.2.1 (iv) we get for the first term on the right hand side:

$$-16\lambda^{-3}\langle\det(x\times a)x\times a,a\rangle=-48\lambda^{-3}\det(x\times a)\langle x,a,a\rangle=$$

$$-16\lambda^{-2}\det(x\times a)\langle x,a\rangle_a=0.$$

With (5.59) we find

$$-24\lambda^{-2}Q_a(x)\langle(x\times a)\times(x\times a),a,a\rangle=$$

$$-6\lambda^{-1}Q_a(x)\langle x,x,a\rangle-18\lambda^{-2}Q_a(x)\langle x,a,a\rangle^2=$$

$$2Q_a(x)^2-6\lambda^{-1}Q_a(x)\langle x,a\rangle_a^2=2Q_a(x)^2.$$

Finally, by (5.5),

$$\frac{1}{2}\langle x^2,a\rangle_a^2=2Q_a(x)^2.$$

Adding up we get $Q_a(x^2)=Q_a(x)^2$ if $\langle x,a\rangle_a=0$. Thus, $A_a$ is a J-algebra.

We now compute $\det_a$. Let $\times_a$ denote the cross product in $A_a$, corresponding to $\langle\ ,\ ,\ \rangle_a$ according to (5.16). By Lemma 5.2.1 (i) and (5.58) we have:

$$x\times_a x=4\lambda^{-1}(x\times a)\times(x\times a)-\langle x,a\rangle_a x\qquad(x\in A).\tag{5.60}$$

Using (5.53), (5.55) and (5.57), we find from (5.60):

$$x\times_a x=-2\lambda^{-1}(x\times x)\times(a\times a)-(Q_a(x)-\frac{1}{2}\langle x,a\rangle_a^2)a\qquad(x\in A).\tag{5.61}$$

So for $x\in A$,

$$3\det_a(x)=\langle x\times_a x,x\rangle_a$$
$$=-6\lambda^{-1}\langle x\times_a x,x,a\rangle+9\lambda^{-2}\langle x\times_a x,a,a\rangle\langle x,a,a\rangle.\tag{5.62}$$

With (5.61) we calculate $\langle x\times_a x,x,a\rangle$ as follows:

$$\langle x\times_a x,x,a\rangle=-2\lambda^{-1}\langle(x\times x)\times(a\times a),x,a\rangle-(Q_a(x)-\frac{1}{2}\langle x,a\rangle_a^2)\langle x,a,a\rangle$$

$$=-2\lambda^{-1}\langle(x\times a)\times(a\times a),x,x\rangle-(Q_a(x)-\frac{1}{2}\langle x,a\rangle_a^2)\langle x,a,a\rangle$$

$$=-\frac{1}{2}\det(x)-\frac{1}{6}\lambda(Q_a(x)-\frac{1}{2}\langle x,a\rangle_a^2),$$

the last equality coming from (5.52), (5.57) and (5.55). Similarly, one finds:

$$\langle x\times_a x,a,a\rangle=-\frac{1}{3}\lambda(Q_a(x)-\frac{1}{2}\langle x,a\rangle_a^2).$$

Plugging in these two expressions one gets from (5.62) that $\det_a(x) = \lambda^{-1} \det(x)$ $(x \in A)$.

In case $a = e$ we have $\lambda = \det(e) = 1$, so $\det_e(x) = \det(x)$ $(x \in A)$. For $x, y \in A$,

$$\langle x, y \rangle_e = -6\langle x, y, e \rangle + 9\langle x, e, e \rangle \langle y, e, e \rangle$$
$$= -2\langle x, y \times e \rangle + \langle x, e \times e \rangle \langle y, e \times e \rangle.$$

By Lemma 5.2.1 (i), $x \times e = -\frac{1}{2}x + \frac{1}{2}\langle x, e \rangle e$. Using this formula, one derives by a straightforward computation that $\langle x, y \rangle_e = \langle x, y \rangle$. In a similar way we find for $x, y \in A$,

$$x._e y = 4(x \times e) \times (y \times e) + \frac{1}{2}\langle x, y \rangle e - \frac{1}{2}\langle x, e \rangle \langle y, e \rangle e$$
$$= x \times y + \frac{1}{2}\langle x, e \rangle y + \frac{1}{2}\langle y, e \rangle x + \frac{1}{2}\langle x, y \rangle e - \frac{1}{2}\langle x, e \rangle \langle y, e \rangle e,$$

which equals $xy$ by Lemma 5.2.1 (i). Thus we see that $A_e = A$.

$A_a$ and $A$ have the same determinant function up to a nonzero factor, so they are simultaneously reduced or proper by Th. 5.5.2 and Cor. 5.4.6, respectively. The last result is a consequence of Th. 5.6.2, as we see by taking for $t : A \to A_a$ the identity map.    □

We call the J-algebra $A_a$ an *isotope* of $A$, and these two J-algebras are said to be *isotopic*.

The next proposition answers the question when two isotopes $A_a$ and $A_b$ are isomorphic. We also obtain a transitivity result for transformations of a J-algebra that leave the determinant invariant. This is the result we hinted at in the introductory paragraph of this section.

An isomorphism $t : A_a \to A_b$ must carry the identity element to the identity element, so $t(a) = b$, and it must preserve the determinants, so $\det_b(t(x)) = \det_a(x)$ $(x \in A)$, provided $\dim_k(A) > 2$ (see Prop. 5.3.10).

**Proposition 5.9.3** *Let $A$ be a proper J-algebra, and let $a, b \in A$, with $\det(a)\det(b) \neq 0$. The following are equivalent.*
*(i)   $A_a \cong A_b$.*
*(ii)  there exists a linear transformation $t : A \to A$ such that $t(a) = b$ and*

$$\langle t(x), t(y), t(z) \rangle = \det(a)^{-1}\det(b)\langle x, y, z, \rangle \quad (x, y, z \in A).$$

*If, moreover, $A$ is reduced, the conditions are also equivalent to*
*(iii) the bilinear forms $\det(a)^{-1}\langle x, y, a \rangle$ and $\det(b)^{-1}\langle x, y, b \rangle$ are equivalent.*

Proof. Recall that $A_a$ and $A_b$ are also proper. We noticed already the implication (i) $\Rightarrow$ (ii). Also, (ii) $\Rightarrow$ (iii) clearly holds in all cases. Because the algebra structure of $A_a$ is completely determined by $a$ (see (5.54) and (5.58)), (ii) implies (i). If $A$ is reduced all $A_a$ have isomorphic associated composition algebras. That in this case (ii) and (iii) are equivalent follows from Th. 5.8.1. □

Finally, we give a characterization of automorphisms of J-algebras.

**Proposition 5.9.4** *If $A$ is a J-algebra of dimension $> 2$, then a linear transformation $t$ of $A$ is an automorphism if and only if $t(e) = e$ and $\det(t(x)) = \det(x)$ $(x \in A)$.*

Proof. The "only if" part is known (see Prop. 5.3.10); we prove the "if" part. Since $A_e = A$,

$$\langle x, y \rangle = -6 \langle x, y, e \rangle + 9 \langle x, e, e \rangle \langle y, e, e \rangle$$

by (5.54). So a linear transformation $t$ that leaves $e$ and det invariant, also leaves $\langle\ ,\ \rangle$ invariant. From

$$\begin{aligned}
\langle t(x \times y), t(z) \rangle &= \langle x \times y, z \rangle \\
&= 3 \langle x, y, z \rangle \\
&= 3 \langle t(x), t(y), t(z) \rangle \\
&= \langle t(x) \times t(y), t(z) \rangle \qquad (x, y, z \in A)
\end{aligned}$$

it follows that

$$t(x \times y) = t(x) \times t(y) \qquad (x, y \in A).$$

Using (i) of Lemma 5.2.1 one derives that $t(xy) = t(x)t(y)$ $(x, y \in A)$.   □

## 5.10 Historical Notes

*Jordan algebras* over the reals were introduced in the early thirties by the physicist P. Jordan, who proposed them in the foundation of quantum mechanics; see [Jo 32], [Jo 33] and the joint paper [JoNW] with J. von Neumann and E. Wigner. The general theory over arbitrary fields of characteristic not two was developed by several people. We just mention A.A. Albert and N. Jacobson, in particular their joint paper [AlJa] where one finds among other things the classification of *Albert algebras* over real closed fields and algebraic number fields.

The definition of *J-algebras*, a limited class of Jordan algebras including the *Albert algebras*, and the whole approach followed in the present chapter originates from T.A. Springer's paper [Sp 59].

The notion of isotopy of Jordan algebras was introduced by Jacobson, see [Ja 68, p.57]. The notion of isotopy introduced in 5.9 is an adaptation to J-algebras.

The characterization of the automorphisms of a J-algebra in Prop. 5.9.4 as the linear transformations that leave the cubic form det invariant and fix the identity element $e$ was earlier proved by N. Jacobson in [Ja 59, Lemma 1]. C. Chevalley and R.D. Schafer [CheSch] gave an equivalent characterization, viz., as transformations that leave the quadratic form $Q$ and the cubic form det invariant. They dealt with Lie algebras $F_4$ and $E_6$ over algebraically closed fields in characteristic zero, so instead of automorphisms they considered derivations; see also H. Freudenthal [Fr 51].

# 6. Proper J-algebras and Twisted Composition Algebras

The study of J-algebras in the previous chapter has yielded a description of all reduced J-algebras. In the present chapter we develop another description of J-algebras which includes all nonreduced ones. For this purpose we make a link between J-algebras and twisted composition algebras. We will see that a J-algebra is reduced if and only if certain twisted composition algebras are reduced. This will lead to the result, already announced at the end of Ch. 5, that every J-algebra over an algebraic number field is reduced (see Cor. 6.3.4).

As in the previous chapter, a field will always be assumed to have characteristic $\neq 2, 3$.

## 6.1 Reducing Fields of J-algebras

Let $A$ be a J-algebra over a field $k$. In Prop. 5.3.8 we saw that for $a \in A$, the minimum polynomial $m_a$ divides the characteristic polynomial $\chi_a$ and has the same roots. The following proposition is an immediate consequence.

**Proposition 6.1.1** *If $\chi_a$ is irreducible over $k$, then $k[a]$ is a cubic extension field of $k$ and $\chi_a$ has a root in this field, viz., $a$ itself. If $\chi_a$ with $a \notin ke$ is reducible over $k$, then $k[a]$ contains an element $x \neq 0$ with $x \times x = 0$. So $A$ is not reduced if and only if $k[a]$ is a cubic extension field of $k$ for all $a \notin ke$.*

Proof. If $\chi_a$ is irreducible, $m_a = \chi_a$. If $\chi_a$ is reducible, it has a root in $k$ since it is of degree 3. If not all roots of $\chi_a$ in a splitting field are equal, $k[a]$ contains a primitive idempotent. If $\chi_a$ has three equal roots in $k$, then $k[a]$ contains a nilpotent element, hence also an $x \neq 0$ with $x^2 = 0$. In either case we find a nonzero $x$ with $x \times x = 0$, so $A$ is reduced (see Th. 5.5.1). The rest is clear. $\qquad\square$

Thus, a nonreduced J-algebra $A$ over $k$ of dimension $> 1$ necessarily contains a cubic extension field $l$ of $k$, viz., any $k[a]$ for $a \notin ke$. In $l \otimes_k l$ there exist idempotents, hence $l \otimes_k A$ is reduced. We call an extension field $l$ of $k$ such that $l \otimes_k A$ is reduced a *reducing field* of $A$. A J-algebra is reduced if and only if the cubic form det represents zero nontrivially by Th. 5.5.2. Hence if $A$ is not reduced and $l$ is a reducing field of $A$, the cubic form det does not represent zero on $A$, but it does represent zero on $l \otimes_k A$; by Lemma 4.2.11 the degree of $l$ over $k$ can not be 2.

If $A$ is not reduced and $l$ is a reducing field, the reduced J-algebra $l \otimes_k A$ is either of quadratic type or proper (cf. Th. 5.4.5). This does not depend on the choice of $l$, for if $l'$ is another reducing field, we pick a common extension $m$ of $l$ and $l'$, then the reduced algebras $l \otimes_k A$ and $l' \otimes_k A$ must be of the same type as $m \otimes_k A$. So it makes sense to call a J-algebra $A$ over $k$ of *quadratic type* or *proper* according to whether $l \otimes_k A$ is of quadratic type or proper for any reducing field $l$ of $k$. $A$ is said to be an *Albert algebra* if and only if $l \otimes_k A$ is an Albert algebra for some (hence any) reducing field $l$. We will see towards the end of this chapter (see Cor. 6.3.3) that a nonreduced J-algebra must necessarily be proper.

In a nonreduced J-algebra the cubic form det does not represent zero nontrivially by Th. 5.5.2, so by Lemma 5.2.3 every nonzero element has a J-inverse. For this reason, a nonreduced J-algebra (Albert algebra) is also called a *J-division algebra* (*Albert division algebra*, respectively).

For given $a \in A$ with $\dim_k k[a] = 3$, we define $F = k[a]^\perp$, so $A = k[a] \oplus F$ as a vector space over $k$. We will in particular be interested in the case that $k[a]$ is a field. From the structure of J-algebra on $A$ we will derive a structure of twisted composition algebra on $F$ over the field $k[a]$ such that $A$ is reduced if and only if $F$ is reduced as a twisted composition algebra. Notice that if $k[a]$ is of degree 3 over $k$ and is not a field, then $A$ is reduced anyway, as we saw above. By way of example, we first consider a simple situation, viz., a reduced J-algebra $H(C; \gamma_1, \gamma_2, \gamma_3)$ and $a = h(\alpha_1, \alpha_2, \alpha_3; 0, 0, 0)$ with three distinct $\alpha_i$, in the notation of (5.1). Then

$$k[a] = \{\, h(\xi_1, \xi_2, \xi_3; 0, 0, 0) \mid \xi_1, \xi_2, \xi_3 \in k \,\}$$

and

$$F = \{\, h(0, 0, 0; c_1, c_2, c_3) \mid c_1, c_2, c_3 \in C \,\}.$$

So in this example $k[a]$ is not a field, but a split cubic extension of $k$, and we will provide $F$ with a structure of twisted composition algebra over a split cubic extension (viz., $k[a]$) of $k$ as treated in the first part of § 4.3. This has sufficient analogy with the field case to exhibit essential phenomena. (If instead of the above split extension $k[a]$ of $k$ we have $k[a] = l$ with $l$ a cubic cyclic field extension of $k$, then $A_l = l \otimes_k A$ will be a J-algebra over $l$ and we may suppose that we are in the situation of the example with $k$ replaced by $l$.) As action of $k[a]$ on $F$ we take

$$b.x = -2b \times x = -2bx + \langle\, b, e\,\rangle x \qquad \left(b \in k[a], \quad x \in F\right).$$

Writing this out explicitly we find

$$h(\xi_1, \xi_2, \xi_3; 0, 0, 0).h(0, 0, 0; c_1, c_2, c_3) = h(0, 0, 0; \xi_1 c_1, \xi_2 c_2, \xi_3 c_3),$$

which is the natural structure of free $k[a]$-module on $F$.

On $k[a]$ we consider an automorphism of order three:

$$\sigma(h(\xi_1, \xi_2, \xi_3; 0, 0, 0)) = h(\xi_2, \xi_3, \xi_1; 0, 0, 0).$$

The group $< \sigma >$ can be considered as a "Galois group" of $k[a]$ over $k$. We compute the cross product of two elements $x = h(0, 0, 0; c_1, c_2, c_3)$ and $y = h(0, 0, 0; d_1, d_2, d_3)$ of $F$; using Lemma 5.2.1 we get:

$$x \times y = -\frac{1}{2} h(\gamma_3^{-1}\gamma_2 \langle c_1, d_1 \rangle, \gamma_1^{-1}\gamma_3 \langle c_2, d_2 \rangle, \gamma_2^{-1}\gamma_1 \langle c_3, d_3 \rangle; 0, 0, 0) +$$

$$+ \frac{1}{2} h(0, 0, 0; \gamma_2^{-1}\gamma_3 \overline{(c_2 d_3 + d_2 c_3)}, \gamma_3^{-1}\gamma_1 \overline{(c_3 d_1 + d_3 c_1)}, \gamma_1^{-1}\gamma_2 \overline{(c_1 d_2 + d_1 c_2)}).$$

If we define

$$N(x, y) = h(\gamma_3^{-1}\gamma_2 \langle c_1, d_1 \rangle, \gamma_1^{-1}\gamma_3 \langle c_2, d_2 \rangle, \gamma_2^{-1}\gamma_1 \langle c_3, d_3 \rangle; 0, 0, 0),$$
$$x * y = h(0, 0, 0; \gamma_2^{-1}\gamma_3 \overline{c_2 d_3}, \gamma_3^{-1}\gamma_1 \overline{c_3 d_1}, \gamma_1^{-1}\gamma_2 \overline{c_1 d_2}),$$

we see that

$$x \times y = -\frac{1}{2} N(x, y) + \frac{1}{2} x * y + \frac{1}{2} y * x.$$

$N( , )$ is a nondegenerate symmetric $k[a]$-bilinear form on the free $k[a]$-module $F$, associated with the quadratic form $N( )$ with $N(x) = \frac{1}{2} N(x, x)$, and $*$ is a $k$-bilinear product in $F$ which is $\sigma$-linear in the first variable and $\sigma^2$-linear in the second one. The conditions (ii) and (iii) of Def. 4.1.1 are easily verified, so $F$ is a twisted composition algebra over the split cubic extension $k[a]$ of $k$.

We now return to the general case.

## 6.2 From J-algebras to Twisted Composition Algebras

We again consider an arbitrary $a \in A$ with $\dim_k k[a] = 3$, and $F = k[a]^{\perp}$, so $A = k[a] \oplus F$. By Lemma 5.2.1 (i), $k[a]$ is closed under the cross product, i.e., $b \times c \in k[a]$ for $b, c \in k[a]$. From (5.5) we infer that $k[a]F = F$ and from Lemma 5.2.1 (i) and (iii) we derive:

$$2b \times (b \times x) = -b^2 \times x \qquad (b \in k[a], \; x \in F). \tag{6.1}$$

As in the example in the previous section, we introduce an action of $k[a]$ on $F$ by $k$-linear transformations:

$$p(b)(x) = -2b \times x = -2bx + \langle b, e \rangle x \qquad (b \in k[a], \; x \in F). \tag{6.2}$$

(For the second equality, see (i) of Lemma 5.2.1.) Then $p : k[a] \to \text{End}_k(F)$ is $k$-linear, $p(e) = \text{id}$, and from equation (6.1) it follows that

$$p(b^2) = p(b)^2 \qquad (b \in k[a]).$$

Linearizing this relation we obtain

$$p(bc) = \frac{1}{2}(p(b)p(c) + p(c)p(b)) \qquad (b, c \in k[a]).$$

Since $p(a^2) = p(a)^2$ it follows that $p(a^3) = p(a)^3$, from which we conclude that

$$p(bc) = p(b)p(c) \qquad (b, c \in k[a]).$$

We can therefore define a structure of $k[a]$-module on $F$ by

$$b.x = p(b)(x) \qquad (b \in k[a], \ x \in F). \tag{6.3}$$

The product $b.x$ written with a dot should be distinguished from the ordinary J-algebra product $bx$ which is written without a dot. Notice that $(\lambda e).x = \lambda x$ for $\lambda \in k$ and $x \in F$, so we may identify $k$ with the subfield $ke$ of $k[a]$, which will usually be done in the sequel. We will write $bc.x$ for $(bc).x$, which equals $b.(c.x)$ $(x \in F, \ b, c \in k[a])$.

*From now on we assume that $a$ is chosen such that $k[a]$ is a field of degree 3 over $k$, which we denote by $l$. As in Ch. 4, $l'$ is the normal closure of $l$ over $k$, so if $l/k$ is not Galois, then $l' = l(\sqrt{D})$ with $D$ a discriminant of $l/k$. Further, $k' = k(\sqrt{D})$. We fix a generator $\sigma$ of $\mathrm{Gal}(l'/k')$, also considered as an element of $\mathrm{Gal}(l'/k)$ and as a $k$-isomorphism of $l$ into $l'$. Finally, $\tau$ is the element of $\mathrm{Gal}(l'/k)$ whose fixed field is $l$ if $l' \neq l$, and $\tau = \mathrm{id}$ if $l' = l$. We first express the cross product in $l$ in terms of the field product and $\sigma$.*

**Lemma 6.2.1** *For $b, c \in l$,*

$$b \times c = \frac{1}{2}(\sigma(b)\sigma^2(c) + \sigma^2(b)\sigma(c)).$$

*We have $\det(b) = N_{l/k}(b)$ and $b \times b = N_{l/k}(b)b^{-1}$ if $b \neq 0$.*

Proof. For $b \in l$, the Hamilton-Cayley equation

$$b^3 - \langle b, e \rangle b^2 - (Q(b) - \frac{1}{2}\langle b, e \rangle^2)b - \det(b) = 0$$

has roots $b$, $\sigma(b)$ and $\sigma^2(b)$ in $l'$, whence

$$b + \sigma(b) + \sigma^2(b) = \mathrm{Tr}_{l/k}(b) = \langle b, e \rangle \tag{6.4}$$

and

$$b\sigma(b)\sigma^2(b) = N_{l/k}(b) = \det(b).$$

Using part (ii) of Lemma 5.2.1 we obtain the last formulas of the lemma. We also see that $b \times b = \sigma(b)\sigma^2(b)$, from which we obtain the first formula.   □

$A$ and $F$ are vector spaces over $l$; recall that $A = l \oplus F$. To prepare for a structure of twisted composition algebra over $l$ on $F$, we define the maps $N : F \times F \to l$ and $f : F \times F \to F$ by

$$x \times y = -\frac{1}{2}N(x, y) + f(x, y) \qquad (x, y \in F). \tag{6.5}$$

**Lemma 6.2.2** *N is a nondegenerate symmetric l-bilinear form on F.*

Proof. Symmetry and $k$-bilinearity are clear. For $b \in l$ and $x, y \in F$ we have:

$$\langle b.x, y \rangle = -2\langle b \times x, y \rangle = -2\langle b, x \times y \rangle = \langle b, N(x, y) \rangle.$$

Using this repeatedly, we derive for $b, c \in l$ and $x, y \in F$:

$$\langle c, N(b.x, y) \rangle = \langle cb.x, y \rangle = \langle cb, N(x, y) \rangle = \langle c, bN(x, y) \rangle.$$

This implies that $N(b.x, y) = bN(x, y)$, so $N$ is a symmetric $l$-bilinear form. If $x \in F$ satisfies $N(x, y) = 0$ for all $y \in F$, then

$$\langle x, y \rangle = \langle e.x, y \rangle = \langle e, N(x, y) \rangle = 0 \qquad (y \in F) \tag{6.6}$$

and hence $x = 0$. Thus, $N$ is nondegenerate. $\qquad\qquad\qquad\qquad$ $\square$

From (6.4) and (6.6) we infer that

$$\langle x, y \rangle = \mathrm{Tr}_{l/k}(N(x, y)) \qquad (x, y \in F). \tag{6.7}$$

Put $N(x) = \frac{1}{2}N(x, x)$.

Now we focus attention on the component $f(x, y)$ of $x \times y$ in $F$. .

**Proposition 6.2.3** *With squaring operation*

$$x^{*2} = f(x, x) \qquad (x \in F)$$

*and norm $N$, $F$ is a twisted composition algebra over $l$ (not necessarily normal).*

Proof. (a) It is obvious that $f$ is symmetric and $k$-bilinear. Linearizing (ii) in Lemma 5.2.1, we find:

$$b(x \times y) + x(b \times y) + y(b \times x) = 3\langle x, y, b \rangle e \qquad (b \in l, \ x, y \in F).$$

The right hand side equals $\langle x \times y, b \rangle e$, in which we replace $x \times y$ by its component in $l$. Thus we get

$$b(x \times y) + x(b \times y) + y(b \times x) = -\frac{1}{2}\langle N(x, y), b \rangle e \qquad (b \in l, \ x, y \in F).$$

From this formula we derive by equating the components in $F$ on either side and using (6.2), (6.5) and Lemma 5.2.1 (i):

$$f(b.x, y) + f(x, b.y) = (\langle b, e \rangle - b).f(x, y) \qquad (x, y \in F, \ b \in l). \tag{6.8}$$

Using the symmetry of $f$ we conclude that

$$f(x, b.x) = \frac{1}{2}(\langle b, e \rangle - b)f(x, x).$$

Now (6.8) gives

$$f(b.x, b.x) = -f(x, b^2.x) + (\langle b, e \rangle - b)f(x, b.x) =$$

$$\frac{1}{2}(-\langle b^2, e \rangle + b^2 + (\langle b, e \rangle - b)^2)f(x, x).$$

From (6.2) we see that $\mathrm{Tr}_{l/k}(b) - b = \sigma(b) + \sigma^2(b)$, and similarly for $b^2$. Inserting these formulas we obtain

$$f(b.x, b.x) = \sigma(b)\sigma^2(b)f(x, x) \quad (b \in l, \ x \in F).$$

This shows that the squaring operation defined by $x^{*2} = f(x, x)$ for $x \in F$ satisfies condition (i) of Def. 4.2.1.

(b) It is obvious that condition (ii) of Def. 4.2.1 is fulfilled, for

$$(x + y)^{*2} - x^{*2} - y^{*2} = 2f(x, y) \quad (x, y \in F)$$

with $f$ as in (6.5), and this is $k$-bilinear.

(c) By Lemma 5.2.1 (iv), we know that

$$(x \times x) \times (x \times x) = \det(x)x. \tag{6.9}$$

On the other hand, we infer from (6.5) that

$$x \times x = -N(x) + x^{*2}. \tag{6.10}$$

This yields

$$(x \times x) \times (x \times x) = N(x) \times N(x) + N(x).x^{*2} - N(x^{*2}) + (x^{*2})^{*2}. \tag{6.11}$$

In (6.9) the right hand side has zero component in $l$, so the same must hold in (6.11). Thus we find with Lemma 6.2.1,

$$N(x^{*2}) = N(x) \times N(x) = \sigma(N(x))\sigma^2(N(x)) \quad (x \in F),$$

which proves (iii) in Def. 4.2.1.

(d) Using Lemma 5.2.1 (i) and (ii) and equation (6.10), we compute:

$$x \times (x \times x) = \det(x) + \frac{1}{2}\langle N(x), e \rangle x - \frac{1}{2}\langle x, x^{*2} \rangle. \tag{6.12}$$

On the other hand, we find with (6.10) and (6.5):

$$x \times (x \times x) = x \times (-N(x) + x^{*2})$$
$$= \frac{1}{2}(N(x).x - N(x, x^{*2})) + f(x, x^{*2}). \tag{6.13}$$

The $l$-component on the right hand side of equation (6.12) is in $k$, so the same must hold for (6.13), i.e., $N(x, x^{*2}) \in k$. This proves (iv) of Def. 4.2.1.    □

Let $u$ be a nonzero element of $l$. So $\lambda = \det(u) = N_{l/k}(u) \neq 0$. Then the isotope $A_u$ is defined. see § 5.9. By (5.58) and (5.61) the product on $A_u$ is given by

$$x._u y = -2\lambda((x \times y) \times (u \times u)) + \frac{1}{2}\lambda^{-1}\langle x, u \times u \rangle y + \frac{1}{2}\lambda^{-1}\langle y, u \times u \rangle x \quad (x, y \in A).$$
(6.14)

Notice that $\lambda^{-1}(u \times u) = u^{-1}$.

For $x, y \in l$ we obtain from part (iii) of Lemma 5.2.1 in a straightforward manner that $x._u uy = u^{-1}(xy)$. It follows that $a$ also generates in the algebra $A_u$ the vector space $l$. Moreover, we see from (5.54) that the orthogonal complement of $l$ relative to $\langle \ , \ \rangle$ is again $F$. We now have on $F$ a new structure of $l$-module

$$(b, x) \mapsto p_u(b)x = -2b \times_u x \quad (b \in l, x \in F).$$

Using the bilinear version of (5.61) we see that this equals $u^{-1}.(b.x) = (u^{-1}b).x$. Likewise, we find for the bilinear function $f_u$ associated to the J-algebra $A_u$ that

$$f_u(x, y) = u^{-1}.f(x, y).$$

Let $F$ be as in Prop. 6.2.3. The preceding results prove imply the following result.

**Proposition 6.2.4** *The twisted composition algebra associated to the isotope $A_u$ is the isotope $F_{u^{-1}}$ of $F$.*

Isotopes of twisted composition algebras were defined in § 4.2.

# 6.3 From Twisted Composition Algebras to J-algebras

Consider a twisted composition algebra $F$ over a cubic extension field $l$ of $k$; let $l'$ be the normal closure of $l$ over $k$ as in the previous section, etc. We construct from $F$ a J-algebra, which will turn out to be proper. Denote the multiplication of elements of $F$ by elements of $l$ by a dot, so $\lambda.x$ for $\lambda \in l$ and $x \in F$; let $N$ be the norm on $F$ and $N(\ ,\ )$ the associated $l$-bilinear form. Take $A = l \oplus F$ as a vector space over $k$. On $l$ we define the quadratic form

$$Q(\lambda) = \frac{1}{2}\mathrm{Tr}_{l/k}(\lambda^2) \quad (\lambda \in l),$$
(6.15)

with associated bilinear form

$$\langle \lambda, \mu \rangle = \mathrm{Tr}_{l/k}(\lambda\mu) \quad (\lambda, \mu \in l).$$
(6.16)

Since $1$, $\sigma$ and $\sigma^2$ are linearly independent over $l'$ by Dedekind's Theorem, $Q$ is nondegenerate. In accordance with (6.7), $Q$ is extended to a nondegenerate quadratic form on $A$ by defining

$$Q(\lambda + x) = Q(\lambda) + \langle N(x), e \rangle$$
$$= \mathrm{Tr}_{l/k}(\tfrac{1}{2}\lambda^2 + N(x)) \qquad (\lambda \in l,\ x \in F). \qquad (6.17)$$

The associated bilinear form is

$$\langle \lambda + x, \mu + y \rangle = \mathrm{Tr}_{l/k}(\lambda\mu + N(x,y)) \qquad (\lambda, \mu \in l,\ x, y \in F). \qquad (6.18)$$

Define the $k$-bilinear

$$f : F \times F \to F,\ f(x,y) = \tfrac{1}{2}((x+y)^{*2} - x^{*2} - y^{*2}).$$

The cross product $\times$ in $A$ is defined by

$$(\lambda + x) \times (\mu + y) = \tfrac{1}{2}\{\sigma(\lambda)\sigma^2(\mu) + \sigma^2(\lambda)\sigma(\mu) - N(x,y)\} +$$
$$f(x,y) - \tfrac{1}{2}(\lambda.y + \mu.x), \qquad (6.19)$$

where $\lambda, \mu \in l$, $x, y \in F$ (cf. Lemma 6.2.1 and equations (6.2) and (6.5)). Define the cubic form det on $A$ by

$$3\det(a) = \langle a, a \times a \rangle$$

(cf. the definition of the cross product as given in (5.16)). An easy computation shows that

$$\det(\lambda + x) = \mathrm{N}_{l/k}(\lambda) - \mathrm{Tr}_{l/k}(\lambda N(x)) + T(x) \qquad (\lambda \in l,\ x \in F), \qquad (6.20)$$

with $T(x) = \langle x^{*2}, x \rangle$ as in Def. 4.2.1 (iv). We define the ordinary product on $A$ as is to be expected from Lemma 5.2.1 (i):

$$ab = a \times b + \tfrac{1}{2}\langle a, e \rangle b + \tfrac{1}{2}\langle b, e \rangle a + \tfrac{1}{2}\langle a, b \rangle e - \tfrac{1}{2}\langle a, e \rangle\langle b, e \rangle e. \qquad (6.21)$$

A straightforward computation yields:

$$(\lambda + x)(\mu + y) = \lambda\mu + \tfrac{1}{2}\{\sigma(N(x,y)) + \sigma^2(N(x,y))\} +$$
$$f(x,y) + \tfrac{1}{2}\{(\sigma(\lambda) + \sigma^2(\lambda)).y + (\sigma(\mu) + \sigma^2(\mu)).x\}. \qquad (6.22)$$

Notice that $\lambda x = \lambda.x$ for $\lambda \in k$ and $x \in F$, but this equality need not hold with arbitrary $\lambda \in l$. The multiplication

$$l \times F \to F,\ (\lambda, x) \mapsto \lambda x$$

does not define a structure of vector space over $l$ on $F$, since $\lambda(\mu x)$ and $(\lambda\mu)x$ are not equal in general.

The cross product and the ordinary product on $A$ are both commutative and $k$-bilinear, and the identity element $e$ of $l$ and $k$ is also identity element for the product in $A$. Thus, $A$ with the ordinary product is a commutative, not necessarily associative, algebra over $k$. We will show that it is a proper J-algebra.

**Definition 6.3.1** For a cubic extension $l$ of $k$ and a twisted composition algebra $F$ over $l$, the $k$- algebra $l \oplus F$ provided with the product as in (6.21) together with the quadratic form $Q$ as defined by (6.17) is denoted by $A(l, F)$ and will be called the *J-algebra associated with $l$ and $F$*.

**Theorem 6.3.2** *If $l$ is a cubic extension of the field $k$ and $F$ a twisted composition algebra over $l$, the algebra $A(l, F)$ is a proper J-algebra over $k$. Every J-algebra over $k$ that contains an element $a$ such that $k[a]$ is a cubic field extension of $k$ is of the form $A(l, F)$, viz., with $l = k[a]$ and $F = k[a]^\perp$.*

*$A(l, F)$ is reduced as a J-algebra if and only if $F$ is reduced as a twisted composition algebra.*

Proof. We first verify the conditions (5.4) - (5.6) for $A(l, F)$; we start with a purely technical result.

(a) If $\mathrm{Tr}_{l/k}(\lambda) = 0$, then $\mathrm{Tr}_{l/k}(\lambda^4) = 2\,\mathrm{Tr}_{l/k}(\lambda^2 \sigma(\lambda)^2)$. This is easily verified by computing the fourth power of $(\lambda + \sigma(\lambda) + \sigma^2(\lambda))$ and equating this to 0.

(b) To prove (5.4), consider $a = \lambda + x$ $(\lambda \in l,\ x \in F)$ with $\langle a, e \rangle = 0$, i.e. with $\mathrm{Tr}_{l/k}(\lambda) = 0$. It is straightforward to compute $Q(a^2)$ and $Q(a)^2$ and to verify that these are equal, using $\mathrm{Tr}_{l/k}(\lambda) = 0$ and the result of (a).

(c) For $a = \lambda + x$, $b = \mu + y$ and $c = \nu + z$ $(\lambda, \mu, \nu \in l,\ x, y, z \in F)$ we find after some computing:

$$\langle ab, c \rangle = \mathrm{Tr}_{l/k}(\dagger),$$

where $\dagger$ stands for

$$\lambda\mu\nu + \frac{1}{2}\{(\sigma(\nu) + \sigma^2(\nu))N(x, y) + (\sigma(\lambda) + \sigma^2(\lambda))N(y, z) +$$

$$(\sigma(\mu) + \sigma^2(\mu))N(x, z)\} + N(f(x, y), z).$$

If we extend $F$ to a normal twisted composition algebra over $l'$ (cf. Prop. 4.2.2), we get

$$N(f(x, y), z) = \frac{1}{2}N(x * y, z) + \frac{1}{2}N(y * x, z).$$

By Def. 4.1.1 (iii), $N(x*y, z) = \sigma(N(y*z, x))$, so $\mathrm{Tr}_{l/k}(N(x*y, z))$ is invariant under cyclic permutations of $x,y$ and $z$, and similarly for $\mathrm{Tr}_{l/k}(N(y * x, z))$. Hence $\langle ab, c \rangle$ is invariant under cyclic permutations of $a,b$ and $c$. This proves (5.5).

(d) From the definition of $Q$ in (6.17) it is immediate that $Q(e) = \frac{3}{2}$, as required in (5.6). This completes the proof that $A(l, F)$ is a J-algebra.

(e) We now want to prove that $A(l, F)$ is proper. To this end, we consider

$$A(l, F)_l = l \otimes_k A(l, F) = l \otimes_k l \oplus l \otimes_k F.$$

This is an algebra over $l$ with $l$ acting on the first factor of the tensor product. The action of $\sigma$ on $l$ is extended to $l \otimes_k l$ as $1 \otimes_k \sigma$. There are three orthogonal primitive idempotents $\varepsilon_1, \varepsilon_2, \varepsilon_3$ in $l \otimes_k l$, which are permuted cyclically by $\sigma$: $\sigma(\varepsilon_i) = \varepsilon_{i-1}$ (indices mod 3).

We consider the product $l \times F \to F$, $(\lambda, x) \mapsto \lambda.x$ as a $k$-bilinear transformation and extend it to an $l$-bilinear product on $(l \otimes_k l) \times (l \otimes_k F)$, also denoted by a dot (.). Similarly, we consider the $l$-bilinear form $N(\,,\,)$ associated with the norm $N$ of the (not necessarily normal) twisted composition algebra $F$ as a $k$-bilinear form and extend it to an $l$-bilinear form on $l \otimes_k F$, denoted by $N(\,,\,)$; this form is nondegenerate.

For $\alpha \in l \otimes_k l$ define the $l$-linear transformation

$$t_\alpha : l \otimes_k F \to l \otimes_k F, \quad x \mapsto \alpha x.$$

From (6.22) one infers that $t_\alpha(x) = \frac{1}{2}(\sigma(\alpha) + \sigma^2(\alpha)).x$ $\left(x \in l \otimes_k F\right)$. Since $N(\,,\,)$ is $l$-bilinear on $F$, $t_\alpha$ is symmetric for all $\alpha \in 1 \otimes_k l$, for

$$N(t_\alpha(x), y) = N(\frac{1}{2}(\sigma(\alpha) + \sigma^2(\alpha)).x, y)$$

$$= N(x, \frac{1}{2}(\sigma(\alpha) + \sigma^2(\alpha)).y)$$

$$= N(x, t_\alpha(y)) \qquad (x, y \in F).$$

Hence $t_\alpha$ is symmetric on $l \otimes_k F$ for all $\alpha \in l \otimes_k l$.

In particular,

$$t_{\varepsilon_i}(x) = \frac{1}{2}(\varepsilon_{i+1} + \varepsilon_{i+2}).x \qquad \left(x \in l \otimes_k F\right)$$

(indices mod 3). Each $t_{\varepsilon_i}$ is a symmetric linear transformation with $t_{\varepsilon_i}^2 = \frac{1}{2}t_{\varepsilon_i}$, so $l \otimes_k F$ is the direct sum of eigenspaces of $t_{\varepsilon_i}$ with eigenvalues 0 and $\frac{1}{2}$. Since $t_{\varepsilon_i}$ and $t_{\varepsilon_j}$ commute, they leave each other's eigenspaces invariant. As $t_{\varepsilon_1} + t_{\varepsilon_2} + t_{\varepsilon_3} = \mathrm{id}$ and $t_{\varepsilon_1}t_{\varepsilon_2}t_{\varepsilon_3} = 0$, the eigenspaces with eigenvalue 0 have dimension over $l$ equal to $\dim_l F$ and the eigenspaces with eigenvalue $\frac{1}{2}$ have dimension $2\dim_l F$.

We pick the idempotent $u = \varepsilon_1$ in $A(l, F)_l$. By the preceding argument, $\dim_l E_0 > 0$ and $\dim_l E_1 > 0$. It follows that $A(l, F)_l$, hence also $A(l, F)$ itself, is proper (see Th. 5.4.5 and its proof).

(f) By Th. 5.5.1, $A(l, F)$ is reduced if and only if there exists $\lambda + x \neq 0$ $\left(\lambda \in l,\right.$ $x \in F)$ such that $(\lambda + x) \times (\lambda + x) = 0$. By (6.19), the latter is equivalent to

$$x^{*2} = \lambda.x \quad \text{and} \quad N(x) = \sigma(\lambda)\sigma^2(\lambda). \qquad (6.23)$$

So, if $A(l, F)$ is reduced, then $F$ is reduced by Th. 4.2.10.

Conversely, if $F$ is reduced, pick a nonzero $x \in F$ such that $x^{*2} = \lambda.x$ for some $\lambda \in l$. By condition (iii) in Def. 4.2.1,

$$\sigma(N(x))\sigma^2(N(x)) = \lambda^2 N(x). \qquad (6.24)$$

By action of $\sigma$ and $\sigma^2$, respectively, on this equation we get two equations:

$$\sigma^2(N(x))N(x) = \sigma(\lambda^2)\sigma(N(x)),$$
$$N(x)\sigma(N(x)) = \sigma^2(\lambda^2)\sigma^2(N(x)).$$

Multiplying these two equations we find:

$$N(x)^2\sigma(N(x))\sigma^2(N(x)) = \left(\sigma(\lambda)\sigma^2(\lambda)\right)^2\sigma(N(x))\sigma^2(N(x)).$$

If $N(x) \neq 0$, this implies $N(x) = \pm\sigma(\lambda)\sigma^2(\lambda)$. From $N(x) = -\sigma(\lambda)\sigma^2(\lambda)$ it would follow that

$$\sigma(N(x))\sigma^2(N(x)) = \lambda^2\sigma(\lambda)\sigma^2(\lambda) = -\lambda^2 N(x),$$

which contradicts (6.24). So $N(x) = \sigma(\lambda)\sigma^2(\lambda)$ if $x^{*2} = \lambda.x$ and $N(x) \neq 0$, hence $x$ satisfies (6.23) and therefore $A(l, F)$ is reduced.

Now let $x^{*2} = \lambda.x$ and $N(x) = 0$. As we saw in step (a) of the proof of Th. 4.1.10, either $x^{*2} = 0$, hence $x$ satisfies (6.23) with $\lambda = 0$, or for $y = x^{*2} \neq 0$ we have $y^{*2} = 0$ and $N(y) = 0$, so $y$ (instead of $x$) satisfies (6.23) with $\lambda = 0$. Again we conclude that $A(l, F)$ is reduced.    $\square$

If the J-algebra $A$ with $\dim_k A > 1$ is not reduced, it certainly contains a cubic extension field $l = k[a]$ of $k$. By the above theorem, $A$ is of the form $A(l, F)$ and therefore proper. Thus we have found:

**Corollary 6.3.3** *If a J-algebra $A$ is not reduced, then it is proper. In particular, a nonreduced J-algebra of dimension 27 is an Albert division algebra.*

The above theorem implies that if a field $k$ has the property that every twisted composition algebra over a cubic extension of $k$ is reduced, then every J-algebra over $k$ is reduced. For fields $k$ with this property, see, e.g., Th. 4.8.3. Examples are the algebraic number fields. We have seen at the end of 4.8 that every twisted composition algebra of dimension 8 over a cubic extension of an algebraic number field is reduced. Together with the above theorem this gives the result we announced in (v) at the end of Ch. 5:

**Corollary 6.3.4** *Over an algebraic number field every Albert algebra is reduced.*

## 6.4 Historical Notes

As remarked in the historical note to Ch. 4, the use of twisted composition algebras for the description of proper J-algebras stems from T.A. Springer (see [Sp 63]). With this device, Springer managed to prove that every Albert algebra over an algebraic number field is reduced, a result originally proved in a quite different way by A.A. Albert (see [Al 58, Th. 10]; in that paper, Albert also proved that Albert algebras over real closed fields are reduced).

# 7. Exceptional Groups

In this chapter we identify two algebraic groups associated with Albert algebras. We first determine the automorphism group; this will be shown to be an exceptional simple algebraic group of type $F_4$. Then we study the group of transformations that leave the cubic form det invariant and show that this is a group of type $E_6$.

As in the previous two chapters, all fields are supposed to have characteristic $\neq 2, 3$. We mainly deal with Albert algebras, but several results hold, more generally, for proper J-algebras.

## 7.1 The Automorphisms Fixing a Given Primitive Idempotent

In this section we study $\mathrm{Aut}(A)_u$, the group of automorphisms of a proper reduced J-algebra $A$ that fix a given primitive idempotent $u$. For some results, we will have to restrict to reduced Albert algebras. Notations being as in § 5.3, an automorphism $s$ that fixes $u$ must leave invariant the zero space $E_0$ and the half space $E_1$ defined by $u$. Since $s$ is orthogonal, it induces orthogonal transformations $t$ in $E_0$ and $v$ in $E_1$; we will see in Th. 7.1.3 that $t$ must even be a rotation. The fact that $s$ is an automorphism implies that $t$ and $v$ satisfy the relation

$$v(xy) = t(x)v(y) \qquad (x \in E_0, \ y \in E_1).$$

This situation will first be analyzed. That analysis will lead to Th. 7.1.3, which identifies $\mathrm{Aut}(A)_u$ as the spin group of the restriction of $Q$ to $E_0$.

**Proposition 7.1.1** (i) Let $A$ be a proper reduced J-algebra. Let $u$ be a primitive idempotent and $E_0$ and $E_1$ the zero and half space, respectively, of $u$ in $A$. For every rotation $t$ of $E_0$ there exists a similarity $v$ of $E_1$ such that

$$v(xy) = t(x)v(y) \qquad (x \in E_0, \ y \in E_1). \tag{7.1}$$

If $t = s_{a_1} s_{a_2} \cdots s_{a_{2h}}$ for certain $a_i \in E_0$, then one may take for $v$ the following map:

$$v(y) = a_1(a_2(\cdots(a_{2h}y)\cdots)) \qquad (y \in E_1).$$

*(ii) Assume that $A$ is an Albert algebra. Then for any rotation $t$ of $E_0$ the similarity $v$ of $E_1$ such that equation (7.1) is satisfied is unique up to multiplication by a nonzero scalar, and the square class of the multiplier of $v$ equals the spinor norm of $t$: $\nu(v) = \sigma(t)$. Moreover, if $t$ is an orthogonal transformation of $E_0$ which is not a rotation, then there does not exist a similarity $v$ of $E_1$ satisfying (7.1).*

Proof. This result is somewhat similar to the Principle of Triality, and the proof resembles the proof we gave for that Principle in Th. 3.2.1.

First consider a reflection $s_a : x \mapsto x - Q(a)^{-1}\langle x, a \rangle a$ in $E_0$. We have

$$
\begin{aligned}
a(xy) &= -x(ay) + \frac{1}{4}\langle a, x \rangle y &&\text{(by Lemma 5.3.3 (ii))} \\
&= -(x - Q(a)^{-1}\langle a, x \rangle a)(ay) &&\text{(by Lemma 5.3.3 (i))} \\
&= -s_a(x)(ay) &&\left(x \in E_0,\ y \in E_1\right).
\end{aligned}
$$

From Lemma 5.3.3 (vi) we infer that $y \mapsto ay$ is a similarity with multiplier $\frac{1}{4}Q(a)$.

It is easily seen that if $(t_1, v_1)$ and $(t_2, v_2)$ satisfy equation (7.1), then so does $(t_1 t_2, v_1 v_2)$. Since every rotation $t$ is a product of an even number of reflections, the existence of a similarity $v$ such that (7.1) holds easily follows. The square class of the multiplier of this $v$ evidently equals the spinor norm of $t$.

Assume that $A$ is an Albert algebra. To prove the uniqueness statement of (ii), it suffices to prove that $t = \mathrm{id}$ implies $v = \lambda.\,\mathrm{id}$ for some $\lambda \in k^*$. So let $v$ be a similarity of $E_1$ with

$$
v(xy) = xv(y) \qquad \left(x \in E_0,\ y \in E_1\right).
$$

Now we saw in Cor. 5.3.4 that there is a representation $\varphi$ of the Clifford algebra $\mathrm{Cl}(Q; E_0)$ in $E_1$ such that $\varphi(x)(y) = 2xy$ $(x \in E_0,\ y \in E_1)$. We have $\dim E_0 = 9$, $\dim E_1 = 16$ and $\dim \mathrm{Cl}(Q; E_0) = 2^9$. Over an algebraic closure $K$ of $k$ it is the sum of two full matrix algebras of dimension $2^8$, so the irreducible representations have dimension 16. Hence $\varphi$ must be absolutely irreducible. The endomorphism $v$ of $E_1$ commutes with the representation $\varphi$, so it follows from Schur's Lemma that $v = \lambda.\,\mathrm{id}$ for some $\lambda \in k$.

Finally, let $t$ be an orthogonal transformation of $E_0$ which is not a rotation and suppose there exists a similarity $v$ of $E_1$ such that (7.1) holds. Write $t = s_a t_1$ with a reflection $s_a$ and a rotation $t_1$, and pick a similarity $v_1$ of $E_1$ such that $t_1^{-1}$ and $v_1$ satisfy (7.1). We saw above that there exists a similarity $v_2$ of $E_1$, viz. $v_2(y) = ay$, such that $v_2(xy) = -s_a(x)v_2(y)$. Then for $w = vv_1v_2$ we find

$$
w(xy) = -xw(y) \qquad \left(x \in E_0,\ y \in E_1\right).
$$

Using again the representation $\varphi$ of the Clifford algebra $\mathrm{Cl}(Q; E_0)$ in $E_1$, we see that $w\varphi(x) = -\varphi(x)w$ for $x \in E_0$. It follows that $w^2$ commutes with the representation $\varphi$, so $w^2 = \lambda^2 \cdot \mathrm{id}$ for some $\lambda \in \bar{k}$. Over $\bar{k}$ we have $E_1 = W_+ \oplus W_-$ with $W_\pm$ the eigenspace of $w$ for the eigenvalue $\pm\lambda$. If $x \in E_0$, then $\varphi(x)$ interchanges the eigenspaces, so they have equal dimension. The even Clifford algebra $\mathrm{Cl}^+(Q; E_0)$ leaves $W_+$ and $W_-$ invariant. In case $\dim C = 8$, $\mathrm{Cl}^+(Q; E_0)$ is a full matrix algebra of dimension $2^8$, so its irreducible representations have dimension 16. This contradicts the fact that the restriction of $\varphi$ to $\mathrm{Cl}^+(Q; E_0)$ has invariant subspaces $W_+$ and $W_-$ of dimension 8. □

To prepare the way for Th. 7.1.3 we need a lemma. Notations as before.

**Lemma 7.1.2** *In a proper reduced J-algebra we have for any rotation $t$ of $E_0$ and similarity $v$ of $E_1$ satisfying equation* (7.1),

$$v(y) \circ v(y) = n(v)t(y \circ y) \qquad (y \in E_1).$$

Proof. In the equation $v(xy) = t(x)v(y)$ we replace $x$ by $y \circ y$; using Lemma 5.3.3 (iv), we then find

$$t(y \circ y)v(y) = \frac{1}{4}Q(y)v(y) \qquad (y \in E_1).$$

Assume $Q(y) \neq 0$. With $x_1 = Q(y)^{-1}t(y \circ y)$ we have $Q(x_1) = \frac{1}{4}$ and $x_1 v(y) = \frac{1}{4}v(y)$, so $v(y) \in E_+$ if we decompose $E_1$ in $E_+$ and $E_-$ with respect to $x_1$, as in § 5.4. By Lemma 5.4.2 (i),

$$v(y) \circ v(y) = Q(v(y))x_1 = n(v)t(y \circ y).$$

Working over an algebraic closure of $k$, this relation holds on all of $E_1$ by Zariski continuity. □

**Theorem 7.1.3** *Let $A$ be a reduced Albert algebra over $k$, $u$ a primitive idempotent and $E_0$ and $E_1$ the zero and half spaces of $u$ in $A$. The restriction mapping*

$$\mathrm{Res}_{E_0} : s \mapsto s|_{E_0} \qquad (s \in \mathrm{Aut}(A)_u)$$

*is a homomorphism of $\mathrm{Aut}(A)_u$ onto the reduced orthogonal group $\mathrm{O}'(Q; E_0)$ with kernel of order two.*

Proof. Let $s$ be an automorphism of $A$ with $su = u$. As we remarked at the beginning of this section, $s$ leaves $E_0$ and $E_1$ invariant, and induces orthogonal transformations $t$ and $v$ in $E_0$ and $E_1$, respectively. Since $s$ is an automorphism, $t$ and $v$ satisfy the relation (7.1). This implies by Prop. 7.1.1 that $t$ is a rotation. Since $v$ is orthogonal, $n(v) = 1$ and hence $\sigma(t) = 1$. Thus we have a homomorphism

$$\mathrm{Res}_{E_0} : \mathrm{Aut}(A)_u \to \mathrm{O}'(Q; E_0), \ s \mapsto s|_{E_0}.$$

Conversely, given a rotation $t$ of $E_0$ with $\sigma(t) = 1$, we choose a similarity $v$ of $E_1$ such that (7.1) holds. Since $\nu(v) = \sigma(t) = 1$, we may choose $v$ such that $n(v) = 1$, so $v$ is an orthogonal transformation. Define the linear transformation $s : A \to A$ by

$$s(e) = e, \quad s(u) = u, \quad s|_{E_0} = t \quad \text{and} \quad s|_{E_1} = v.$$

With Lemma 7.1.2 one easily verifies that $s(z^2) = s(z)^2$ for $z \in A$; since the product in $A$ is commutative, $s$ is an automorphism of $A$. This proves surjectivity of $\text{Res}_{E_0} : \text{Aut}(A)_u \to O'(Q; E_0)$.

If $t = \text{id}$, then $v = \lambda.\text{id}$ by Prop. 7.1.1; but since $v$ is orthogonal in the present situation, $v = \pm \text{id}$. Hence the kernel of $\text{Res}_{E_0}$ consists of two elements.     $\square$

In Prop. 7.1.1 we found a relation between spinor norms of rotations in $E_0$ and square classes of multipliers of similarities in $E_1$. We are now going to show that the group of spinor norms in $E_0$ coincides with the group of square classes of multipliers in $E_1$. First a lemma.

**Lemma 7.1.4** *Let $A$ be a proper reduced J-algebra. For any $y, z \in E_1$ satisfying $Q(y)Q(z) \neq 0$ there exist elements $a_1, a_2, \ldots, a_l \in E_0$ such that*

$$z = a_1(a_2(\cdots (a_l y) \cdots)).$$

*It is always possible to do this with an even number $l$ of multiplications.*

Proof. The notations are as in § 5.4. We first consider the case that $y \circ y = \lambda z \circ z$ for some $\lambda \in k^*$; this implies that $Q(y)^2 = \lambda^2 Q(z)^2$, so $Q(y) = \pm \lambda Q(z)$. Take $x_1 = Q(z)^{-1} z \circ z$. Then $Q(x_1) = \frac{1}{4}$ and $x_1 z = \frac{1}{2} z$, so $z \in E_+$. Further, $x_1 y = \pm \frac{1}{4} y$, so $y \in E_\pm$. If $y \in E_-$, there is $c \in C$ such that $z = cy$ (by Lemma 5.4.3, with $a_- = y$ and $a_+ = z$). If $y \in E_+$, we pick any $c \in C$ with $Q(c) \neq 0$, then $y' = cy \in E_-$, so we can find $c' \in C$ such that $c'y' = z$.

Now assume that $y \circ y$ and $z \circ z$ are linearly independent. For $\alpha = Q(y)Q(z)^{-1}$ we have $Q(y \circ y) = \alpha^2 Q(z \circ z) = Q(\alpha z \circ z)$. By Witt's Theorem there exists an orthogonal transformation $t$ of $E_0$ such that $t(y \circ y) = \alpha z \circ z$; since $\dim(E_0) > 1$, we may assume that $t$ is a rotation. Write $t$ as a product of reflections, $t = s_{a_1} s_{a_2} \cdots s_{a_l}$. By Prop. 7.1.1, the similarity

$$v : E_1 \to E_1, \quad w \mapsto a_1(a_2(\cdots (a_l w) \cdots))$$

satisfies relation (7.1). By Lemma 7.1.2,

$$v(y) \circ v(y) = n(v)t(y \circ y) = \lambda z \circ z.$$

So by the first part of the proof, $v(y)$ can be transformed into $z$ by one or two multiplications by elements of $E_0$.

If in this way we end up with an odd number of multiplications, we can multiply in addition by $4x_1$ with $x_1 = Q(z)^{-1} z \circ z$ as in the first part of the proof, since $4x_1 z = z$.     $\square$

**Proposition 7.1.5** *Let $A$ be a proper reduced $J$-algebra, $u$ a primitive idempotent and $E_0$ and $E_1$ the zero and half spaces, respectively, of $u$ in $A$. The group of spinor norms of rotations of $E_0$ with respect to $Q|_{E_0}$ coincides with the group of square classes of multipliers of similarities of $E_1$ with respect to $Q|_{E_1}$.*

Proof. In view of Prop. 7.1.1 it suffices to show that for every similarity $v$ of $E_1$ there exists a rotation $t$ of $E_0$ such that $\sigma(t) = \nu(v)$. Pick $y \in E_1$ with $Q(y) \neq 0$. Applying the preceding lemma to $y$ and $z = v(y)$, we obtain an even number of elements $a_1, a_2, \ldots, a_l$ of $E_0$ such that

$$v(y) = a_1(a_2(\cdots (a_l y) \cdots)).$$

Then $n(v) = Q(a_1)Q(a_2)\cdots Q(a_l)$. So for the rotation

$$t = s_{a_1} s_{a_2} \cdots s_{a_l}$$

one has indeed $\sigma(t) = \nu(v)$.  □

More explicitly, this proposition says that the group of spinor norms of the quadratic form $\xi^2 + \alpha N(c)$ in dim $C + 1$ variables coincides with the group of square classes of multipliers of similarities with respect to the quadratic form $N(c) + \alpha N(c')$ in $2 \dim C$ variables (for those $\alpha \in k^*$ whose norm class $\kappa(\alpha)$ is the norm class of a primitive idempotent $u \in A$; cf. § 5.7).

We next discuss the algebraic groups occurring of the situation of the present section. As in previous chapters we denote algebraic groups by boldface letters. Let again $K$ denote an algebraic closure of $K$ and put $A_K = K \otimes_k A$ etc. View $A$ as a subset of $A_K$.

The group $\mathrm{Aut}(A_K)_u$ is an algebraic group, denoted by $\mathbf{G}_u$. Further, we have the spin group $\mathbf{Spin}(Q; E_0)$.

**Proposition 7.1.6** $\mathbf{G}_u$ *is isomorphic to* $\mathbf{Spin}(Q, E_0)$.

Proof. The spin group $\mathbf{Spin}(Q; E_0)$ is the subgroup of the even Clifford algebra $\mathrm{Cl}^+(Q; E_0)_K$ consisting of the products $s = a_1 \circ \ldots \circ a_{2h}$ with $a_i \in E_0$, $Q(a_i) = 1$, see § 3.1. For such an $s$ define $\psi(s) \in \mathbf{G}_u$ to be the linear map of $A_K$ fixing $e$, $u$, $E_0$ and $E_1$, such that $\psi(s)|_{E_0} = s_{a_1} \cdots s_{a_{2h}}$ and

$$\psi(s)|_{E_1}(y) = a_1(a_2(\cdots (a_{2h}y)\cdots)) \qquad (y \in E_1).$$

Then $\psi$ is a homomorphism of algebraic groups. It follows from Th. 7.1.3 that it is bijective.

Let $\pi : \mathbf{Spin}(Q; E_0) \to \mathbf{SO}(Q; E_0)$ be the canonical homomorphism. Then $\pi = \mathrm{Res}_{E_0} \circ \psi$. Since the characteristic is not 2, $\pi$ is a separable homomorphism. Hence the Lie algebra homomorphism $d\pi$ is bijective (see [Sp 81, 4.3.7 (ii)]). But then $d\psi$ also must be bijective. It follows that $\psi$ is an isomorphism (see [loc.cit., 5.3.3]).  □

**Corollary 7.1.7** *The Lie algebra* $L = L(\mathbf{G}_u)$ *is the space of derivations $d$ of* $A_K$ *with $du = 0$.*

Proof. By [Hu, p. 77], $L$ is contained in the space $S$ of these derivations (cf. 2.4.5). If $d \in S$ then $de = du = 0$, and $d(E_i) \subset E_i$ $(i = 0, 1)$. Put $d_i = d|_{E_i}$. From Lemma 5.3.2 we see that

$$\langle d_i(x), y \rangle + \langle x, d_i(y) \rangle = 0$$

for $x, y \in E_i$. In particular, for $x \in E_0$ we have that $d_0(x)$ is skew symmetric relative to $Q|_{E_0}$. Using [Sp 81, 7.4.7 (3)] it follows that $d \mapsto d_0$ is a linear map of $S$ to $L(\mathbf{SO}(Q; E_0))$. If $x \in E_0$ and $d_0(x) = 0$ then $d_1(xy) = xd_1(y)$ for $y \in E_1$. Arguing as in the proof of the uniqueness part of Prop. 7.1.1 we see that $d_1 = \lambda \, \mathrm{id}$. But since $\lambda \langle y, y \rangle = \langle y, d_1(y) \rangle = 0$ $(y \in E_1)$ we must have $\lambda = 0$. Hence the map $d \mapsto d_0$ of $S$ is injective. It follows that

$$\dim L \leq \dim S \leq \dim L(\mathbf{SO}(Q; E_0)) = \dim \mathbf{SO}(Q; E_0) = \dim \mathbf{G}_u = \dim L.$$

It follows that $S = L$, as asserted.  $\square$

## 7.2 The Automorphism Group of an Albert Algebra

We are now going to show that the automorphisms of an Albert algebra form an exceptional simple algebraic group of type $F_4$. Let $A$ be an Albert algebra over a field $k$, and $K$ an algebraic closure of $k$. We keep the notations of the previous section. Then $\mathbf{G} = \mathrm{Aut}(A_K)$ is an algebraic group.

**Theorem 7.2.1** $\mathbf{G}$ *is a connected simple algebraic group of type $F_4$ which is defined over $k$.*

Proof. (a) We first show that $\mathbf{G}$ is a connected algebraic group of dimension 52. We do this by considering its action on the variety $V$ of primitive idempotents in $A_K$. By Cor. 5.8.2, this action is transitive. We claim that $V$ is an irreducible variety of dimension 16. Consider the orthogonal primitive idempotents $u_1 = u$, $u_2 = \frac{1}{2}(e - u) + x_1$ and $u_3 = \frac{1}{2}(e - u) - x_1$ (cf. the beginning of the proof of Th. 5.4.5), and let $V_i$ be the Zariski open subset of $V$ defined by

$$V_i = \{ t \in V \,|\, \langle t, u_i \rangle \neq 0 \}.$$

$V_1$ consists of the primitive idempotents of type (i) in Prop. 5.5.3, i.e., the elements

$$t = (Q(y) + 1)^{-1}(u + \frac{1}{2}Q(y)(e - u) + y \circ y + y) \qquad (y \in E_1, \ Q(y) \neq -1).$$

Since the $y \in E_1$ with $Q(y) \neq -1$ form a Zariski open set in $E_1$, $V_1$ is an irreducible variety of dimension 16. Similarly for $V_2$ and $V_3$. If $t \in V$, $t \notin V_1$, it is of type (ii) as in Prop. 5.5.3:

$$t = \frac{1}{2}(e - u) + x + y \qquad \left(x \in E_0, \ Q(x) = \frac{1}{4}, \ y \in E_1, \ Q(y) = 0\right).$$

Then $\langle t, u_2 \rangle = \frac{1}{2} + \langle x_1, x \rangle$ and $\langle t, u_3 \rangle = \frac{1}{2} - \langle x_1, x \rangle$. At least one of these two must be $\neq 0$, so $t \in V_2$ or $t \in V_3$. Hence $V = V_1 \cup V_2 \cup V_3$. Moreover, we see that $V_2 \cap V_3 \neq \emptyset$, and similarly for $V_1 \cap V_2$ and $V_1 \cap V_3$. Hence $V_i \cap V_j$ is an open dense subset of $V_i$. We conclude that $V_1 \cap V_2 \cap V_3$ is an irreducible open subset of each $V_i$. Hence its closure contains all $V_i$, and must be $V$. Then $V$, being the closure of an irreducible subset, is itself irreducible. As $V = \bar{V_1}$ it has dimension 16.

By Prop. 7.1.6, the stabilizer $\mathbf{G}_u$ of $u$ in $\mathbf{G}$ is isomorphic to $\mathbf{Spin}(Q; E_0)$. The latter is a connected quasisimple algebraic group of type $B_4$, which has dimension 36. It follows that $\mathbf{G}$ is connected (cf. [Sp 81, 5.5.9 (1)] and has dimension 52.

(b) $\mathbf{G}$ leaves $e^{\perp} = K(e - 3u) \oplus E_0 \oplus E_1$ invariant, since it leaves the quadratic form $Q$ invariant (see Prop. 5.3.10); we show that this is an irreducible representation.

The stabilizer $\mathbf{G}_u$ of $u$ in $\mathbf{G}$ leaves $K(e - 3u)$, $E_0$ and $E_1$ invariant. In $E_0$ it induces the rotation group by Th. 7.1.3, so there it acts irreducibly. $\mathbf{G}_u$ induces in $E_1$ the spin representation of $\mathbf{Spin}(Q; E_0)$, which is irreducible. (This irreducibility also follows from Prop. 7.1.1 and Lemma 7.1.4.) So the representation of $\mathbf{G}_u$ in $e^{\perp}$ is the sum of the three inequivalent irreducible representations in $K(e - 3u)$, $E_0$ and $E_1$.

Every $\mathbf{G}$-invariant subspace of $e^{\perp}$ contains a $\mathbf{G}_u$-invariant subspace. Since $\mathbf{G}$ is transitive on the primitive idempotents, it can move $u$ to $u_2$, so $e - 3u$ to $e - 3u_2 = -\frac{1}{2}(e - 3u) - 3x_1$ and $E_1$ to $E_+ + C$ (where $C = x_1^{\perp} \cap E_0$, cf. § 5.4). It follows that the representation of $\mathbf{G}$ in $e^{\perp}$ is irreducible.

(c) Since $\mathbf{G}$ has a faithful irreducible representation, it is reductive (see the proof of Th. 2.3.5). A central element of $\mathbf{G}$ must induce $\lambda.\mathrm{id}$ in $e^{\perp}$ for some $\lambda \in K^*$; since its restriction to $E_0$ is a rotation (cf. Th. 7.1.3) we must have $\lambda = 1$. So $\mathbf{G}$ has trivial center and is semisimple.

(d) Since $\mathbf{G}$ has trivial center, $\mathbf{G} = \mathbf{G}_1 \times \cdots \times \mathbf{G}_s$, each $\mathbf{G}_i$ being simple and $\mathbf{G}_i$ centralizing $\mathbf{G}_j$ for $i \neq j$; see [Sp 81, Th. 8.1.5]. Notice that $\mathbf{G}_i \cap \prod_{j \neq i} \mathbf{G}_j = \mathrm{id}$, since a finite normal subgroup must be central. Let $\pi_i : \mathbf{G} \to \mathbf{G}_i$ be the projection on the $i$-th component.

By Th. 7.1.6, $\mathbf{G}_u$ is a quasisimple algebraic group of type $B_4$, so $\dim \mathbf{G}_u = 36$. We have $\pi_i(\mathbf{G}_u) \neq \mathrm{id}$ for some $i$, say, for $i = 1$. Since $\mathbf{G}_u$ is quasisimple, the kernel of the restriction of $\pi_1$ to $\mathbf{G}_u$ is finite, so $\dim \pi_1(\mathbf{G}_u) = 36$. Hence $\dim \mathbf{G}_1 \geq 36$. Since $\mathbf{G}$ has dimension 52, we must have $\pi_i(\mathbf{G}_u) = \mathrm{id}$ for $i > 1$, so $\mathbf{G}_u \subseteq \mathbf{G}_1$.

$\mathbf{G}_2, \ldots, \mathbf{G}_s$ centralize $\mathbf{G}_1$, hence also $\mathbf{G}_u$. If $t \in \mathbf{G}$ normalizes $\mathbf{G}_u$, then $t(u)$ is fixed under the action of $\mathbf{G}_u$; since $\mathbf{G}_u$ leaves no other idempotents fixed than $u$, it follows that $t(u) = u$. Hence $N(\mathbf{G}_u) = \mathbf{G}_u$. This implies that $\mathbf{G}_i \subseteq \mathbf{G}_u \subseteq \mathbf{G}_1$ for $i > 1$, hence $s = 1$. We conclude that $\mathbf{G}$ is a simple group of dimension 52. A simple algebraic type of classical type has dimension $l^2 - 1$

or $\frac{1}{2}l(l-1)$ for some integer $l$. Such a dimension cannot equal 52, hence $\mathbf{G}$ must be of exceptional type, and can only be of type $F_4$ (see e.g. [Bour, Planches]).

(e) To prove that $\mathbf{G}$ is defined over $k$, we proceed as in the proof of Prop. 2.4.6. In the present case it suffices to show that the Lie algebra $L(\mathbf{G})$ coincides with the space of derivations $D = \mathrm{Der}(A_K)$, or that $\dim D \leq 52$.

Let $u$ be a primitive idempotent of $A_K$ and let $E_1$ be its half space, as usual. If $d \in D$ we have $d(u) = d(u^2) = 2u.d(u)$, which shows that $d(u) \in E_1$. On the other hand, the subspace of $D$ of derivations $d$ with $d(u) = 0$ has dimension 36, by Cor. 7.1.7. Since $\dim E_1 = 16$, we see that $\dim D \leq 16+36 = 52$, as desired. $\hfill\square$

We state explicitly a result mentioned in the proof.

**Corollary 7.2.2** $L(\mathbf{G}) = \mathrm{Der}(A_K)$.

## 7.3 The Invariance Group of the Determinant in an Albert Algebra

$A$ is as in the previous section. In this section we consider the algebraic group $\mathbf{H}$ of linear transformations of $A_K$ leaving det invariant, and prove that this is an exceptional simple algebraic group of type $E_6$. To a large extent the proof goes along the same lines as that of the previous theorem. At the end of the proof, however, we need an extra argument, viz., that $\mathbf{H}$ has an outer automorphism; we first give a proof of that fact.

For a bijective linear transformation $t$ of any J-algebra $A$, define $\tilde{t} : A \to A$ by

$$\langle t(x), \tilde{t}(y) \rangle = \langle x, y \rangle \qquad (x, y \in A),$$

the *contragredient* of $t$. It is clear that $\tilde{\tilde{t}} = t$ and that $\widetilde{tu} = \tilde{t}\tilde{u}$ for bijective linear transformations $t$ and $u$ of $A$.

**Proposition 7.3.1** *Let $A$ be a J-algebra and let $H$ be the group of bijective linear transformations $t$ of $A$ leaving the cubic form* det *invariant. Then $t \in \mathrm{GL}(A)$ lies in $H$ if and only if*

$$t(x) \times t(y) = \tilde{t}(x \times y) \qquad (x, y \in A).$$

*If $t$ lies in $H$ then so does $\tilde{t}$. The mapping*

$$\tilde{\ } : H \to H, \, t \mapsto \tilde{t},$$

*is an outer automorphism of order 2 of $H$.*

Proof. It suffices to give the proof for algebraically closed $k = K$. If $t$ leaves det invariant, then,

$$\langle \tilde{t}(x \times y), t(z) \rangle = \langle x \times y, z \rangle = 3\langle x, y, z \rangle = 3\langle t(x), t(y), t(z) \rangle$$
$$= \langle t(x) \times t(y), t(z) \rangle \quad (x, y, z \in A),$$

so $\tilde{t}(x \times y) = t(x) \times t(y)$. The argument may be reversed, which proves the first statement of the proposition.

If $t \in H$ then

$$\tilde{t}(x \times x) \times \tilde{t}(x \times x) = (t(x) \times t(x)) \times (t(x) \times t(x))$$
$$= \det(t(x))t(x) = \det(x)t(x) \quad (x \in A)$$

by Lemma 5.2.1 (iv). Replacing $x$ by $x \times x$ and using (iv) and (vii) of the same lemma, we arrive at

$$\det(x)^2 \tilde{t}(x) \times \tilde{t}(x) = \det(x \times x)t(x \times x) = \det(x)^2 t(x \times x) \quad (x \in A).$$

Hence $\tilde{t}(x) \times \tilde{t}(x) = t(x \times x)$ if $\det(x) \neq 0$; by Zariski continuity it holds for all $x \in A$. Linearization yields $\tilde{t}(x) \times \tilde{t}(y) = t(x \times y)$ $(x, y \in A)$, which implies that $\tilde{t} \in H$.

So $t \mapsto \tilde{t}$ is an automorphism of $H$. Suppose it were inner. Then there would be $u \in H$ such that $\tilde{t} = utu^{-1}$ for all $t \in H$. Let $\varepsilon$ be a primitive third root of unity in $k$, and take $t = \varepsilon \cdot \mathrm{id}$. Then $\tilde{t} = utu^{-1} = \varepsilon \cdot \mathrm{id}$. But from the definition of $\tilde{t}$ we infer that $\tilde{t} = \varepsilon^{-1} \cdot \mathrm{id}$. Thus we arrive at a contradiction. $\square$

Now we come to the main result of this section.

**Theorem 7.3.2** $H$ *is a connected, quasisimple, simply connected algebraic group of type* $E_6$ *which is defined over* $k$.

Proof. (a) By Cor. 5.4.6 the polynomial function det on $A_K$ is irreducible. Hence

$$W = \{ x \in A_K \mid \det(x) = 1 \}$$

is a 26-dimensional irreducible algebraic variety. $H$ acts on it. For $a \in W$, let $B_a$ be the bilinear form on $A_K$ with $B_a(x, y) = \langle x, y, a \rangle$ for $x, y \in A_K$. This form is nondegenerate, for

$$\langle x, y, a \rangle = 0 \quad (x \in A_K)$$

implies that $y \times a = 0$, hence $y = 0$ by Lemma 5.9.1. Over the algebraically closed field $K$ any two nondegenerate symmetric bilinear forms of the same dimension are equivalent. So by Prop. 5.9.3, $H$ acts transitively on $W$.

(b) The stabilizer $H_e$ of $e$ in $H$ is the automorphism group $G$, by Prop. 5.9.4. By Th. 7.2.1 this is a connected algebraic group of dimension 52. Together with (a) this implies that $H$ is a connected algebraic group of dimension $52 + 26 = 78$.

(c) $\mathbf{H}_e = \text{Aut}(A_K)$, acting in $A_K$, has as irreducible subspaces $Ke$ and $e^\perp$, as we saw in part (b) of the proof of Th. 7.2.1. $\mathbf{H}$ leaves neither of these subspaces invariant, so its faithful representation in $A_K$ is irreducible. Hence $\mathbf{H}$ is reductive (see the proof of Th. 2.3.5). A central element of $\mathbf{H}$ must be of the form $\lambda.\text{id}$ with $\lambda^3 = 1$, so the center of $\mathbf{H}$ has order 3. It follows that $\mathbf{H}$ is semisimple.

(d) To prove that $\mathbf{H}$ is quasisimple we argue as in the proof of part (d) of Th. 7.2.1. There are some complications, however; one is that $\mathbf{H}$ has non-trivial center, so it need not be a direct product of quasisimple groups. Let $\mathbf{H} = \mathbf{H}'/D$, where $\mathbf{H}' = \mathbf{H}_1 \times \cdots \times \mathbf{H}_s$, each $\mathbf{H}_i$ being quasisimple and $\mathbf{H}_i$ centralizing $\mathbf{H}_j$ for $i \neq j$, and where $D$ is a finite central subgroup. Let $\varrho$ be the projection of $\mathbf{H}'$ onto $\mathbf{H}$, and $\pi_i$ that of $\mathbf{H}'$ onto its $i$-th component $\mathbf{H}_i$. Let $\mathbf{H}'_e$ be the identity component of $\varrho^{-1}(\mathbf{H}_e)$. Then $\varrho(\mathbf{H}'_e) = \mathbf{H}_e$ and $\mathbf{H}'_e$ is a simple group of type $F_4$, so $\dim \mathbf{H}'_e = 52$. We have $\pi_i(\mathbf{H}'_e) \neq \text{id}$ for some $i$, say, for $i = 1$. The kernel of the restriction of $\pi_i$ to $\mathbf{H}'_e$ is trivial since $\mathbf{H}'_e$ is simple, so $\dim \pi_1(\mathbf{H}'_e) = 52$. Hence $\dim \mathbf{H}_1 \geq 52$. Since $\mathbf{H}'$ has dimension 78, we must have $\pi_i(\mathbf{H}'_e) = \text{id}$ for $i > 1$, so $\mathbf{H}'_e \subseteq \mathbf{H}_1$ and therefore $\mathbf{H}_e \subseteq \varrho(\mathbf{H}_1)$.

$\mathbf{H}_2, \ldots, \mathbf{H}_s$ centralize $\mathbf{H}_1$, so $\varrho(\mathbf{H}_2), \ldots, \varrho(\mathbf{H}_s)$ centralize $\mathbf{H}_e$. Now consider the normalizer $N(\mathbf{H}_e)$ of $\mathbf{H}_e$ in $\mathbf{H}$. If $t \in \mathbf{H}$ normalizes $\mathbf{H}_e$, then $t(e)$ is fixed under the action of $\mathbf{H}_e$. Since the representation of $\mathbf{H}_e$ in $e^\perp$ is irreducible, the elements of $W$ that are fixed under $\mathbf{H}_e$ lie in $ke \cap W = \{\varepsilon e \,|\, \varepsilon^3 = 1\}$. It follows that the identity component of $N(\mathbf{H}_e)$ is $\mathbf{H}_e$ itself, so $\varrho(\mathbf{H}_i) \subseteq \mathbf{H}_e \subseteq \varrho(\mathbf{H}_1)$ for $i > 1$. This implies that $s = 1$, since $\varrho$ has finite kernel. Hence $\mathbf{H}$ is a quasisimple group of dimension 78. The argument of the proof of Th. 7.2.1 now gives that there are three possible types for such a group, viz., $B_6$, $C_6$ and $E_6$ Since $\mathbf{H}$ has an outer automorphism by Prop. 7.3.1, it can not be of type $B_6$ or $C_6$, so it is of type $E_6$; see [Hu, § 27.4], or [Sp 81, Th. 9.6.2]. Since its center has order 3, $\mathbf{H}$ must be the simply connected group of that type (see [Sp 81, 8.1.11] and [Bour, Planches]).

(e) The proof that $\mathbf{H}$ is defined over $k$ is similar to part (e) of the proof of Th. 7.2.1. The Lie algebra $L(\mathbf{H})$ is contained in the space $S$ of linear maps $t$ of $A_K$ such that

$$\langle t(x), x, x \rangle = 0 \quad (x \in A_K),$$

as follows for example by an argument using dual numbers. From (5.14) we see that for $t \in S$ we have $\langle t(e), e \rangle = 0$, hence $t(e)$ lies in a hyperplane of $A_K$, which has dimension 26. Using Cor. 7.2.2 we conclude that $\dim S \leq 78$, from which one deduces that $S$ coincides with $L(\mathbf{H})$. An application of [Sp 81, 12.1.2] proves that $\mathbf{H}$ is defined over $k$.    □

## 7.4 Historical Notes

C. Chevalley and R.D. Schafer [CheSch] discovered that the automorphism group of an Albert algebra over $\mathbb{R}$ or $\mathbb{C}$ is a simple Lie group of type $F_4$; see also [Fr 51]. The characterization of $E_6$ as the stabilizer of the cubic form det goes back to H. Freudenthal [Fr 51] for the real case; Chevalley and Schafer gave a different description of $E_6$. Notice that Chevalley and Schafer as well as Freudenthal all worked with Lie algebras, so with derivations rather than automorphisms, and with linear transformations $t$ such that $\langle t(x), x, x \rangle = 0$ for all $x$.

It was again L.E. Dickson who in 1901 considered the analog of the complex Lie group $E_6$ over an arbitrary field, as a linear group in 27 variables that leaves a certain cubic form invariant; see [Di 01b] and [Di 08]. N. Jacobson, inspired by Dickson and by Chevalley's Tôhoku paper [Che 55], studied the automorphism group of an Albert algebra and the stabilizer of the cubic form det over arbitrary fields of characteristic not two or three in a series of papers [Ja 59], [Ja 60], [Ja 61]. He proved, for instance, that these groups are simple (quasisimple, respectively) if the Albert algebra contains nilpotent elements (or is reduced, respectively).

The result in Th. 7.1.3 that the automorphisms of an Albert algebra over an algebraically closed field that leave a primitive idempotent invariant form a group isomorphic to Spin(9) is found in [Ja 60].

# 8. Cohomological Invariants

In this final chapter we discuss a number of more recent developments in the theory of octonion and Albert algebras. Specifically, we deal with some cohomological invariants. At the end we make the connection with the Freudenthal-Tits construction (or first Tits construction) of Albert division algebras. The presentation will be more sketchy than in the preceding chapters. The invariants we will discuss are elements of certain Galois cohomology groups. In the first section we therefore give a rudimentary exposition of some notions from Galois cohomology, mainly referring to the literature for definitions and proofs.

## 8.1 Galois Cohomology

Let $k$ be a field and $k_s$ a separable closure. Denote the (topological) Galois group $\mathrm{Gal}(k_s/k)$ by $\Gamma$. Let $A$ be a finite abelian group on which $\Gamma$ acts continuously, i.e., via a finite quotient by an open subgroup. We then have the cohomology groups $H^i(\Gamma, A)$, also written as $H^i(k, A)$. See [Se 64, Ch. I, § 2 and Ch. II, § 1]. The group operation in $H^i(k, A)$ is usually written as addition. For generalities about homological algebra and, in particular, cohomology of groups, see also [Ja 80, Ch. 6].

Assume $B$ is another finite abelian group with continuous $\Gamma$-action. There are cup product maps

$$H^i(k, A) \times H^j(k, B) \to H^{i+j}(k, A \otimes B), \ (c, d) \mapsto c \cup d.$$

Here $A \otimes B$ is the tensor product of $A$ and $B$ considered as $\mathbb{Z}$-modules, and the $\Gamma$-action on it is defined by $\gamma(a \otimes b) = \gamma(a) \otimes \gamma(b)$. For an $i$-cocycle $f$, and a $j$-cocycle $g$, the cup product of their cohomology classes $[f] \in H^i(k, A)$ and $[g] \in H^j(k, A)$ is $[f] \cup [g] = [h]$ with the $(i+j)$-cocycle $h$ defined by

$$h(\sigma_1, \ldots, \sigma_i, \tau_1, \ldots, \tau_j) = f(\sigma_1, \ldots, \sigma_i) \otimes \sigma_1 \cdots \sigma_i(g(\tau_1, \ldots, \tau_j))$$

$$(\sigma_1, \ldots, \sigma_i, \tau_1, \ldots, \tau_j \in \Gamma) \tag{8.1}$$

(see [CaEi, Ch. 11, § 7] or [Br, Ch. 5, § 3]; in the latter the definition is slightly different, with a factor $(-1)^{ij}$ inserted). We have

$$c \cup d = (-1)^{ij} d \cup c \qquad \left(c \in H^i(k, A),\ d \in H^j(k, B)\right). \qquad (8.2)$$

Let $n$ be an integer prime to $\text{char}(k)$. We denote by $\mu_n$ the group of $n$-th roots of unity in $k_s$ with the natural $\Gamma$-action. $\text{Br}(k)$ denotes the Brauer group of $k$; it may be identified with $H^2(\Gamma, k_s^*)$ (see [Ja 80, § 4.7 and § 8.4]). If $A$ is a central simple algebra of dimension $n^2$ over its center $k$, then its class $[A]$ has order dividing $n$ in $\text{Br}(k)$. We denote the subgroup of $\text{Br}(k)$ of elements whose order divides $n$ by $_n\text{Br}(k)$. The following facts are well known.

**Proposition 8.1.1** *Let $n$ be prime to* $\text{char}(k)$.
*(i)* $H^1(k, \mu_n) \cong k^*/(k^*)^n$.
*(ii)* $H^2(k, \mu_n) \cong {_n\text{Br}(k)}$.

Proof. We have an exact sequence

$$1 \to \mu_n \to k_s^* \xrightarrow{n} k_s^* \to 1,$$

the third arrow being the $n$-th power map. This gives rise to a long exact sequence

$$1 \to H^0(\Gamma, \mu_n) \to H^0(\Gamma, k_s^*) \xrightarrow{n} H^0(\Gamma, k_s^*) \to H^1(\Gamma, \mu_n) \to H^1(\Gamma, k_s^*) \xrightarrow{n}$$
$$\xrightarrow{n} H^1(\Gamma, k_s^*) \to H^2(\Gamma, \mu_n) \to H^2(\Gamma, k_s^*) \xrightarrow{n} H^2(\Gamma, k_s^*).$$

If $A$ is a $\Gamma$-module, $H^0(\Gamma, A)$ is the subgroup of $\Gamma$-invariant elements in $A$, so $H^0(\Gamma, k_s^*) = k^*$. By Hilbert's Theorem 90, $H^1(\Gamma, k_s^*) = 1$. Hence we find an exact sequence

$$k^* \xrightarrow{n} k^* \to H^1(\Gamma, \mu_n) \to 1,$$

which implies (i). Since $H^2(\Gamma, k_s^*) \cong \text{Br}(k)$, the last four terms of the long exact sequence yield the exact sequence

$$1 \to H^2(\Gamma, \mu_n) \to \text{Br}(k) \xrightarrow{n} \text{Br}(k).$$

This implies (ii). $\qquad\qquad\qquad\qquad\qquad\qquad\qquad\qquad\qquad\qquad\qquad$ □

The isomorphism of (i) can be made explicit. For $\alpha \in k^*$, let $\xi \in k_s^*$ be an $n$-th root of $\alpha$. The map

$$\Gamma \to \mu_n, \ \sigma \mapsto \xi^{-1}\sigma(\xi), \qquad\qquad\qquad\qquad (8.3)$$

is a 1-cocycle of $\Gamma$ with values in $\mu_n$ whose cohomology class $[\alpha]$ depends only on the coset $\alpha(k^*)^n$. The map $\alpha(k^*)^n \mapsto [\alpha]$ is an isomorphism $k^*/(k^*)^n \xrightarrow{\sim} H^1(k, \mu_n)$. We will consider this as an identification, so $[\alpha]$ stands for the 1-cohomology class as well as for the class of $\alpha \bmod (k^*)^n$ in $k^*/(k^*)^n$. As before, we assume that $n$ is prime to $\text{char}(k)$. Let $\mathbb{Z}/n\mathbb{Z}$ be the cyclic group of order $n$ with trivial $\Gamma$-action. If $\mu_n \subset k$, the $\Gamma$-action on it is trivial, so then $\mu_n \cong \mathbb{Z}/n\mathbb{Z}$ as a $\Gamma$-module. In that situation, choose a primitive $n$-th root of unity $\zeta \in k$. Then we have an isomorphism

$$\mathbb{Z}/n\mathbb{Z} \to \mu_n, \; i + n\mathbb{Z} \mapsto \zeta^i. \tag{8.4}$$

We identify $\mathbb{Z}/n\mathbb{Z} \otimes \mathbb{Z}/n\mathbb{Z}$ with $\mathbb{Z}/n\mathbb{Z}$ via the isomorphism

$$(i + n\mathbb{Z}) \otimes (j + n\mathbb{Z}) \mapsto ij + n\mathbb{Z}. \tag{8.5}$$

Via the isomorphism of (8.4) this yields an isomorphism

$$\phi_\zeta : \mu_n \otimes \mu_n \to \mu_n. \tag{8.6}$$

As the notation indicates, this isomorphism depends on $\zeta$: if $\omega$ is another primitive $n$-th root of unity and $\zeta = \omega^a$, then

$$\phi_\omega(\lambda) = \phi_\zeta(\lambda)^a \qquad (\lambda \in \mu_n \otimes \mu_n),$$

as is readily checked. We also obtain an isomorphism

$$\phi'_\zeta : \mu_n \otimes \mu_n \to \mathbb{Z}/n\mathbb{Z}. \tag{8.7}$$

If $\omega$ is as above, we have $\phi'_\omega(\lambda) = a^2 \phi'_\zeta(\lambda)$.

We continue to assume that $\mu_n \subset k$ with $n$ prime to char($k$). Let $\alpha, \beta \in k^*$, and let $\zeta \in k^*$ be a primitive $n$-th root of unity. Define the *cyclic algebra* $A_\zeta(\alpha, \beta)$ to be the associative algebra over $k$ generated by elements $x$ and $y$ subject to the relations

$$x^n = \alpha, \quad y^n = \beta, \quad xyx^{-1} = \zeta y.$$

This is a central simple algebra over $k$ of dimension $n^2$. If $\alpha$ or $\beta$ is an $n$-th power in $k$, then $A_\zeta(\alpha, \beta)$ is isomorphic to the matrix algebra $M_n(k)$ (see [Mi, § 15, pp. 143-144]). Cyclic crossed products (see [Al 61, Ch. V], [ArNT, Ch. VIII, §§ 4 and 5] or [Ja 80, §§ 8.4 and 8.5]) of dimension $n^2$ over $k$ can be viewed as cyclic algebras. For let $l$ be a cyclic field extension of $k$ of degree $n$, and let $\sigma$ be a generator of the Galois group $\mathrm{Gal}(l/k)$. We can write $l = k(\eta)$ with $\eta^n = \beta \in k^*$. Then $\sigma(\eta) = \zeta y$ for some primitive $n$-th root of unity $\zeta \in k$. For $\alpha \in k^*$, the cyclic crossed product $(l, \sigma, \alpha)$ is the vector space over $l$ with basis $1, u, u^2, \ldots, u^{n-1}$ such that $u^n = \alpha$ and $u\varrho = \sigma(\varrho)u$ for$\varrho \in l$. Taking $x = u$ and $y = \eta$ as generators, one sees that $(l, \sigma, \alpha) = A_\zeta(\alpha, \beta)$.

The result in the following lemma is well-known, see e.g. [KMRT, p. 415]. For the convenience of the reader we sketch a proof. We first explain a notation. For $\alpha, \beta \in k^*$ we have their cohomology classes $[\alpha]$ and $[\beta]$ in $H^1(k, \mu_n)$ as defined by (8.3). The cup product $[\alpha] \cup \beta]$ lies in $H^2(k, \mu_n \otimes \mu_n)$. The isomorphism $\phi_\zeta$ of (8.6) induces an isomorphism of cohomology groups

$$\phi_\zeta^* : H^2(k, \mu_n \otimes \mu_n) \to H^2(k, \mu_n),$$

sending $[f]$ to $[\phi_\zeta \circ f]$.

**Lemma 8.1.2** *Assume that $n$ is prime to* char$(k)$ *and that* $\mu_n \subset k$. *Let* $\alpha, \beta \in k^*$, *and let* $\zeta \in k^*$ *be a primitive $n$-th root of unity. Then the class of* $A_\zeta(\alpha, \beta)$ *in* Br$(k)$ *lies in* $_n$Br$(k)$ *and equals* $\phi_\zeta^*([\alpha] \cup [\beta])$.

Proof. $A_\zeta(1,1) \cong M_n(k)$ is generated by elements $X$ and $Y$ with relations $X^n = Y^n = 1$ and $XYX^{-1} = \zeta Y$. Now $A_\zeta(\alpha, \beta)$ is a $k$-form of $A_\zeta(1,1)$, that is, the $k_s$-algebras $k_s \otimes_k A_\zeta(\alpha, \beta)$ and $k_s \otimes_k A_\zeta(1,1)$ are isomorphic. We give an explicit isomorphism of the former algebra onto the latter. Let $\xi, \eta \in k_s$ be $n$-th roots of $\alpha$ and $\beta$, respectively. Then

$$\psi(1 \otimes x) = \xi \otimes X, \quad \psi(1 \otimes y) = \eta \otimes Y,$$

defines such an isomorphism.

The Galois group $\Gamma$ acts on $k_s \otimes_k A_\zeta(\alpha, \beta)$ and on $k_s \otimes_k A_\zeta(1,1)$ via the first factor. For $\sigma \in \Gamma$,

$$c_\sigma = \psi \circ \sigma \circ \psi^{-1} \circ \sigma^{-1} = \psi \circ {}^\sigma \psi^{-1}$$

is an automorphism of $k_s \otimes_k A_\zeta(1,1) \cong M_n(k_s)$. It defines a noncommutative 1-cocycle of $\Gamma$ in the automorphism group of $k_s \otimes_k A_\zeta(1,1)$ (see [Se 64, Ch. I, § 5]); this automorphism group is isomorphic to PGL$_n(k_s)$ by the Skolem-Noether Theorem.

Define functions $a$ and $b$ on $\Gamma$ with values in $\{0, 1, \ldots, n-1\}$ by

$$\sigma(\xi) = \zeta^{a(\sigma)}\xi, \quad \sigma(\eta) = \zeta^{b(\sigma)}\eta.$$

Notice that $a(\sigma) + a(\tau) - a(\sigma\tau) = 0$ or $n$ for $\sigma, \tau \in \Gamma$ and similarly for $b$. A direct check shows that $c_\sigma$ is the inner automorphism Inn$(g_\sigma)$, where

$$g_\sigma = X^{-b(\sigma)}Y^{a(\sigma)} \quad (\sigma \in \Gamma).$$

Put

$$c_{\sigma,\tau} = g_\sigma \sigma(g_\tau)g_{\sigma\tau}^{-1} \quad (\sigma, \tau \in \Gamma).$$

We compute this explicitly:

$$c_{\sigma,\tau} = X^{-b(\sigma)}Y^{a(\sigma)}X^{-b(\tau)}Y^{a(\tau)}Y^{-a(\sigma\tau)}X^{b(\sigma\tau)}$$
$$= \zeta^{a(\sigma)b(\tau)}X^{-b(\sigma)-b(\tau)}Y^{+a(\sigma)+a(\tau)-a(\sigma\tau)}X^{b(\sigma\tau)}.$$

Since $a(\sigma) + a(\tau) - a(\sigma\tau) = 0$ or $n$ and similarly for $b$, and since further $X^n = Y^n = 1$, we have $Y^{+a(\sigma)+a(\tau)-a(\sigma\tau)} = 1$ and $X^{-b(\sigma)-b(\tau)+b(\sigma\tau)} = 1$, whence

$$c_{\sigma,\tau} = \zeta^{a(\sigma)b(\tau)} \quad (\sigma, \tau \in \Gamma).$$

This function $c$ is a 2-cocycle of $\Gamma$ with values in $\mu_n$, and it is well known that its cohomology class in $H^2(k, \mu_n) = {}_n$Br$(k)$ is the class of $A_\zeta(\alpha, \beta)$ (see [Sp 81, § 12.3.5 (1)] and [Se 64, Ch. I,§§ 5.6 and 5.7]).

On the other hand, by (8.3) $[\alpha] = [f]$ with

$$f(\sigma) = \xi^{-1}\sigma(\xi) = \zeta^{a(\sigma)} \qquad (\sigma \in \Gamma),$$

and similarly for $[\beta] = [g]$. By (8.1) the cup product is $[\alpha] \cup [\beta] = [h]$ with

$$h(\sigma, \tau) = \zeta^{a(\sigma)} \otimes \zeta^{b(\tau)} \qquad (\sigma, \tau \in \Gamma).$$

Then $\phi_\zeta^*([\alpha] \cup [\beta]) = [\phi_\zeta \circ h]$, and

$$(\phi_\zeta \circ h)(\sigma, \tau) = \zeta^{a(\sigma)b(\tau)}.$$

This proves the Lemma. □

Recall that $\phi_\zeta$ as defined by (8.6) depends on the choice of $\zeta$, hence so does $\phi_\zeta^*$. If $\omega = \zeta^a$ is another primitive $n$-th root of unity, then the class of $A_\zeta(\alpha, \beta)$ in the Brauer group equals $a\phi_\omega^*([\alpha] \cup [\beta])$.

Only for $n = 2$ is the isomorphism $\phi_\zeta$ canonical, viz., $\phi_{-1}$. In that case $A_{-1}(\alpha, \beta)$ is a quaternion algebra, whose class in $_2\mathrm{Br}(k)$ is $\phi_{-1}^*([\alpha] \cup [\beta])$; we simply write $[\alpha] \cup [\beta]$ for this class in the sequel.

For $n = 3$ we have a canonical isomorphism $\phi_\zeta' : \mu_3 \otimes \mu_3 \to \mathbb{Z}/3\mathbb{Z}$ as in (8.7), since there is only one other root of unity besides $\zeta$, viz., $\zeta^2$, and $\phi_{\zeta^2}' = 2^2\phi_\zeta' = \phi_\zeta'$. Hence the class $\phi_\zeta'^*([\alpha] \cup [\beta])$ in $H^2(k, \mathbb{Z}/3\mathbb{Z})$ is uniquely determined. By abuse of notation, we write $[\alpha] \cup [\beta]$ for this class and call it "the cup product of $[\alpha]$ and $[\beta]$ in $H^2(k, \mathbb{Z}/3\mathbb{Z})$".

## 8.2 An Invariant of Composition Algebras

The first cohomological invariant we deal with is an invariant of composition algebras.

We will use a theorem of Merkuryev-Suslin [MeSu]. Let $D$ be a division algebra with center $k$, of degree $n$ over $k$. Assume $n$ to be prime to $\mathrm{char}(k)$. Then the class $[D]$ in the Brauer group of $k$ is an element of $H^2(k, \mu_n)$. For $\alpha \in k^*$ denote by $[\alpha]$ its class in $k^*/(k^*)^n = H^1(k, \mu_n)$. The cup product $[\alpha] \cup [D]$ lies in $H^3(k, \mu_n \otimes \mu_n)$.

**Theorem 8.2.1 (Merkuryev-Suslin)** *Assume that $n$ is prime to $\mathrm{char}(k)$ and not divisible by a square. Let $\alpha \in k^*$, and let $D$ be a division algebra with center $k$, of degree $n$ over $k$. Then $[\alpha] \cup [D] = 0$ if and only if $\alpha$ is the reduced norm of an element of $D$.*

For a proof, see [MeSu, 12.2]. The difficult part of the theorem is the "only if" part. We will need the theorem for $n = 2, 3$. In these cases we have $\mu_n \otimes \mu_n \cong \mathbb{Z}/n\mathbb{Z}$ (see the end of the previous section).

Assume that $\mathrm{char}(k) \neq 2$. Let $C$ be a composition algebra of dimension 4 or 8. In $C$ we choose an orthogonal basis of the form $e, a, b, ab$ or $e, a, b, ab, c, ac, bc, (ab)c$ as in Cor. 1.6.3. Recall that in the octonion case we call such elements $a, b, c$ a basic triple. If $C$ is a quaternion algebra, it determines an element of order 1 or 2 in the Brauer group, so an element $[C] \in H^2(k, \mathbb{Z}/2\mathbb{Z})$. Taking as generators $a$ and $b$ as above, we see that $C$ is the cyclic algebra $A_{-1}(-N(a), -N(b))$, so $[C]$ equals the cup product $[-N(a)] \cup [-N(b)]$ (see Lemma 8.1.2 and the remark following it). Hence that cup product is independent of the choice of the elements $a$ and $b$ that provide the orthogonal basis. Conversely, this cup product determines the class $[C]$ in the Brauer group; since $C$ is the only 4-dimensional algebra in its class, it is determined up to isomorphism by $[-N(a)] \cup [-N(b)]$. $C$ is split if and only if $[C] = 0$, that is, if $[-N(a)] \cup [-N(b)] = 0$.

We now exhibit a similar invariant in the case that $C$ is an octonion algebra.

**Theorem 8.2.2** *Let $C$ be an octonion algebra over $k$, with $\mathrm{char}(k) \neq 2$, and let $a, b, c$ be a basic triple in $C$.*
*(i) The cup product $\langle a, b, c \rangle = [-N(a)] \cup [-N(b)] \cup [-N(c)] \in H^3(k, \mathbb{Z}/2\mathbb{Z})$ does not depend on the choice of the basic triple $a, b, c$.*
*(ii) $\langle a, b, c \rangle = 0$ if and only if $C$ is split.*

Proof. Let $D$ be the subalgebra of $C$ with basis $e, a, b, ab$. It is a quaternion algebra over $k$ and $[D] = [-N(a)] \cup [-N(b)]$. This is 0 if and only if $D$ is split, in which case $C$ is also split.

Assume that $D$ is a division algebra. The element $c$ is anisotropic and orthogonal to $D$. It follows from Prop. 1.5.1 that the class of $N(c)$ modulo the group $N_D(D^*)$ of reduced norms of nonzero elements of $D$ is uniquely determined. (Recall that the reduced norm of $D$ coincides with the composition algebra norm $N$.) By the "if" part of Th. 8.2.1, $\langle a, b, c \rangle = \langle a, b, c' \rangle$ if the anisotropic elements $c$ and $c'$ are both orthogonal to $D$. It also follows from Th. 8.2.1 that $\langle a, b, c \rangle = 0$ if and only if $-N(c) \in N_D(D^*)$. Using Prop. 1.5.1 we see that this is so if and only if $C$ is split. Now (ii) follows. If $C$ is split (i) also follows.

Assume that $C$ is a division algebra. Let $a', b', c'$ be another basic triple; we have to prove that $\langle a', b', c' \rangle = \langle a, b, c \rangle$. Denote by $D'$ the quaternion subalgebra generated by $a'$ and $b'$. The 4-dimensional subspaces $D^\perp$ and $D'^\perp$ of $e^\perp$ have an intersection of dimension $\geq 1$. Taking $d \neq 0$ in that intersection, we have

$$\langle a, b, c \rangle = \langle a, b, d \rangle = \langle d, a, b \rangle.$$

(Notice that the cup products are symmetric since the coefficient group is $\mathbb{Z}/2\mathbb{Z}$.) We have, similarly,

$$\langle a', b', c' \rangle = \langle d, a', b' \rangle.$$

Hence, in order to prove (i) we may assume that $a = a'$. Then a similar argument yields that we may assume that $c = c'$, or by symmetry of the cup products, $b = b'$. But then we are in the case already dealt with.    □

We write now $\langle a, b, c \rangle = f(C)$. This is an invariant of the octonion algebra $C$, lying in $H^3(k, \mathbb{Z}/2\mathbb{Z})$. In fact, $f(C)$ completely characterizes the $k$-isomorphism class of $C$, see [Se 94, Th. 9].

In characteristic 2 there also exists a cohomological invariant that characterizes octonion algebras up to $k$-isomorphism; see [Se 94, § 10.3].

## 8.3 An Invariant of Twisted Octonion Algebras

In this section we introduce an invariant of twisted octonion algebras, which will be used in the next section to obtain an invariant of Albert algebras. We first define it in a special case and will afterwards handle the general situation.

From now on, all fields are assumed to have characteristic not 2 or 3. Let $l$ be a cubic cyclic field extension of $k$ and $F$ a normal twisted octonion algebra over $l$. We assume that $k$ contains the third roots of unity. There is $\alpha \in k$ such that $l = k(\xi)$ with $\xi^3 = \alpha$. Fix a generator $\sigma$ of $\mathrm{Gal}(l/k)$ and a primitive third root of unity $\zeta \in k$ such that $\sigma(\xi) = \zeta\xi$. We further assume that $F$ is isotropic and we choose $a \in F$ with $N(a) = 0$ and $T(a) = \lambda \neq 0$. Decompose $F$ with respect to $a$:

$$F = la \oplus la * a \oplus E_1(a) \oplus E_2(a)$$

(see §§ 4.5 and 4.9). Let $D$ be the $k$-algebra generated by $l$ and the transformation $t$ with $t^3 = -\lambda$ that we introduced in the first paragraph of § 4.7. $D$ is isomorphic to the cyclic crossed product $(l, \sigma, -\lambda)$ (see Lemma 4.7.1), so to the cyclic algebra $A_\zeta(-\lambda, \alpha)$. The class of $D$ in $_3\mathrm{Br}(k) = H^2(k, \mathbb{Z}/3\mathbb{Z})$ is $[D] = [-\lambda] \cup [\alpha]$ (see Lemma 8.1.2 and the remark at the end of § 8.1). In $H^1(k, \mathbb{Z}/3\mathbb{Z}) = k^*/(k^*)^3$ we have $[-\lambda] = [\lambda] = [T(a)]$, so by (8.2) we find $[D] = -[\alpha] \cup [T(a)]$.

Choose $v \in E_1(a)$ with $T(v) \neq 0$; the existence of such a $v$ follows from Lemma 4.7.2. Consider the cup product

$$g(a, v) = [\alpha] \cup [T(a)] \cup [T(v)] = -[D] \cup [T(v)] \in H^3(k, \mathbb{Z}/3\mathbb{Z} \otimes \mathbb{Z}/3\mathbb{Z}).$$

Identifying $\mathbb{Z}/3\mathbb{Z} \otimes \mathbb{Z}/3\mathbb{Z}$ with $\mathbb{Z}/3\mathbb{Z}$ by the isomorphism of (8.5), we get $g(a, v)$ in $H^3(k, \mathbb{Z}/3\mathbb{Z})$. If we replace $v$ by another element $w \in E_1(a)$ with $T(w) \neq 0$, $T(v)$ gets replaced by $T(v)\nu$, where $\nu$ is a nonzero reduced norm of an element of $D$, according to Lemma 4.7.6. This does not change the cup product $[D] \cup [T(v)]$ by the "if" part of Th. 8.2.1. Hence $g(a, v)$ depends only on $a$, and we may write $g(a)$ instead.

In a similar way one sees that the cup product $[\alpha] \cup [T(a)] \cup [T(v')]$ with $v' \in E_2(a)$, $T(v') \neq 0$, is independent of the particular choice of $v'$.

**Lemma 8.3.1** *There exists $v' \in E_2(a)$ with $\langle v, v' \rangle = T(v)$ and $T(v') = T(v)^2$.*

Proof. By Th. 4.6.2 we may assume that $F = \mathcal{F}(V, t)$, with $V = E_1(a), V' = E_2(a)$. We have the $\sigma$-linear map $t$ of $V$ with $t(x) = x * a$ $(x \in V)$. Take $v' = t(v) \wedge t^{-1}(v)$. By (4.76),

$$T(v) = \langle v, t(v) \wedge t^{-1}(v) \rangle = \langle v, v' \rangle.$$

Further, using (4.64) we see that

$$t'(v') = t^{-1}(v) \wedge v, \ t'^{-1}(v') = v \wedge t(v).$$

Using (4.70) we conclude that

$$t'(v') \wedge t'^{-1}(v') = T(v)v,$$

whence by (4.76)

$$T(v') = \langle t'(v') \wedge t'^{-1}(v'), v' \rangle = T(v)\langle v, v' \rangle = T(v)^2.$$

$\square$

In $H^1(k, \mathbb{Z}/3\mathbb{Z}) \cong k^*/(k^*)^3$ we find

$$[T(v')] = [T(v)^2] = [T(v)^{-1}] = -[T(v)].$$

Thus we have also:

$$g(a) = -[\alpha] \cup [T(a)] \cup [T(v')] = [D] \cup [T(v')]$$

for $v' \in E_2(a), T(v') \neq 0$.

Now we are going to prove that $g(a)$ is independent of the particular choice of $a$, and that it is zero if and only if $F$ is reduced.

**Proposition 8.3.2** *Assume that $k$ contains third roots of unity, that $l$ is a cubic cyclic extension of $k$ and $F$ an isotropic twisted octonion algebra over $l$. If $a, b \in F$ are isotropic with $T(a)T(b) \neq 0$, then $g(a) = g(b)$. $F$ is reduced if and only if $g(a) = 0$.*

Proof. If $F$ is reduced, then there exist a nonzero $v \in E_1(a)$ and $u \in D$ such that $T(v) = N_D(u)$ by Th. 4.8.1. If $T(v) \neq 0$, then

$$g(a) = -[D] \cup [T(v)] = -[D] \cup [N_D(u)] = 0$$

by Th. 8.2.1. Assume now that $T(v) = 0$. Pick $v_0 \in E_0$ with $T(v_0) \neq 0$. By Lemma 4.7.6, $T(v) = T(v_0)N_D(w)$ for some nonzero $w \in D$. Then $N_D(w) = 0$, so $D$ is not a division algebra. Hence $D \cong M_3(k)$, so $[D] = 0$ and therefore $g(a) = 0$. If, conversely, $g(a) = 0$, then $T(v) \in N_D(D)$ by Th. 8.2.1. This implies by Th. 4.8.1 that $F$ is reduced.

Now assume that $F$ is not reduced. Then $D$ is a division algebra by Cor. 4.8.2, hence $T(v) \neq 0$ for all nonzero $v \in E_2(a)$ by Lemma 4.7.6. First assume that $a * b = 0$. By Lemma 4.9.2, $E_2(a) \cap E_1(b) \neq 0$. Pick a nonzero $v \in E_2(a) \cap E_1(b)$; then $v$ is isotropic and $T(v) \neq 0$. According to Lemma 4.9.1, $a \in E_1(v)$ and $b \in E_2(v)$. Then

$$\begin{aligned} g(a) &= -[\alpha] \cup [T(a)] \cup [T(v)] \\ &= [\alpha] \cup [T(v)] \cup [T(a)] \qquad \text{(by (8.2))} \\ &= g(v) \qquad \bigl(\text{since } a \in E_1(v)\bigr). \end{aligned}$$

Similarly, $g(b) = g(v)$. Hence $g(a) = g(b)$. Finally, let $a * b = d \neq 0$. By condition (ii) of Def. 4.1.1, $d$ is isotropic, and by (4.4) and (4.6) we have $b * d = d * a = 0$. According to what we already proved, $g(b) = g(d) = g(a)$. $\square$

From the Proposition we see that $g(a)$ is, in fact, an invariant of $F$. We denote it by $g(F)$ or $g(F, k)$. To define $g(F, k)$ we assumed that $k$ contains the third roots of unity and that $F$ is isotropic. We now want to get rid of these restrictions and we also want to include nonnormal twisted octonion algebras.

We first recall some results from Galois cohomology which we shall use. Let $k$, $k_s$ and $A$ be as in the beginning of § 8.1. If $m$ is a finite separable extension of $k$, the Galois group $\text{Gal}(k_s/m)$ is a subgroup of $\Gamma = \text{Gal}(k_s/k)$. The cohomology groups $H^i(m, A)$ are defined and we have a restriction homomorphism

$$\text{Res}_{m/k} : H^i(k, A) \to H^i(m, A)$$

which is induced by the restriction of cocycles of $\text{Gal}(k_s/k)$ with values in $A$ to the subgroup $\text{Gal}(k_s/m)$. If $m/k$ is a Galois extension, the Galois group $\text{Gal}(m/k)$ acts on $H^i(m, A)$ (see [Se 64, Ch. I, p. 12-13]) and the image of $\text{Res}_{m/k}$ is fixed elementwise by $\text{Gal}(m/k)$ ([loc.cit., p. 11]). If, moreover, the order of $A$ is prime to the degree $[m : k]$, then $\text{Res}_{m/k}$ defines an isomorphism of $H^i(k, A)$ onto the subgroup of $\text{Gal}(m/k)$-invariant elements of $H^i(m, A)$ (as follows from [CaEi, Cor. 9.2, p. 257]).

$m$ being arbitrary, let $m'$ be a finite separable extension of $m$. Then

$$\text{Res}_{m'/k} = \text{Res}_{m'/m} \circ \text{Res}_{m/k}.$$

Finally, $\text{Res}_{m/k}$ is compatible with cup products.

Let now $k$ be any field with $\text{char}(k) \neq 2, 3$, and let $F$ be a twisted octonion algebra over a cubic field extension $l$ of $k$. We shall define an invariant $g(F) = g(F, k)$ of $F$, lying in $H^3(k, \mu_3 \otimes \mu_3) = H^3(k, \mathbb{Z}/3\mathbb{Z})$ (for this identification see the end of § 8.1). We proceed in several steps.

(a) $F$ is an isotropic normal octonion algebra over $l$ and $\sigma$.

We use the notations of the beginning of this section, but we do not assume that $k$ contains the third roots of unity. $D$ is defined as before. The class $[D]$ now lies in $H^2(k, \mu_3)$. Choose again $v \in E_1(a)$ with $T(v) \neq 0$. We have $[T(v)] \in H^1(k, \mu_3)$. Define $g(F, k)$ to be the element $-[D] \cup [T(v)]$

of $H^3(k, \mu_3 \otimes \mu_3) = H^3(k, \mathbb{Z}/3\mathbb{Z})$. We have to show that this is independent of choices. Let $k' = k(\mu_3)$ and put $l' = k' \otimes_k l$, $F' = l' \otimes_l F$. Then
$F'$ carries an obvious structure of normal twisted octonion algebra over $l'$
and $\sigma$. We have the invariant $g(F', k')$ dealt with above, and it is clear that
$\mathrm{Res}_{k'/k}(g(F, k)) = g(F', k')$. To prove that $g(F, k)$ is independent of choices,
we use the injective homomorphism $\mathrm{Res}_{k'/k}$ to pass to $k'$, over which field we
have already proved independence (in Prop. 8.3.2).

(b) Let $F$ be as in (a) and let $F_\mu$ be an isotope of $F$ (see § 4.1). We claim
that $g(F_\mu, k) = g(F, k)$.

$F_\mu$ and $F$ have the same underlying vector space and proportional quadratic
forms. The cubic form of $F_\mu$ is $\mathrm{N}_{l/k}(\mu)T$, where $T$ is the cubic form of $F$. An
isotropic vector $a$ for $F$ can also serve for $F_\mu$. The cyclic crossed product which
it defines for $F_\mu$ is $(l, \sigma, -\mathrm{N}_{l/k}(\mu)\lambda)$, which is isomorphic to $D = (l, \sigma, -\lambda)$
(the notations being as before). Since the space $E_1(a)$ is the same for $F$ and
$F_\mu$, we conclude that

$$g(F_\mu, k) = -[D] \cup [\mathrm{N}_{l/k}(\mu)T(v)].$$

As $\mathrm{N}_{l/k}(\mu)$ is a reduced norm of an element of $D$ we conclude that $g(F_\mu, k) =
g(F, k)$, establishing our claim.

(c) $F$ is an arbitrary normal octonion algebra over $l$ and $\sigma$.

We may assume that $F$ is anisotropic. Replacing $F$ by an isotope we may
assume that we are in the situation of Case (A) in § 4.11. We then have the
quadratic extension $k_1$ of $k$ and the cyclic extension $l_1 = k_1 \otimes_k l$ of $k_1$, whose
Galois group is is generated by $\sigma$. Moreover, we have the isotropic normal
twisted composition algebra $F_1$ over $l_1$ and $\sigma$. Then $g(F_1, k_1)$ is defined.
Denote again by $\tau_1$ the nontrivial automorphism of $k_1/k$. It acts on $l_1$ and
commutes with $\sigma$. We have a $\tau_1$-linear automorphism $v$ of $F_1$. This can be
viewed as an isomorphism of $F_1$ onto the twisted algebra $^{\tau_1}(F_1)$ (i.e., $F_1$ with
the scalar action of $l_1$ twisted by $\tau_1$). It follows that $g(^{\tau_1}(F_1), k_1) = g(F_1, k_1)$.
But $g(^{\tau_1}(F_1), k_1) = \tau_1.g(F_1, k_1)$. Hence $g(F_1, k_1)$ is fixed by $\mathrm{Gal}(k_1/k)$ and
there exists a unique $g(F, k) \in H^3(k, \mathbb{Z}/3\mathbb{Z})$ with

$$g(F_1, k_1) = \mathrm{Res}_{k_1/k}(g(F, k)).$$

As in step (a), $g(F, k)$ is independent of choices. Also, if $F'$ is an isotope of
$F$ we have $g(F', k) = g(F, k)$ (by step (b)),

(d) Let $F$ be a normal twisted octonion algebra over $l$ and $\sigma$. The *opposite*
$F^o$ of $F$ is a normal twisted composition algebra over $l$ and $\sigma^2$. It has the
same underlying vector space and quadratic form as $F$, but its product $x *^o y$
is the reversed product $y * x$ $(x, y \in F)$. The cubic form of $F^o$ is the same as
the form $T$ of $F$. We claim that $g(F^o, k) = g(F, k)$.

To show this we may perform quadratic extensions, to reduce to the situation
that $F$ is isotropic over $k$ and $\mu_3 \subset k$. Then choose $a$ as before, with $T(a) =
\lambda \neq 0$. The cyclic crossed product defined by $a$ for $F^o$ is $(l, \sigma^2, -\lambda)$, which is
isomorphic to the opposite $D'$ of $D = (l, \sigma, -\lambda)$. Then

$$g(F^\circ, k) = -[D'] \cup [T(v')] = [D] \cup [T(v')],$$

where $v'$ lies in the space like $E_1(a)$ relative to $F^\circ$ and $T(v') \neq 0$. But from § 4.9 we see that this space coincides with the space $E_2(a)$ relative to $F$. Using Lemma 8.3.1, we conclude that $g(F^\circ, k) = g(F, k)$, as claimed.

(e) $F$ is arbitrary.

We may assume that $F$ is nonnormal. Let $k', l'$ and $F'$ be as in Prop. 4.2.4. Then $F'$ is a normal twisted octonion algebra over $l'$ and $\sigma$. So $g(F', k')$ is defined. $\tau$ being the nontrivial automorphism of $k'/k$, we have the $\tau$-linear antiautomorphism $u$ of $F'$ of $F'$. This can be viewed as an isomorphism of the twisted algebra $^\tau(F')$ onto $(F')^\circ$. Proceeding as in step (c) we see that $\tau.g(F', k') = g((F')^\circ, k')$. By step (d) this equals $g(F', k')$. It follows that there exists a unique $g(F, k) \in H^3(k, \mathbb{Z}/3\mathbb{Z})$ with $g(F', k') = \mathrm{Res}_{k'/k}(g(F, k))$.

We have now defined $g(F, k)$ for any twisted octonion algebra $F$. It follows from our constructions that this invariant can be defined in the following manner. Let $m/k$ be a tower of quadratic extensions such that $m$ contains third roots of unity and that $F_m = m \otimes_k F$ is an isotropic normal twisted octonion algebra. Let $g(F_m, m)$ be as in Prop. 8.3.2. Then $g(F, k)$ is the unique element of $H^3(k, \mathbb{Z}/3\mathbb{Z})$ such that

$$\mathrm{Res}_{m/k}(g(F, k)) = g(F_m, m).$$

If $m'/m$ is a finite tower of quadratic extensions, then it follows from the definitions that

$$\mathrm{Res}_{m'/m}(g(F_m, m)) = g(F_{m'}, m').$$

From this one concludes that $g(F, k)$ does not depend on the particular choice of $m$. By Th. 5.5.2 and Lemma 4.2.11, $F_m$ is reduced if and only if $F$ is so. Hence $F$ is reduced if and only $g(F, k) = 0$.

Thus, $g(F)$ is an invariant which detects whether $F$ is reduced or not. The answer to the following question is not known: assuming that $F$ is isotropic, is its isomorphism class uniquely determined by $g(F)$? This is related to a similar question for Albert algebras; see the end of the next section.

## 8.4 An Invariant of Albert Algebras

Let $A$ be an Albert algebra over a field $k$, $\mathrm{char}(k) \neq 2, 3$. We will attach to it an invariant $g(A) \in H^3(k, \mathbb{Z}/3\mathbb{Z})$. Consider $a \in A$, $a \notin ke$. If $k[a]$ is not a cubic field extension of $k$, then we set $g_a(A) = 0$; by Prop. 6.1.1, $A$ is reduced in this case. Assume now that $k[a] = l$ is a cubic field extension of $k$. As in § 6.2, we take $F = l^\perp$ and make this a twisted octonion algebra over $l$. We define

$$g_a(A) = g(F).$$

It is obvious that this depends only on the field $l$ and not on the particular choice of $a$ in $l$. It is our purpose to show that it is even independent of $l$, in other words, that $g_a(A)$ is an invariant of $A$.

By Prop. 8.3.2, $g(F) = 0$ if and only if $F$ is reduced; by Th. 6.3.2, this is the case if and only if $A$ is reduced. So, in particular, if $F$ is reduced, $g(F)$ does not depend on $l$.

Assume now that $A$ is a division algebra, so $F$ is not reduced. We recall from Th. 5.5.2 that $A$ is a division algebra if and only if the cubic form det does not represent 0 nontrivially over $k$, and according to Lemma 4.2.11 the property of a cubic form of not representing 0 nontrivially is not affected by quadratic extensions of the base field. We have seen in the previous section that $g(F)$ is not affected either by quadratic extensions of $k$, so we may assume that $\mu_3 \subset k$. Then $a$ may be chosen in $l$ such that $a^3 = \alpha \in k$. By the Hamilton-Cayley equation (5.7) this is equivalent to $Q(a) = \langle a, e \rangle = 0$ and $\det(a) = \alpha$. Moreover, $l$ is a cyclic extension of $k$. As in § 8.3, we fix a third root of unity $\zeta \in k$, and denote by $\sigma$ the unique generator of $\mathrm{Gal}(l/k)$ with $\sigma(a) = \zeta a$. Since $l/k$ is cyclic, we may consider $F$ as a normal twisted octonion algebra over $l$ with a product which is $\sigma$-linear in the first factor and $\sigma^2$-linear in the second one. After another quadratic extension, if necessary, we may assume that the norm $N$ of $F$ is isotropic. Recall that the cubic form $T$ associated with $F$ does not represent zero nontrivially if $F$ is not reduced (see Th. 4.1.10).

We saw in the previous section that $g_a(A) = g(F) = [\alpha] \cup [T(b)] \cup [T(c)]$, where $b \in F$, $b \neq 0$ (and hence $T(b) \neq 0$), $Q(b) = 0$, and $c \in E_1(b)$, $c \neq 0$. This can also be written as

$$g_a(A) = [\det(a)] \cup [\det(b)] \cup [\det(c)],$$

since $T(x) = \det(x)$ for $x \in F$ by (6.20). Recall from the previous section that we may replace $c \in E_1(b)$ by a nonzero $c' \in E_2(b)$, provided we put a minus sign in front of the cup product:

$$g_a(A) = -[\det(a)] \cup [\det(b)] \cup [\det(c')].$$

The restriction to $F$ of $Q$ is related to $N$ by

$$Q(x) = \mathrm{Tr}_{l/k}(N(x)) \qquad (x \in F)$$

(see (6.7) or (6.17)). It follows that for an $l$-subspace of $F$ the orthogonal complement with respect to $\langle \ , \ \rangle$ (the $k$-bilinear form associated with $Q$) coincides with the orthogonal complement with respect to $N(\ , \ )$ (the $l$-bilinear form associated with $N$). From (6.22) we get that $x^{*2} = x^2$ if $x \in F$, $N(x) = 0$. The action of $l$ on $F$ is given by (6.3) and (6.2); recall that this action is denoted with a dot to distinguish it from the J-algebra product in $A$, so we write $d.x$ for $d \in l$ and $x \in F$. In particular, we have

$$e.x = x, \quad a.x = -2ax \qquad (x \in F), \tag{8.8}$$

since $\langle a, e \rangle = 0$. All this together leads to the following conclusion.

**Lemma 8.4.1** $E(b)$, *the orthogonal complement of* $lb \oplus lb*b$ *in* $F$ *with respect to* $N(\ ,\ )$, *is the orthogonal complement in* $A$ *with respect to* $\langle\ ,\ \rangle$ *of*

$$E_0(b) = ke \oplus ka \oplus ka^2 \oplus kb \oplus kab \oplus ka^2b \oplus kb^2 \oplus kab^2 \oplus ka^2b^2.$$

Next we characterize $E_1(b)$ and $E_2(b)$ within $E(b)$ in terms of the product in $A$.

**Lemma 8.4.2** *For* $i = 1, 2$ *we have*

$$E_i(b) = \{\, w \in E(b) \mid (ab)w = \zeta^i b(aw) = \zeta^{2i} a(bw)\,\}.$$

Proof. From (6.22) we know:

$$v * w + w * v = 2vw \qquad (v, w \in F,\ N(v, w) = 0). \tag{8.9}$$

Replacing $v$ by $a.v$ we get

$$(a.v) * w + w * (a.v) = 2(a.v)w \qquad (v, w \in F,\ N(v, w) = 0),$$

which can be written as

$$\zeta a.(v * w) + \zeta^2 a.(w * v) = 2(a.v)w \qquad (v, w \in F,\ N(v, w) = 0). \tag{8.10}$$

Similarly,

$$\zeta^2 a.(v * w) + \zeta a.(w * v) = 2v(a.w) \qquad (v, w \in F,\ N(v, w) = 0). \tag{8.11}$$

By (4.55) and (4.56), $E_1(b) = \{\, w \in E_0(b)^\perp \mid b*w = 0\,\}$. Let $w \in E_1(b)$. From (8.9) we see that $w * b = 2bw$. From (8.10) and (8.11) with $v$ replaced by $b$ we obtain, using (8.8),

$$-4\zeta^2 a(bw) = -4(ab)w,$$
$$-4\zeta a(bw) = -4b(aw),$$

whence $(ab)w = \zeta b(aw) = \zeta^2 a(bw)$.

Conversely, let $w \in E_0(b)^\perp$ satisfy $(ab)w = \zeta b(aw)$. From (8.11) and (8.8) we obtain

$$\zeta^2 a.(b * w) + \zeta a.(w * b) = -4b(aw) = -4\zeta^2(ab)w.$$

On the other hand, we have by (8.10),

$$\zeta a.(b * w) + \zeta^2 a.(w * b) = -4(ab)w.$$

Dividing this by $\zeta$ and then subtracting it from the previous equation, we get $(\zeta^2 - 1)a.(b * w) = 0$. Hence $b * w = 0$, so $w \in E_1(b)$.

This proves the Lemma for $E_1(b)$. The case of $E_2(b)$ is similar.     $\square$

As we see from the proof, the condition $(ab)w = \zeta^i b(aw)$ already suffices to characterize $E_i(b)$, but for the application that follows it is convenient to have a condition in which $a$ and $b$ appear symmetrically.

We now interchange the roles of $a$ and $b$. Let $l' = k[b]$, $F' = k[b]^\perp$. We take the generator $\sigma'$ of $\mathrm{Gal}(l'/k)$ such that $\sigma'(b) = \zeta^2 b$, so we also interchange the roles of $\zeta$ and $\zeta^2$. We have the subspaces $E'(a)$, $E'_1(a)$ and $E'_2(a)$ in $F'$, and also $E'_0(a)$.

**Lemma 8.4.3** $E_1(b) = E'_2(a)$.

Proof. Since $a$ and $b$ play a symmetric role in the set of generators of $E_0(b)$ (see Lemma 8.4.1), $E'_0(a) = E_0(b)$ and hence $E'(a) = E(b)$. Now the result follows from Lemma 8.4.2.  □

After these preparations we can prove that $g_a(A)$ is independent of the choice of $a$. Replacing $k$ by a tower of quadratic extensions (which we are allowed to do) we may assume that

$$a \in A,\ a \notin ke,\ Q(a) = \langle a, e \rangle = 0;\ b \in k[a]^\perp, b \neq 0, Q(b) = 0;$$
$$c \perp e, a, a^2, b, ab, a^2 b, b^2, ab^2, a^2 b^2, c \neq 0, (ab)c = \zeta b(ac).$$

The independence of $g_a(A)$ follows from the following theorem.

**Theorem 8.4.4** *Let $A$ be an Albert division algebra over $k$ and let $a, b, c$ be as above. Then*

$$g(A) = [\det(a)] \cup [\det(b)] \cup [\det(c)]$$

*is a nonzero element of $H^3(k, \mathbb{Z}/3\mathbb{Z})$ that is independent of the particular choice of the elements $a$, $b$ and $c$.*

Proof. Let $a' \in A$ satisfy the same conditions as $a$. We have to prove that $g_a(A) = g_{a'}(A)$. After performing quadratic extensions we may assume that $k$ contains the third roots of unity and that $k[a]$ and $k[a']$ are cyclic over $k$. Choose an isotropic element $b \neq 0$ in $A$ orthogonal to $e, a, a^2, a', (a')^2$ (which may require another quadratic extension of $k$).

Pick a nonzero $c \in E'_2(a)$. Then

$$
\begin{aligned}
g_b(A) &= -[\det(b)] \cup [\det(a)] \cup [\det(c)] \\
&= [\det(a)] \cup [\det(b)] \cup [\det(c)] \quad \text{(by (8.2))} \\
&= g_a(A) \quad \text{(by Lemma 8.4.3).}
\end{aligned}
$$

Similarly, $g_b(A) = g_{a'}(A)$. Hence $g_a(A) = g_{a'}(A)$.  □

With the aid of [PeRa 96, § 4.2] and the results of the next section one can identify $g(A)$ with plus or minus the Serre-Rost invariant. (Due to choices that have to be made, there is a sign ambiguity in the definition of the latter invariant, anyway.) J.-P. Serre has raised the question whether $g(A)$ together with two invariants of $A$ in $H^*(k, \mathbb{Z}/2\mathbb{Z})$ characterizes $A$ up to isomorphism (see [Se 94, § 9.4] or [PeRa 94, p. 205, Q. 1]).

## 8.5 The Freudenthal-Tits Construction

In this section we briefly indicate how the decomposition of an Albert division algebra $A$ into subspaces $E_0(b)$, $E_1(b)$ and $E_2(b)$ is related to the *Freudenthal-Tits construction* (or *first Tits construction*) of $A$ (see [Ja 68, Ch. IX,§ 12], [PeRa 94, p. 200], [PeRa 96, § 2.5] or [PeRa 97, § 6]). We continue to consider the situation of § 8.4, assuming that $\mu_3 \subset k$. Let $D$ again be the $k$-algebra generated by $l$ and the transformation $t$ with $t^3 = -\lambda$ that we introduced in the first paragraph of § 4.7. Since $A$ is a division algebra $F$ is not reduced, hence $D$ is a division algebra by Cor. 4.8.2. $E_0(b)$ is a 3-dimensional vector space over $l = k[a]$, on which $t$ acts $\sigma$-linearly by

$$t(a^i b^j) = \sigma(a)^i b^{j+1} = \zeta^i a^i b^{j+1},$$

with $b^3 = \det(b)$. This provides $E_0(b)$ with a structure of 1-dimensional vector space over $D$.

$D$ acts on $V = E_1(b)$ as in § 4.7, which makes it a 1-dimensional vector space over $D$. The opposite algebra $D'$ acts on $V' = E_2(b)$ (see § 4.7). Since $D'$ and $D$ are anti-isomorphic by Lemma 4.7.5, $E_2(b)$ is also a 1-dimensional vector space over $D$.

Take $a_0 = e$, then $E_0(b) = Da_0$. Pick $a_1 \neq 0$ in $E_1$, then $E_1(b) = Da_1$. In $E_2(b)$ we take $a_2 = T(a_1)^{-1} t(a_1) \wedge t^{-1}(a_1)$. By Lemma 8.3.1, $T(a_2) = T(a_1)^{-1}$ and $\langle a_1, a_2 \rangle = 1$. Further, $\langle t^i(a_1), a_2 \rangle = 0$ for $i = 1, 2$. We now have a decomposition

$$A = Da_0 \oplus Da_1 \oplus Da_2.$$

Besides the reduced norm $\mathrm{N}_D$, we have on $D$ the *reduced trace* $\mathrm{T}_D$. Over the dual numbers $k[\varepsilon]$ $\left(\varepsilon \neq 0, \, \varepsilon^2 = 0\right)$ one has

$$\mathrm{N}_{k[\varepsilon] \otimes_k D}(1 + \varepsilon u) = 1 + \varepsilon \mathrm{T}_D(u),$$

so from Lemma 4.7.4 we get that

$$\mathrm{T}_D(\xi_0 + \xi_1 t + \xi_2 t^2) = \mathrm{Tr}_{l/k}(\xi_0).$$

Put $\alpha = T(a_1)$; then $T(a_2) = \alpha^{-1}$.

For $u = \xi_0 + \xi_1 t + \xi_2 t^2$ let $u'$ be as in Lemma 4.7.5 and put $\tilde{u} = \xi_0 - \sigma(\xi_1)t + \sigma^2(\xi_2)t^2$. A tedious but straightforward calculation, using the results of Chapter 7, yields that for $z = \tilde{d}_0 a_0 + d_1 a_1 + d_2' a_2 \in A$ we have

$$\det(z) = \mathrm{N}_D(d_0) + \alpha \mathrm{N}_D(d_1) + \alpha^{-1} \mathrm{N}_D(d_2) - \mathrm{T}_D(d_0 d_1 d_2). \tag{8.12}$$

This is precisely the cubic form that plays the role of det in the Freudenthal-Tits construction, also called first Tits construction (see[PeRa 94, (14)], [PeRa 96, § 2.5] or [PeRa 97, § 6]). Starting with a central simple 9-dimensional algebra $D$ over $k$ and an element $\alpha \in k^*$ the construction produces a structure of Albert algebra on $D \oplus D \oplus D$, whose identity element is $(1, 0, 0)$ and whose cubic form is given by (8.12). One can verify, using Prop. 5.9.4, that the Albert algebra obtained from our $D$ and $\alpha$ is isomorphic to $A$.

## 8.6 Historical Notes

The invariant of octonion algebras dealt with in § 8.2 stems from J.-P. Serre; see [Se 94, § 8 and§ 10.3]. The invariant mod 3 of Albert algebras in § 8.4 has been introduced by M. Rost [Ro], following a suggestion by Serre [Se 91]. H.P. Petersson and M.L. Racine gave a simpler proof for its existence [PeRa 96] and named it the *Serre-Rost invariant*. Their proof is valid in all characteristics except three, and in [PeRa 97] they show that with certain modifications their approach works in characteristic three as well.

Our construction of the invariant of twisted octonion algebras in § 8.3 was inspired by the work of Petersson and Racine. It is, in fact, the Serre-Rost invariant in disguise. The approach to the Serre-Rost invariant via the twisted composition algebras, given here, works only in characteristic not two or three. It would be interesting to extend this to the remaining characteristics.

What is nowadays usually called the first Tits construction is already found in H. Freudenthal's paper [Fr 59, § 26] for the special case that $D$ is the $3 \times 3$ matrix algebra over the reals. This is why we use the name *Freudenthal-Tits construction*. J. Tits communicated the construction in its present general form and a second construction which is closely related to the first one to N. Jacobson, who published them in his book [Ja 68, Ch. IX, § 12]

# References

[Al 58] A.A. Albert: A construction of exceptional Jordan division algebras. Ann. of Math. 67 (1958), 1-28.

[Al 61] A.A. Albert: Structure of Algebras. Amer. Math. Soc. Coll. Publ. 24. Providence, 1961.

[AlJa] A.A. Albert and N. Jacobson: On reduced exceptional simple Jordan algebras. Ann. of Math. 66 (1957), 400-417.

[Ar] E. Artin: Geometric Algebra. Interscience, New York, 1957, 1988.

[ArNT] E. Artin, C.J. Nesbitt and R.M. Thrall: Rings with Minimum Condition. University of Michigan Press, Ann Arbor, 1944.

[Ba] E. Bannow: Die Automorphismengruppen der Cayley-Zahlen. Abh. Math. Sem. Univ. Hamburg 13 (1940), 240-256.

[Bl] F. van der Blij: History of the octaves. Simon Stevin 34 (1961), 106-125.

[BlSp 59] F. van der Blij and T.A. Springer: The arithmetics of octaves and the group $G_2$. Proc. Kon. Nederl. Akad. Wetensch. A 62 (=Indag. Math. 21) (1959), 406-418.

[BlSp 60] F. van der Blij and T.A. Springer: Octaves and Triality. Nieuw Arch. Wisk. (3) 8 (1960), 158-169.

[Bor] A. Borel: Linear Algebraic Groups. Benjamin, New York, Amsterdam, 1969. Second ed.: Graduate Texts in Math. 126. Springer, Berlin, New York etc., 1991.

[Bour] N. Bourbaki: Groupes et Algèbres de Lie, Chapitres 4, 5 et 6. Hermann, Paris, 1968, Seconde éd. Masson, Paris, 1981.

[Br] K.S. Brown: Cohomology of Groups. Graduate Texts in Math. 87. Springer, Berlin, New York etc., 1982.

[Ca 14] E. Cartan: Les groupes réels simples, finis et continus. Ann. Sci. École Norm. Sup. (3) 31 (1914), 263-355. Oeuvres I,1, 399-491.

[Ca 25] E. Cartan: Le principe de dualité et la théorie des groupes simples et semi-simples. Bull. Sci. Math. 49 (1925), 361-374. Œuvres I,1, 555-568.

[CaEi] H. Cartan and S. Eilenberg: Homological Algebra. Princeton University Press, Princeton, 1956.

[Che 35] C. Chevalley: Démonstration d'une hypothèse de M. Artin. Abh. Math. Sem. Univ. Hamburg 11 (1935), 73-75.

[Che 54] C. Chevalley: The Algebraic Theory of Spinors. Columbia University Press, New York, 1954. Reprinted in: The Algebraic Theory of Spinors and Clifford Algebras. Collected Works, Vol. 2. Springer, Berlin, New York etc., 1997.

[Che 55] C. Chevalley: Sur certains groupes simples. Tôhoku Math. J. (2) 7 (1955), 14-66.

[CheSch] C. Chevalley and R.D. Schafer: The exceptional simple Lie algebras $F_4$ and $E_6$. Proc. Nat. Acad. Sci. U.S.A. 36 (1950), 137-141.

[De] V.B. Demyanov: On cubic forms in discretely valued fields. Dokl. Akad. Nauk. SSSR (N.S.) 74 (1950), 889-891 (Russian).

[Di 01a] L.E. Dickson: Theory of linear groups in an arbitrary field. Trans. Amer. Math. Soc. 2 (1901), 363-394. Math. papers II, 43-74.

[Di 01b] L.E. Dickson: A class of groups in an arbitrary realm connected with the configuration of 27 lines on a cubic surface. Quart. J. Pure Appl. Math. 33 (1901), 145-173. Math. papers V, 385-413.

[Di 05] L.E. Dickson: A new system of simple groups. Math. Ann. 60 (1905), 400-417.

[Di 08] L.E. Dickson: A class of groups in an arbitrary realm connected with the configuration of 27 lines on a cubic surface (second paper). Quart. J. Pure Appl. Math. 39 (1908), 205-209. Math. papers VI, 145-149.

[Dieu] J. Dieudonné: La géométrie des groupes classiques. Ergebnisse der Math. und ihrer Grenzgebiete, Neue Folge, Band 5. Springer, Berlin etc., 1955, Second ed. 1963.

[Eb] H.-D. Ebbinghaus et al.: Zahlen. Grundwissen Math. 1. Springer, Berlin, New York etc., 1983.

[Fr 51] H. Freudenthal: Oktaven, Ausnahmegruppen und Oktavengeometrie. Mimeogr. notes, Math. Inst. Utrecht, 1951, 1960. Reprinted in Geom. Dedicata 19 (1985), 7-63.

[Fr 59] H. Freudenthal: Beziehungen der $E_7$ und $E_8$ zur Oktavenebene. VIII. Proc. Kon. Nederl. Akad. Wetensch. A 62 (=Indag. Math. 21) (1959), 447-465.

[Gre] M. Greenberg: Lectures on Forms in Many Variables. Benjamin, New York, Amsterdam, 1969.

[Gro] H. Gross, Quadratic Forms in Infinite Dimensional Vector Spaces. Progress in Math. 1. Birkhäuser, Boston, Basel etc., 1979.

[Ho 78] G.M.D. Hogeweij: Ideals and Automorphisms of Almost-classical Lie Algebras. Dissertation, Utrecht University, 1978.

[Ho 82] G.M.D. Hogeweij: Almost-classical Lie algebras. Proc. Kon. Nederl. Akad. Wetensch. A 85 (=Indag. Math. 44) (1982), 441-460.

[Hu] J.E. Humphreys: Linear Algebraic Groups. Graduate Texts in Math. 21. Springer, Berlin, New York etc., 1975.

[Ja 39] N. Jacobson: Cayley numbers and normal simple Lie algebras of type G. Duke Math. J. 5 (1939), 775-783.

[Ja 58] N. Jacobson: Composition algebras and their automorphisms. Rend. Circ. Mat. Palermo (2) 7 (1958), 55-80.

[Ja 59] N. Jacobson: Some groups of transformations defined by Jordan algebras. I. J. Reine Angew. Math. 201 (1959), 178-195.

[Ja 60] N. Jacobson: Some groups of transformations defined by Jordan algebras. II, Groups of type $F_4$. J. Reine Angew. Math. 204 (1960), 74-98.

[Ja 61] N. Jacobson: Some groups of transformations defined by Jordan algebras. III, Groups of type $E_{6I}$. J. Reine Angew. Math. 207 (1961), 61-85.

[Ja 62] N. Jacobson: Lie Algebras. Interscience, New York, 1962.

[Ja 64a] N. Jacobson: Lectures in Abstract Algebra III – Theory of Fields and Galois Theory. D. Van Nostrand, Princeton etc., 1964. Second pr. Graduate Texts in Math. 32. Springer, Berlin, New York etc., 1975.

[Ja 64b] N. Jacobson: Triality and Lie algebras of type $D_4$. Rend. Circ. Mat. Palermo (2) 13 (1964), 129-153.

[Ja 68] N. Jacobson: Structure and Representations of Jordan Algebras. Amer. Math. Soc. Coll. Publ. 39. Providence, 1968.

[Ja 69] N. Jacobson: Lectures on Quadratic Jordan Algebras. Tata Institute of Fundamental Research, Bombay, 1969.

[Ja 71] N. Jacobson: Exceptional Lie Algebras. Lecture Notes in Pure and Applied Mathematics, Vol. 1. Marcel Dekker, New York, 1971.

[Ja 74]  N. Jacobson: Basic Algebra I. Freeman, San Francisco, 1974, Second ed. Freeman, New York, 1985.

[Ja 80]  N. Jacobson: Basic Algebra II. Freeman, San Francisco, 1980, Second ed. Freeman, New York, 1989.

[Ja 81]  N. Jacobson: Structure Theory of Jordan Algebras. The Univ. of Arkansas Lecture Notes in Math., Vol. 5. Fayetteville, 1981.

[Jo 32]  P. Jordan: Über eine Klasse nichtassoziativer hyperkomplexer Algebren. Nachr. Ges. Wiss. Göttingen (1932), 569-575.

[Jo 33]  P. Jordan: Über Verallgemeinerungsmöglichkeiten des Formalismus der Quantenmechanik. Nachr. Ges. Wiss. Göttingen (1933), 209-214.

[JoNW]  P. Jordan, J. von Neumann and E. Wigner: On an algebraic generalization of the quantum mechanical formalism. Ann. of Math. (2) 36 (1934), 29-64.

[KMRT]  M.-A. Knus, A. Merkurjev, M. Rost, J.-P. Tignol, The Book of Involutions, Amer. Math. Soc. Coll. Publ. 44. Providence, 1998.

[Lam]  T.Y. Lam: The Algebraic Theory of Quadratic Forms. Benjamin, Reading, Mass., 1973.

[Lang]  S. Lang: Algebra. Addison Wesley, Reading, Mass. etc., 1965.

[Le]  D.J. Lewis: Cubic homogeneous polynomials over p-adic number fields. Ann. of Math. 56 (1952), 473-478.

[McC]  K. McCrimmon: A general theory of Jordan rings. Proc. Nat. Acad. Sci. U.S.A. 56 (1966), 1072-1079.

[MeSu]  A.S. Merkuryev and A.A. Suslin: K-cohomology of Severi-Brauer varieties and the norm residue homomorphism. Izv. Akad. Nauk. SSSR 46 (1982), 1011-1046 (Russian). English translation: Math. USSR Izv. 21 (1983), 307-340.

[Mi]  J. Milnor, Introduction to Algebraic K-theory. Annals of Math. Studies Nr. 72. Princeton University Press, Princeton, 1971.

[O'M]  O.T. O'Meara: Introduction to Quadratic Forms. Grundlehren math. Wissenschaften 117. Springer, Berlin, New York etc., 1963.

[PeRa 94]  H.P. Petersson and M.L. Racine: Albert algebras. In: W. Kaup, K. McCrimmon and H.P. Petersson (ed.): Jordan Algebras (Proc. Conf. Oberwolfach, August 9–15, 1992), pp. 197-207. W. de Gruyter, Berlin, New York, 1994.

[PeRa 96]  H.P. Petersson and M.L. Racine: An elementary approach to the Serre-Rost invariant of Albert algebras. Indag. Math., N.S., 7 (1996), 343-365.

[PeRa 97]  H.P. Petersson and M.L. Racine: The Serre-Rost invariant of Albert algebras in characteristic three. Indag. Math., N.S., 8 (1997), 543-548.

[Ro]  M. Rost: A mod 3 invariant for exceptional Jordan algebras. C.R. Acad. Sci. Paris Sér. I Math. 315 (1991), 823-827.

[Schaf]  R.D. Schafer: An Introducion to Nonassociative Algebras. Academic Press, New York, London, 1966.

[Schar]  W. Scharlau: Quadratic and Hermitian Forms. Springer, Berlin, New York etc., 1985.

[Se 64]  J.-P. Serre: Cohomologie Galoisienne. Lecture Notes in Math. 5. Springer, Berlin etc., 1964; 5-e éd. 1994. English translation, Galois Cohomology, with new material and an Appendix added, 1997.

[Se 70]  J.-P. Serre: Cours d'arithmétique. Presses Universitaires de France, Paris, 1970.

[Se 73]  J.-P. Serre: A Course in Arithmetic. (English translation of [Se 70].) Graduate Texts in Math. 7. Springer, New York etc., 1973.

[Se 91]  J.-P. Serre: Résumé des cours de l'année 1990-91. Annuaire du Collège de France (1991), 111-121.

[Se 94]  J.-P. Serre: Cohomologie galoisienne: progrès et problèmes. Sém. Bourbaki, no. 783, 1994. Astérisque 227 (1995), 229-257.

[Sp 52]  T.A. Springer: Sur les formes quadratiques d'indice zéro. C.R. Acad. Sci. Paris 234 (1952), 1517-1519.

[Sp 55]  T.A. Springer: Some properties of cubic forms over fields with a discrete valuation. Proc. Kon. Nederl. Akad. Wetensch. A 58 (=Indag. Math. 17) (1955), 512-516.

[Sp 59]  T.A. Springer: On a class of Jordan algebras. Proc. Kon. Nederl. Akad. Wetensch. A 62 (=Indag. Math. 21) (1959), 254-264.

[Sp 63]  T.A. Springer: Oktaven, Jordan-Algebren und Ausnahmegruppen. Vorlesungsausarbeitung von P. Eysenbach. Mimeogr. notes, Math. Inst. Göttingen, 1963.

[Sp 73]  T.A. Springer: Jordan Algebras and Algebraic Groups. Ergebnisse der Math. und ihrer Grenzgebiete, Band 75. Springer, Berlin, New York etc., 1973.

[Sp 81]  T.A. Springer: Linear Algebraic Groups. Progress in Math., Vol. 9. Birkhäuser, Boston,Basel etc., 1981, second ed., 1998.

[Stei]  R. Steinberg, Lectures on Chevalley Groups. Mimeogr. notes, Yale University Math. Dept., 1967.

[Stu]  E. Study: Grundlagen und Ziele der analytischen Kinematik. Sitzungsber. Berliner Math. Ges. 12 (1913), 36-60.

[Ti]  J. Tits: Classification of algebraic semisimple groups. In: A. Borel and G.D. Mostow (eds): Algebraic Groups and Discontinuous Subgroups, pp. 33-62. Amer. Math. Soc. Proc. of Symposia in Pure Math., Vol. 9. Providence, R.I., 1966.

[Va]  F. Vaney: Le parallélisme absolu dans les espaces elliptiques réels à 3 et à 7 dimensions et le principe de trialité. Thèse, Paris, 1929.

[Weil]  A. Weil: Basic Number Theory. Grundlehren math. Wissenschaften, Band 144. Springer, Berlin, New York etc., 1967, Third ed. 1974.

[Weiss]  E.A. Weiss: Oktaven, Engelscher Komplex, Trialitätsprinzip. Math. Z. 44 (1938), 580-611.

[Zo 31]  M. Zorn: Theorie der alternativen Ringe. Abh. Math. Sem. Univ. Hamburg 8 (1931), 123-147.

[Zo 33]  M. Zorn: Alternativkörper und quadratische Systeme. Abh. Math. Sem. Univ. Hamburg 9 (1933), 395-402.

# Index

Printing: Weihert-Druck GmbH, Darmstadt
Binding: Buchbinderei Schäffer, Grünstadt